D1064571

PLANT DISEASES:

Epidemics and Control

PLANT DISEASES:

Epidemics and Control

By

J. E. VAN DER PLANK

Plant Protection Research Institute
Department of Agricultural Technical Services
Pretoria, South Africa

1963

ACADEMIC PRESS

New York and London

ACADEMIC PRESS, INC.
111 Fifth Avenue, New York, New York 10003

United Kingdom Edition published by
ACADEMIC PRESS, INC. (LONDON) LTD.
Berkeley Square House, London W1X 6BA

LIBRARY OF CONGRESS CATALOG CARD NUMBER: 63–16978

Second Printing, 1969

PRINTED IN THE UNITED STATES OF AMERICA

PREFACE

This book describes new methods of epidemiological analysis based largely on infection rates and on the relation between the amount of inoculum and the amount of disease it produces.

Much information has been published over the years on the progress of disease in crops from one date of examination to the next. This information on the amount of disease at successive dates is the material that can be analyzed to estimate various types of infection rates. There is also a substantial amount of information, much of it gathered recently, on the relation between inoculum and disease. All this information is enough to allow a system of epidemiology to be developed and shown to be closely related to experimental observation.

This system can be applied to the control of disease. The main forms of control—by sanitation, resistant varieties, or fungicides—reduce either the inoculum or the infection rate or both, and are appropriate subjects for epidemiological analysis. Particularly in relation to the use of resistant varieties, epidemiological analysis has revealed some principles that remained hidden so long as thinking about breeding for resistance was guided mainly by genetics.

Epidemiological analysis has come to stay. It welds together widely different observations into a coherent whole, and far more than any other way of thinking makes plant pathology a quantitative science. I believe that within a few years it will be taken for granted that it is as essential for a plant pathologist to be trained in epidemiology as it is for him to be taught mycology, virology, or genetics.

This book is primarily for plant pathologists and for plant breeders concerned with breeding for resistance against disease. Exercises at the end of chapters and mathematical tables will help students, teachers, and research workers alike. Others who will find much to interest them are those concerned with defense against plant disease used as a weapon in war. This topic has some unusual features; and although only one chapter is written specifically about defense it links directly with many other chapters. The last chapter in the book is for those who have to design field experiments. It is about the technique of field experimentation when the pathogen (or insect pest) moves freely from plot to plot despite the presence of guard rows in between.

J. E. VAN DER PLANK

Pretoria, South Africa

June, 1963

CONTENTS

CHAPTER 4

How to Plot the Progress of an Epidemic

CHAPTER 5

The Basic Infection Rate

CHAPTER 6

The Latent Period

CHAPTER 7

Average Values of Infection Rates. Increase of Populations of Lesions and of Foci. Independent Action of Propagules

CHAPTER 8

Corrected Infection Rates

CHAPTER 9

Stochastic Methods in Epidemiology

CHAPTER 10

A Guide to the Chapters on Control of Disease

CHAPTER 11

Sanitation with Special Reference to Potato Blight

CHAPTER 12

Sanitation with Special Reference to Wheat Stem Rust

CHAPTER 13

Sanitation and Two Systemic Diseases. Sanitation when Other Things Are Not Equal

CHAPTER 14

Vertical and Horizontal Resistance against Potato Blight

CHAPTER 15

A Note on the History of Stem Rust Epidemics in Spring Wheat in North America

CHAPTER 16

Plant Disease in Biological Warfare

CHAPTER 17

The Bases of Vertical Resistance

CHAPTER 18

General Resistance against Disease

CHAPTER 19

The Choice of Type of Resistance

CHAPTER 20

The Quantitative Effect of Horizontal Resistance

CHAPTER 21

Control of Disease by Fungicides

CHAPTER 22

How Disease Spreads as It Increases

CHAPTER 23

The Cryptic Error in Field Experiments

CHAPTER 1

The Control of Plant Disease Studied as Part of Epidemiology

1.1. The Population of Pathogens

In order to control rust in fields of wheat one must stop, or at least retard, the growth of millions or billions of rust pustules on thousands or millions of wheat plants. The fate of a single wheat plant and the growth of a single pustule are relatively trivial details in a large picture.

As it is with rust in wheat, so it is for most other plant diseases. Control is aimed at the whole population of the pathogen: at the pathogen often in its millions and billions. In this sense, control of plant disease is part of the study of populations of pathogens. Chemical protectants retard the rate of increase of the population of the pathogen. Chemical eradicants reduce the population. Resistant varieties retard the increase of the population, or reduce the population that initiates the epidemics. Sanitation—the use of crop rotation, clean seed, isolation, etc.—reduces the population that reaches the crop. The chemical and physical properties of the fungicide, the apparatus to apply it, the genetics of host and parasite, the technique of plant breeding, and sanitation are all relevant to the control of disease only as they reduce the initial population of the pathogen or retard its subsequent increase.

Horsfall and Dimond saw this when they planned and edited their treatise on Plant Pathology. They called Volume 3, "The Diseased Population Epidemics and Control." But when this volume was written, in 1957 and 1958, too little was known about how to study the increase of pathogen populations. More is now known, though much is still unpublished; and the time has come to take up this task anew, and write about disease control as part of epidemiology.* Hence this book.

Chemical industry and plant breeders forge fine tactical weapons; but only epidemiology sets the strategy. We hope this book will convince readers of the need to review the strategy, as distinct from the actual tactics in battle, against plant disease.

1.2. Epidemics and Biological Warfare

This book is about defense against epidemics. Most of it is about epidemics that arise naturally. But epidemics are potential weapons in war. Sometimes we are a little slow in realizing the threat, although we are reminded of it by the splendid work on the epidemiology of plant diseases that has been published by the United States Army Chemical Corps.

Defense against disease in war is tied with how crops are defended in peace. When we work to control disease in an agricultural crop of great national importance, we should pause and ask ourselves whether the defenses we are building could easily be breached in time of war.

Consider stem rust of wheat. About 50 years ago a method of control was evolved that has since become generally applied. Resistant varieties are used. These varieties are resistant against all the prevalent races of the stem rust fungus. As long as the races stay unchanged, the resistance stays unchanged. But the fungus is mutable; and new races arise. If these can attack a variety and become abundant, the variety is destroyed. This happened in the United States and Canada in 1935, when race 56 ruined the fields of spring wheat. It happened again in 1953 and 1954, when the newer varieties that were resistant to race 56 went down to race 15B. Still newer varieties resist both race 56 and race 15B. But races are now known that can attack these varieties, too. The peacetime question is whether the new races will become abundant enough to do what race 56 did in 1935 and race 15B in 1953 and 1954. If one thinks of war, the question is whether an enemy could release enough inoculum of the appropriate races to undo 50 years' wheat breeding. Present commercial varieties would become hostages to an enemy who could prepare the inoculum and deliver it. And neither preparation nor delivery are technologically inherently difficult with modern skills.

* Epidemiology is the science of disease in populations. Sometimes one needs to distinguish between the study of disease in populations and the study of populations of pathogens, but not in this book.

Fortunately, good defense is possible. It is discussed in Chapter 16. But it is not enough to know how to build defenses; it is necessary also to see that they are indeed built in good time.

The great corn crop seems safe at present from enemy attack with disease, for which a long line of plant breeders must be thanked. But for some years now new methods of breeding for resistance have been proposed and pursued, which could yet make corn as vulnerable as wheat.

It cannot be too strongly stressed that breeding for resistance against disease is not a matter of genetical convenience but of epidemiological safety; and that policy about control measures must be decided at the highest level on all the evidence that can be presented.

1.3. The Language of Epidemiology. Some Expressions

Most of what we discuss about epidemics can be stated in plain language, without mathematical expressions. We aim at plain statement in all except Chapters 5 to 9, and these can be skipped at the first reading.

But the use of these four expressions makes writing conciser and reading easier: r (or r_l); x; $\log [x/(1-x)]$; and $\log [1/(1-x)]$. We introduce them forthwith, at the price of a little repetition later.

The book is largely about the infection rate r. It is the rate at which the population of the pathogen increases.

Between 1940 and 1950 the human population of the United States increased at an average rate of 135 (persons) per 10,000 (persons) per year. One can vary how one puts this information. One can say the average rate was 1.35% per year, or the average rate was 0.0135 per unit (i.e., per person) per year. In a similar way one can state that the population of the potato blight fungus *Phytophthora infestans* increased in unsprayed fields of the susceptible variety Bintje in the sand areas of the Netherlands at an estimated rate of $r = 0.42$ per unit per day during July 1953.

The infection rate r tells in a single figure much about an epidemic. A rate $r = 0.5$ per unit per day is a fast rate even for the potato blight fungus or the wheat stem rust fungus. A rate as fast as 0.5 per unit per day means that the race of the pathogen is aggressive, the variety of the host plant is susceptible, and the weather favors infection. By measuring the rate of increase of the pathogen's population, one measures also the condition of the host plants and the environment. Thus during July 1953 in the sand areas of the Netherlands the estimated infection rate r of potato blight was 0.42 per unit per day on the variety Bintje but only

0.11 per unit per day on the less susceptible variety Voran. The rate r reflects the field resistance of the variety.

When r is estimated during the early (logarithmic) phase of an epidemic we put the subscript l after it. The distinction is sometimes important, but in most chapters one can think of r and r_l as alike.

The rate r, or, to be precise, r_l, is like the rate of increase of money at compound interest. The next chapter deals with the arithmetic of interest on money.

The rate of increase of a human population is estimated as an average that is determined from the results of successive censuses. So too r is an average estimated from successive assessments of the population of the pathogen. We assess the population as the proportion x of infected plants if the disease is systemic, or of infected susceptible tissue if the pathogen causes local lesions. Thus, if 18% of the foliage is infected with a leafspot disease, we write $x = 0.18$. We use x here for the quantity that increases. In the next chapter this is money; but in the course of Chapter 3, x settles down as the estimated proportion of disease, and keeps that meaning thereafter.

When one plots x against time one commonly gets a disease progress curve that is roughly S-shaped. Early in the epidemic, at the bottom of the S, the absolute rate of increase of disease is small because only a little inoculum is present. Toward the end of the epidemic, at the top of the S, the rate is also slow, because little tissue remains to be infected. In the middle of the S, where x is around 0.5, the absolute rate is relatively fast, because it is little restricted either by lack of inoculum or lack of tissue that can still be infected.

An S-shaped curve is awkward to analyze. To help straighten it, we plot log $[x/(1-x)]$, instead of x, against time. We define r so that if r is constant log $[x/(1-x)]$ plotted against time gives a straight line. Log $[x/(1-x)]$ is a line straightener; and any deviations from straightness show that r has varied.

What we have just written is for pathogens that increase in a form similar to compound interest on money. At times pathogens increase in a form similar to simple interest on money. Then we use log $[1/(1-x)]$ instead of log $[x/(1-x)]$.

1.4. A Suggested Order of Reading the Chapters

The next three chapters (Chapters 2 to 4) are meant to introduce r, x, log $[x/(1-x)]$, and log $[1/(1-x)]$ more fully. It is suggested that the reader then goes straight on to Chapter 10, which starts the part about the control of disease.

Chapter 10 outlines how epidemiology is used to study control. Chapters 11, 12, and 13 are about sanitation. Chapters 14 to 19 are about the use of resistant varieties; for the convenience of plant breeders, agronomists, and geneticists the chapters on resistance against disease are kept in a block. For Chapters 10 to 19 it is enough to be familiar with r, x, $\log [x/(1-x)]$, and $\log [1/(1-x)]$. The few references in Chapters 10 to 19 to matters in Chapters 5 to 8 can be accepted for the time being and checked later.

Chapters 5 to 8, which, with Chapter 9, we suggest can be skipped at a first reading, are mainly about infection rates. When it is written that the rate of increase of the population of the United States between 1940 and 1950 was 1.35% per year, the figure is the barest summary of the censuses. The rate can be analyzed without end. It is affected by births, deaths, and marriages, fertility rates and ages, immigration and emigration, differences between town and country, and so on. So too an infection rate can be analyzed to see how various factors affect it.

An analysis of the infection rate is needed if we are to investigate some aspects of disease control. Spraying fields of Bintje potatoes from 3 to 5 times during the season with fungicides reduced the estimated average value of r for potato blight from 0.42 to 0.082 per unit per day in July 1953. These two rates by themselves tell much: they tell what the degree of control was like. But if one needs more information, for example on the proportion of spores that got through the fungicide barrier, one must break the figures down in a way that gives the needed information. How to break the figures down is the topic of much of Chapters 5 to 8.

These chapters have a little more mathematics than the others. There is a well tried device about any equations or expressions that give difficulty: accept them, and carry on. (How many of us understand all the mathematics in the statistical tests of significance we use without a qualm?)

1.5. The Spread of Disease and Its Bearing on the Technique of Field Experiments

As disease multiplies, it tends to spread. As the population of the pathogen increases, it tends to cover more plants and more fields. Chapter 22 touches on this. The spread of disease is very relevant to control; but in this book we stress the role of multiplication rather than of spread.

The benefit to farmers of resistant varieties or fungicides is first determined in experiments with plots laid out in some approved statistical design. These experiments have an important part in the control of plant disease, and the second longest chapter in the book, Chapter 23, is about them.

In any experiment with plots designed to give information to farmers, whether it is about chemical fertilizers, fungicidal sprays, or anything else, the plots must not interfere with one another. This is a basic principle universally acknowledged. Guard strips left between plots attest the importance of the principle.

In plant pathology and entomology one must reckon with the fact that pathogens and insects move. The problem is epidemiological. Continental epidemics of wheat stem rust spread annually from Mexico and Texas to Canada. How long will miniature epidemics of wheat stem rust take to spread between experimental plots separated by guard strips a few feet wide? The answer, obtained experimentally, is what one would expect: the stem rust fungus swiftly crosses the guard strips and moves from plot to plot.

With foliage diseases in particular, masses of spores can move from plot to plot, despite the presence of guard strips. The experimental evidence for this is, we believe, now incontrovertible. Superficial visual impressions are misleading. Disease is too often pictured as moving like lava, engulfing as it moves, and showing clearly where it has moved. The picture is not inept for some diseases that move slowly. But fast moving pathogens can move between plots in a replicated experimental layout without showing in clear gradients that they have moved. The faster they move and multiply, the more they obliterate the visual traces of their having moved.

When spores move, they cause plots to interfere with each other and undermine the foundations of field experiments. But, although one cannot wholly prevent plots from interfering with each other, one can at least greatly reduce either the interference itself or the effects of the interference, by methods that are statistically sound and reasonably simple.

Mutual interference among plots affects experiments in plant pathology and entomology directly, and sometimes other experiments indirectly. For example, plots might interfere with one another when sprinkle irrigation is compared with flood irrigation of a pasture susceptible to a disease favored by wet foliage.

For 40 years and more statistical methods have been actively applied to problems of technique in agricultural experiments, with rich rewards. But there is still a fine field for study of interference among plots and its effects, and results are likely to be as exciting and as fresh as any found before. The field is for statisticians, plant pathologists, and entomologists alike.

CHAPTER 2

About Interest on Money. Logarithmic Increase

SUMMARY

This chapter is mainly to refresh the memory about systems of simple and compound interest on money. Of special importance is the hypothetical system in which interest is added to capital continuously as it is earned, and instantly begets further interest. At any moment of time, the rate of increase of the total amount of capital plus interest is proportional to the total amount itself at that moment. This is logarithmic increase. Examples are worked out in order to make calculations familiar.

2.1. Interest is Proportional to Initial Capital

If 1000 dollars yield 100 dollars in interest, then 2000 dollars, invested in the same way, yield 200. In any given investment, the interest that is earned is proportional to the initial capital that earns it.

Suppose that the fungus in lesions of a disease releases spores which move away and germinate elsewhere in the field to start daughter lesions. If the fungus in 1000 parent lesions starts N daughter lesions, and the fungus in 2000 parent lesions in the same field at the same time starts $2N$ daughter lesions, the fungus is increasing in the same way as money at interest; and one can apply to the increase of the fungus the same arithmetic as one applies to the increase of money at interest. The test is simple. Subsequent chapters will show that the increase of disease is a complicated process, with incubation periods and all that. But so long as there is valid reason to expect that twice as many parent lesions will produce (with allowance for sampling errors) twice as many daughter lesions, one can handle the population problem arithmetically as one would handle problems of interest on money.

So consider interest on money.

2.2. Interest per Cent and Interest per Unit

If 100 dollars of capital earns 10 dollars of interest in a year, that is interest at the rate of 10% per year. At the same rate 1 dollar earns 1/10 dollar per year. Instead of dollar, one can substitute pound or franc or lira or any other unit of money, without altering the rate of interest. The rate is 1/10 unit per unit of money per year, or, simply, 1/10 per unit per year.

In this book the rate of interest (which becomes the infection rate when one deals with disease instead of money) is given per unit and not per cent. This will be a little unfamiliar at first, but more convenient in the long run.

2.3. Simple Interest

At simple interest the interest-bearing capital stays unchanged. Interest does not earn interest. If 1000 dollars is invested at simple interest at the rate of 1/10 per unit per year, the interest after 1 year is $1000 \times 1/10 = 100$ dollars. The total amount in dollars of capital plus interest is then $1000(1+\frac{1}{10})$. After 2 years it is $1000(1+\frac{2}{10})$. After 10 years it is

$$1000\left(1+\frac{10}{10}\right) = 1000 \times 2$$

The bottom line of Fig. 2.1 illustrates increase of money at simple interest, paid yearly. The total amount of capital plus interest increases by successive steps, all steps being equal in height.

2.4. Discontinuous Compound Interest

Compound interest itself earns interest as soon as it is added to capital. As before, the total amount in dollars of capital plus interest at the end of the first year is $1000(1+\frac{1}{10})$. All of this now bears interest during the second year, and after 2 years the total amount is

$$[1000(1+\tfrac{1}{10})](1+\tfrac{1}{10}) = 1000(1+\tfrac{1}{10})^2$$

At the end of 10 years the total amount is

$$1000(1+\tfrac{1}{10})^{10} = 1000 \times 2.5937$$

Increase of money at discontinuous compound interest paid yearly is illustrated by the middle line of Fig. 2.1. The total amount of capital plus interest increases in steps, each step being higher than the previous one.

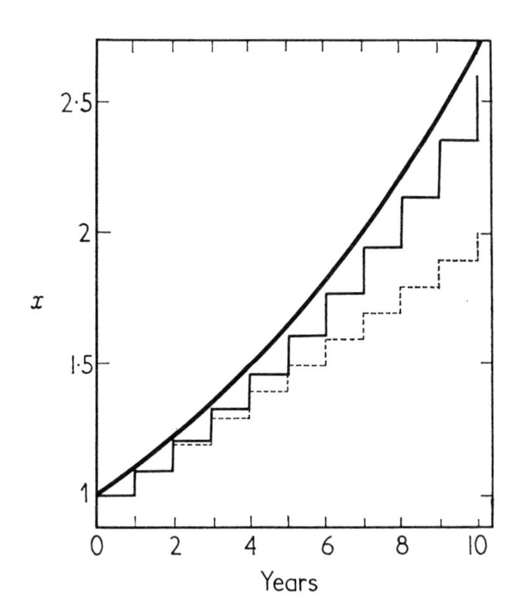

FIG. 2.1. The increase of the amount x of capital plus interest with time. The bottom (broken) line with equal steps, the middle line with progressively higher steps, and the top line without steps represent, respectively, simple interest added once a year, discontinuous compound interest added once a year, and compound interest added continuously. The rate of interest is 1/10 per unit per year, and the initial capital is 1. These lines illustrate the numerical examples in Sections 2.3, 2.4, and 2.5, except that the 1000 has been omitted for convenience.

2.5. Continuous Compound Interest

If the money is being put to use all the time, it is unfair to wait a whole year before adding the interest. If the interest could be reinvested once a month, the total amount of capital plus interest after a month would be

$$1000\{1+[1/(10\times 12)]\}$$

At the end of 10 years ($=$ 120 months) it is

$$1000\left(1+\frac{1}{10\times 12}\right)^{10\times 12} = 1000\times 2.7070$$

It would be fairer still for interest to be earned on the daily balance of the total amount of capital plus interest. If one forgets about leap years, the total amount at the end of 10 years is

$$1000\left(1+\frac{1}{10\times 365}\right)^{10\times365} = 1000\times 2.7181$$

One can carry on the process indefinitely. If one ignores the realities of investment, one can visualize reinvestment of interest every second instead of every day. The total amount at the end of 10 years is then

$$1000\left(1+\frac{1}{10\times 365\times 24\times 60\times 60}\right)^{10\times365\times24\times60\times60} = 1000\times 2.71828$$

At an interest rate of 1/10 per unit per year, 1 sec is a short enough interval for practical purposes, because the answer it gives (1000×2.71828) tallies, to at least six digits, with 1000×2.7182818, etc., which is the figure when interest is reinvested after an infinitesimally small period of time. That is, it is the figure when money increases at interest added continuously.

Continuous compound interest is illustrated by the top curve in Fig. 2.1. There are now no steps.

The number 2.7182818, etc., is always written as e or ε. It is defined as

$$\left(1+\frac{1}{n}\right)^{n}$$

where n is very large.

One uses e for calculations of continuous compound interest. If x_0 is the capital at the start, the total amount x of capital plus interest added continuously for t years at the rate of r per unit per year is given by the equation

$$x = x_0\, e^{rt} \qquad (2.1)^*$$

In the previous example, $x_0 = 1000$ (dollars), $r = 1/10$ per unit per year, and $t = 10$ years.

$$\begin{aligned} x &= 1000e^{(1/10)\times 10} \\ &= 1000e \\ &= 1000\times 2.71828 \end{aligned}$$

which is the same answer as before.

* The clue to Eq. (2.1) is that, when n is very large,

$$\begin{aligned} \left(1+\frac{r}{n}\right)^{nt} &= \left(1+\frac{1}{n}\right)^{nrt} \\ &= e^{rt} \end{aligned}$$

If $t = 19$, with x_0 and r as before,

$$\begin{aligned} x &= 1000e^{(1/10)\times 19} \\ &= 1000e^{1.9} \\ &= 1000 \times 6.6869 \end{aligned}$$

For calculations of this sort, one can use tables or logarithms. A table of e^{rt} is given in the Appendix with rt written as a. A few numerical examples are worked out at the end of this chapter.

2.6. Continuous Compound Interest Seen in Another Way

In mathematics d is shorthand for "a very small bit of." It is applied to quantities that vary. If the total amount x of capital plus interest varies with time t, then dt means a very small interval of time, and dx is the very small bit that x increases in that interval. We make dt small enough for dx to be negligible by comparison with x. Then during the interval dt we can consider interest to be on a fixed amount and therefore simple. (For example, suppose the rate of simple interest to be 1/10 per unit per year, as in our original example. Suppose $dt = 1/1000$ second and $x = 1{,}000{,}000$ dollars. In 1/1000 sec, 1,000,000 dollars would earn in interest 3/10,000 cent, approximately, which is dx. Because 3/10,000 cent is insignificant in comparison with 1,000,000 dollars, one can consider x as staying constant over the interval dt, and that interest is therefore simple over this interval.) Simple interest is the capital $\times$ the rate of interest $\times$ the time during which interest is earned. It is $xrdt$, where r is the rate of interest. But the interest is also dx, the amount that x increases in the interval dt. Therefore

$$dx = xrdt \tag{2.2}$$

Although dx is insignificantly small in comparison with x, it is not insignificantly small in comparison with xdt, because dt is itself very small. (To continue the previous example, at simple interest 3/10,000 cent in 1/1000 sec is approximately 100,000 dollars in 1 year.)

By a process that can be found in books on the infinitesimal calculus, one can collect all the dx's together and all the dt's in Eq. (2.2). The equation then becomes

$$x = x_0\, e^{rt}$$

which is Eq. (2.1) as before.

2.7. Equivalent Rates of Continuous and Discontinuous Compound Interest

This is important when one comes to infection rates in epidemiology. Consider an example. With discontinuous compound interest, added yearly, at a rate of 1/10 per unit per year 1000 becomes $1000(1+\frac{1}{10})^{10}$

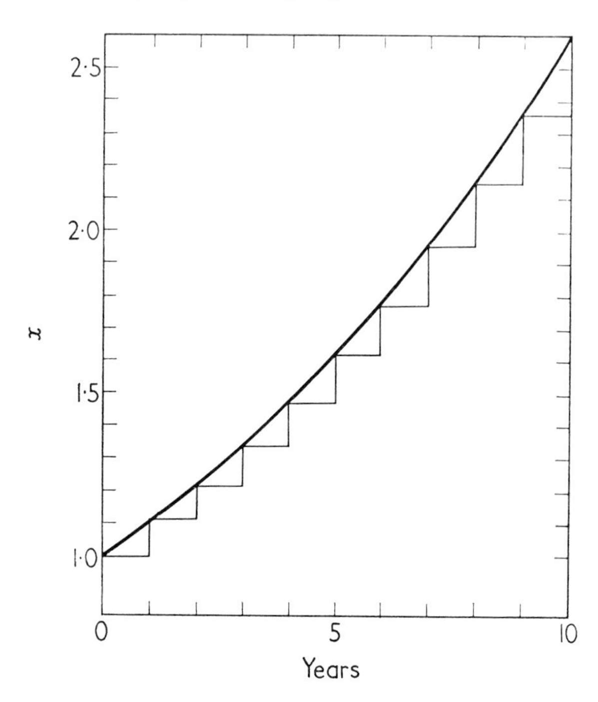

FIG. 2.2. The increase of the amount x of capital plus interest with time. The line with steps represents discontinuous compound interest at a rate of 1/10 per unit per year and added once a year. The smooth upper line represents continuous compound interest at a rate of 0.09531 per unit per year. In both instances the initial capital is 1.

in 10 years. At what rate r of continuous compound interest would 1000 become $1000(1+\frac{1}{10})^{10}$ in 10 years?

$$1000(e^r)^{10} = 1000(1+\tfrac{1}{10})^{10}$$

$$e^r = 1.1$$

$$r = 0.09531$$

(This calculation will be given as an example at the end of the chapter.)

In Fig. 2.2 the stepped line for discontinuous compound interest is the same as the corresponding line in Fig. 2.1. The smooth curve for continuous compound interest at the rate of 0.09531 per unit per year interest touches this line at the end of every year. Provided that one is interested in the amount after a whole number of years (without fractions of a year), it does not matter whether one receives discontinuous compound interest at the rate of 1/10 per unit per year payable yearly, or continuous compound interest at the rate of 0.09531 per unit per year. The amount is the same.

2.8. Variable Rates of Continuous Compound Interest

This, too, is important when one comes to infection rates in epidemiology. It will be discussed further in Chapter 7.

When the rate r is not constant over the whole period t, the final amount is that given by the average value of r during the period. For example, suppose that for 5 years $r = 0.04$ per unit per year. At the end of those 5 years

$$\begin{aligned} x &= x_0\, e^{0.04 \times 5} \\ &= x_0\, e^{0.20} \\ &= x_0 \times 1.2214 \end{aligned}$$

Suppose that for a further 5 years $r = 0.16$ per unit per year. At the end of those years

$$\begin{aligned} x &= x_0 \times 1.2214 \times e^{0.16 \times 5} \\ &= x_0 \times 1.2214 \times 2.2255 \\ &= x_0 \times 2.718 \end{aligned}$$

This is what we would have got by taking $r = (0.04 + 0.16)/2 = 0.1$ per unit per year for the whole 10 years.

2.9. Consistent Units of Time

If r is given per unit per year, t must be given in years. If r is given per unit per day (which it often is for disease infection rates), t must be given in days.

2.10. Natural Logarithms

These are to the base e, and are written $\log_e$. Just as $\log_{10} 10^a = a$, so $\log_e e^a = a$.

Written as logarithms, Eq. (2.1) becomes

$$\log_e x = \log_e x_0 + rt \tag{2.3}$$

or

$$r = \frac{1}{t}(\log_e x - \log_e x_0) \tag{2.4}$$

$$= \frac{1}{t} \log_e \frac{x}{x_0} \tag{2.5}$$

To find natural logarithms from tables of common logarithms (to the base 10), multiply the common logarithm by 2.30259. Equation (2.5) then becomes

$$r = \frac{2.30259}{t} \log_{10} \frac{x}{x_0} \tag{2.6}$$

With data for plant disease, it is usually accurate enough to cut the factor down to three digits, i.e., to 2.30 or 2.3.

2.11. Logarithmic Increase

In Fig. 2.3 we use natural logarithms to plot the increase of money at continuous compound interest from $x_0 = 1$ to $x = 2.7183 (= e)$ in 10 years at a rate of 1/10 per unit per year. These are the same data as those for the top curve of Fig. 2.1. The line in Fig. 2.3 is straight, which means that r stayed unchanged over the whole 10-year period.

If we had not known the value of r, we could easily have got it from Fig. 2.3. By taking details from the graph and using them in Eq. (2.4), we find

$$r = \frac{1-0}{10} = \frac{1}{10} \text{ per unit per year}$$

All interest, simple or compound, earned over an appropriate period of time, for instance 1 year, is proportional to the initial capital, other things being equal. With continuous compound interest, the rate of increase of money at any moment of time is also proportional to the total amount of capital plus interest at that moment. Thus dx is the increase of x (the total amount) in the interval dt; and dx/dt, which is the rate of increase of x, is proportional to x. This can be seen in Eq. (2.2).

When the rate of increase of a quantity at any moment is proportional to the quantity itself at that moment, this is known as logarithmic

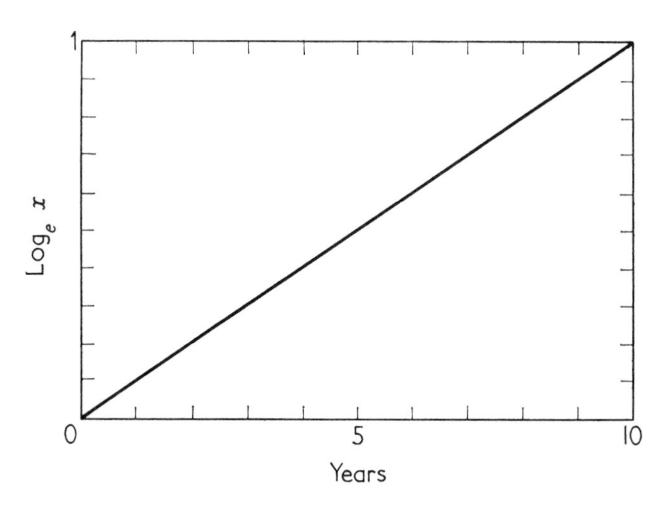

FIG. 2.3. The increase of the amount x of capital plus interest with time. The line represents compound interest added continuously at a rate of 1/10 per unit per year. The initial capital is 1. The line is the logarithmic form of the top line in Fig. 2.1. It suggests why increase at continuous compound interest is called logarithmic increase.

increase. Money at continuous compound interest increases logarithmically. These words have this precise meaning, and will be used only with this meaning.

This definition of logarithmic increase does not imply that the rate stays constant at all times. That is, it does not imply that $\log x$ plotted against time necessarily gives a straight line as in Fig. 2.3.

EXERCISES

1. If $e^r = 1.1$, find r. (Answer: 0.0953.) First write down the natural logarithms of both sides of the equation, thus

$$r = \log_e 1.1$$

Then convert to common logarithms

$$\begin{aligned} r &= 2.3026 \log_{10} 1.1 \\ &= 2.3026 \times 0.04139 \\ &= 0.0953 \end{aligned}$$

2. If the rate of continuous compound interest is r per unit per day, what is it per unit per year? (Answer: $365r$)

From Eq. (2.1), the amount at the end of 365 days is $x = x_0 e^{365r}$. Here $t = 365$ because the unit of time in r is the day.

Let s be the rate per unit per year, $x = x_0 e^s$. Here $t = 1$, because the unit of time in s is the year.

$$x_0 e^{365r} = x_0 e^s$$

$$s = 365r \text{ per unit per year}$$

3. If the rate of interest is r per unit per year, what is it per cent per year? (Answer: $100r$)

4. $e^{0.693} = 2$. If the compound interest is continuous and $r = 0.03$ per unit per year, how long will money take to double itself? (Answer: $0.693/0.03 = 23.1$ years)

5. If money, increasing logarithmically, doubles itself in 15 years, what is r? (Answer: $0.693/15 = 0.0462$ per unit per year)

6. If money increases logarithmically from 1244 units to 2172 units in 8 years, what is r? From Eq. (2.6)

$$r = \frac{2.3026}{8} \log_{10} \frac{2172}{1244}$$

$$= 0.0697 \text{ per unit per year}$$

Alternatively, from Eq. (2.1),

$$e^{8r} = \frac{2172}{1244} = 1.746$$

From Table 4 in the Appendix, $8r = 0.56$ and $r = 0.07$ per unit per year. This method is useful for a quick check.

CHAPTER 3

The Logarithmic and the Apparent Infection Rates

SUMMARY

The change is made from increase of money to increase of disease. Rates ot interest become infection rates. Infection proceeds intermittently. Nevertheless one may determine a logarithmic infection rate if time is measured in whole years or days, whichever are the natural units of time. Logarithmic rates are applicable only when the proportion x of tissue that is diseased is small, i.e., when, say, $x < 0.05$. When the proportion is not small, one may determine an apparent infection rate. For this, the rate of increase of disease at any time is related to the proportion, $1-x$, of susceptible tissue remaining to be infected at that time.

3.1. Statement of the General Problem of This Book

We must find a system of dealing with populations of pathogens. It must be substantially valid, and shown to be valid. It must apply to problems of sanitation, fungicides, and the resistance to disease. It must apply to all infectious plant diseases. It must not become bogged down in intractable algebra.

Sometimes pathogen populations increase in a form similar to simple interest on money. Examples appear in Sections 4.4 and 5.4, but are omitted from this chapter.

Usually, the increase is not simple, and will be dealt with by the mathematics of continuous compound interest, i.e., of logarithmic increase or modifications of it. The analogy between the increase of populations of pathogens and the logarithmic increase of money is faint. There are four major breaks in the analogy. Infection occurs intermittently, not continuously. There is a limit to the amount of disease that can develop (a field cannot become more than 100% infected), whereas money can increase without limit. These two matters are discussed in the present chapter and again later. Newly infected tissue is not immediately infectious; there is nothing comparable in the logarithmic increase of money. Disease tends to occur in foci (i.e., areas in which there is more than an average concentration of disease), and in these foci pathogen populations increase less than logarithmically. These two matters are discussed in Chapters 5, 6, and 7. There are other breaks in the analogy as well. Nevertheless the mathematics of logarithmic increase, appropriately modified, gives a workable and valid method of dealing with pathogen populations (apart, of course, from the instances of increase at simple interest).

What was written in the previous paragraph explains why much of the book is about general epidemiology. Many chapters must pass before we come to discuss the control of disease.

3.2. The Intermittent Increase of Disease

Spores of *Phytophthora infestans* germinate only when the host plant is wet. Most pathogens have some requirement that stops infection from proceeding continuously, month in and month out, at all times of day and night. Infection proceeds intermittently.

Consider a fairly extreme example. Potato leaf roll virus spreads from plant to plant in summer. The virus enters the tubers; and when the same stock of potatoes is planted year after year without the removal of diseased plants there is a steady increase in the proportion of plants infected (except in rare vector-free areas). During a single season newly infected plants are seldom sources of virus for further spread to other plants (Broadbent *et al.*, 1960; Heathcote and Broadbent, 1961). The accumulation of disease from year to year is almost the same as that of money at discontinuous compound interest, with interest added only once a year. Broadbent *et al.* (1960) recorded an increase in the number of plants with leaf roll from 0.3 in 1956 to 2.1 in 1957 to 13.3% in 1958. Consider the sevenfold increase in 1 year, from 0.3 in 1956 to 2.1% in 1957.

$$1+6=e^{1.95}=7$$

One can consider the sevenfold increase to mean a rate of either 6 per unit per year at discontinuous compound interest added once a year or of 1.95 per unit per year for logarithmic increase. These two rates both state that 1 will increase to 7 in 1 year. Of the two, a rate of 6 per unit per year at discontinuous compound interest probably gives a better picture of the biological process of leaf roll infection. It gives a picture of each diseased plant in the field infecting an average of 6 other plants during the year. But we are not looking for a picture of the process of infection. We are looking for a consistent method of calculating rates of infection. The arithmetic of logarithmic increase, with modification where necessary, provides such a method.

One justifies the model of logarithmic increase on two grounds.

First, it was shown in Section 2.6 and Fig. 2.2 that, by appropriate change in the interest rate, one can fit a curve for continuous compound interest, i.e., for logarithmic increase, to the increase of money at discontinuous compound interest. This holds too for the increase of disease. The same condition applies as before: for accurate calculations one must use a whole number of years, if years are the appropriate units, or a whole number of days, if days are the appropriate units.

Second, for calculating logarithmic increase of money at a variable rate one may use the mean rate of interest over the whole period. This was pointed out by means of an example in Section 2.7. The matter will be taken up again in Chapter 7. It will be shown there that one may use the mean infection rate to calculate the increase of disease, and that this is true generally and not just for a logarithmic increase.

To illustrate this theorem we shall return to the example of leaf roll in potatoes. Disease increased sevenfold in 1 year. Suppose, by way of hypothesis, that the whole of this sevenfold increase took place on 1 single day in the year (i.e., suppose that the vectors were active on 1 day only). If we base our calculations on this 1 day, we have $t=1$ day, $x/x_0=7$. Hence on this 1 day, by Eq. (2.6), $r=1.95$ per unit per day. But this rate, if maintained, is $r=1.95\times 365$ per unit per year. (See Exercise 2 of Chapter 2.) On the other 364 days of the year, $r=0$ by hypothesis. Hence the average rate for the whole year, with each day given equal weight, is $r=1.95$ per unit per year. But this is what r would have been if the sevenfold increase had been brought about at a uniform rate throughout the year. For this, $t=1$ year; $x/x_0=7$, as before; and, from Eq. (2.6), $r=1.95$ per unit per year. If one's purpose is simply to calculate a logarithmic infection rate for the year, a sevenfold increase during the year means $r=1.95$ per unit per year, irrespective of whether the increase took place all on 1 day or was spread evenly over the whole year (assuming this to be biologically possible).

The theorem is also illustrated by Exercise 3, for logarithmic increase, and 4, for increase that is not logarithmic, at the end of this chapter.

3.3. The Percentage and the Proportion of Disease

We shall in future ordinarily use the proportion x of disease instead of the percentage. The percentage is $100x$. Thus, $x = 0.021$ if 2.1% of the potato plants in a field have leaf roll.

This is in keeping with stating the infection rate per unit instead of per cent.

3.4. Logarithmic Increase of Disease and Increase That Is Not Logarithmic

Money at interest can increase without limit. Disease can increase only until all susceptible tissue is infected, i.e., until $x = 1$.

At very low proportions of disease this distinction is unimportant. Almost all susceptible tissue is healthy and still available for infection. The pathogen can spread practically unhindered by lack of susceptible tissue which it can infect. We can write, as we did in Eq. (2.2),

$$\frac{dx}{dt} = rx$$

The rate of increase of disease dx/dt is proportional to x. That is, the increase is logarithmic. To make this clear always, the equation will in future be written

$$\frac{dx}{dt} = r_l x \tag{3.1}$$

The subscript l denotes logarithmic increase. It is used only for increase at very low proportions of disease. We shall interpret this as $x < 0.05$. This limit for logarithmic increase may seem somewhat high. But there is nothing to stop one from using a lower limit, say $x < 0.005$, if one believes the data are accurate enough to justify the refinement.

The increase in the proportion x of potatoes with leaf roll from 0.003 to 0.021 was treated in Section 3.2 as logarithmic because $x < 0.05$.

As x increases, the proportion, $1 - x$, of susceptible tissue still healthy and available for infection decreases. To allow for this we replace Eq. (3.1) by

$$\frac{dx}{dt} = rx(1 - x) \tag{3.2}$$

Except at very low values of x, when $1-x$ is so close to 1 that it can be ignored, the increase of disease is no longer logarithmic, and r is written without a subscript. We call r the apparent infection rate. Large (1945) used an equation such as Eq. (3.2) for potato blight caused by *Phytophthora infestans*. (Large's equation was for logistic increase, which is discussed in an appendix at the end of this book.)

Equation (3.2) is only a definition. It defines r as $dx/[x(1-x)\,dt]$. The interpretation of estimates of r obtained from experimental data is left to later chapters.

To estimate r_l and r one estimates the proportion of disease in the field on two dates. If x_1 and x_2 are the proportions on dates t_1 and t_2, Eq. (3.1) becomes

$$r_l = \frac{1}{t_2 - t_1} \log_e \frac{x_2}{x_1} \tag{3.3}$$

or

$$r_l = \frac{2.3}{t_2 - t_1} \log_{10} \frac{x_2}{x_1} \tag{3.4}$$

Similarly Eq. (3.2) becomes

$$r = \frac{1}{t_2 - t_1} \log_e \frac{x_2(1-x_1)}{x_1(1-x_2)}$$

or

$$r = \frac{2.3}{t_2 - t_1} \log_{10} \frac{x_2(1-x_1)}{x_1(1-x_2)}$$

More conveniently we use log $[x/(1-x)]$ and rewrite these equations as:

$$r = \frac{1}{t_2 - t_1} \left(\log_e \frac{x_2}{1-x_2} - \log_e \frac{x_1}{1-x_1} \right) \tag{3.5}$$

and

$$r = \frac{2.3}{t_2 - t_1} \left(\log_{10} \frac{x_2}{1-x_2} - \log_{10} \frac{x_1}{1-x_1} \right) \tag{3.6}$$

See Exercises 2.b and 2.c at the end of this chapter.

3.5. The Meaning of Infection Rates

If rain makes the pathogen multiply faster, rain increases the infection rate. A more susceptible host, a more aggressive pathogen, and weather more favorable to disease all increase the rate. Every factor that affects the rate of increase of disease affects the logarithmic and the apparent infection rates, irrespective of whether the factor is contributed by host, pathogen, environment, or fungicide. Every contribution, whatever its source, is pooled in the one single comprehensive figure that estimates the rate, and in the pool loses its identity.

There will be more about this later in Section 5.3.

3.6. The Words, "per Unit," Again

When we hear that the rate of interest on money is 10% per year, we do not ask, 10% of what? We take it for granted that the interest and the capital that earns it are both in the same currency: dollars or pounds, for example. When (as in Section 3.2) we calculate that a sevenfold increase per year in the number of potato plants infected with leaf roll means an infection rate of 1.95 per unit per year, we need not say what unit. It can be taken for granted that the new infections during the year and the old inoculum that produced them were both measured alike.

The words, per unit, are not strictly necessary. Instead of writing that the rate was 1.95 per unit per year, one could write that the rate was 1.95 per year or 1.95/year or 1.95 year^{-1}. Or one could write that the annual rate was 1.95.

If per unit is not strictly necessary, why use it? The answer is that it shows the rate to be relative. In this book we do not ordinarily estimate the absolute rate dx/dt, but, for example, the relative rate, dx/xdt, which is r_l. The words, "per unit, per cent, per thousand," all tag a rate as relative. For this reason we think per unit worth the space it takes. Others may not. One may choose as one will.

Examples in which the infection rate is absolute and *per unit* is omitted are given in Exercises 7 and 8 at the end of Chapter 4.

EXERCISES

1. Broadbent *et al.* (1960) sprayed potato fields with DDT to control the vectors of leaf roll virus. In the sprayed fields 0.2% of the plants had leaf roll in 1956, 0.7% in 1957, and 1.5% in 1958. Take 1956 and 1958 as t_1 and t_2, respectively. Then $t_2-t_1 = 2$ years, $x_1 = 0.002$ and $x_2 = 0.015$. The proportions are small enough for us to take the increase as logarithmic and apply Eq. (3.4):

$$r_l = \frac{2.3}{2} \log_{10} \frac{0.015}{0.002}$$

$$= 1.01 \text{ per unit per year}$$

This compares with $r_l = 1.95$ per unit per year for unsprayed fields, calculated earlier in this chapter.

2. Asai (1960) started an epidemic by inoculating a plot of wheat with spores of the stem rust fungus, *Puccinia graminis tritici*. On July 13 there were an average 64 pustules of rust per culm of wheat. On July 18 there were 350 pustules per culm. Thus $t_2 - t_1 = 18 - 13 = 5$ days. To change from pustules per culm to proportions, one notes that Asai's data show that infection rose to a maximum of 1000 pustules per culm when the epidemic was complete, i.e., when $x = 1$. So for 64 and 350 pustules per culm, we write $x_1 = 0.064$ and $x_2 = 0.35$.* Find r.

From Eq. (3.6)

$$r = \frac{2.3}{t_2 - t_1} \log_{10} \frac{x_2(1 - x_1)}{x_1(1 - x_2)}$$

$$= \frac{2.3}{5} \log_{10} \frac{0.35 \times 0.936}{0.064 \times 0.65}$$

$$= 0.41 \text{ per unit per day}$$

One can perform this calculation in several ways.

a. If a calculating machine is at hand, first find

$$\frac{0.35 \times 0.936}{0.064 \times 0.65} = 7.88$$

and

$$\log_{10} 7.88 = 0.90$$

Then,

$$r = \frac{2.3 \times 0.90}{5}$$

$$= 0.41 \text{ per unit per day}$$

b. In Eq. (3.6) instead of $\log_{10} \{[x_2(1 - x_1)]/[x_1(1 - x_2)]\}$ write

$$\log_{10} \frac{x_2}{1 - x_2} - \log_{10} \frac{x_1}{1 - x_1}$$

A table of $\log_{10} [x/(1 - x)]$ is given as Appendix Table 1 at the end of the book. From this table, if $x = 0.35$, $\log_{10} [x/(1 - x)] = \bar{1}.73$. That is, if $x_2 = 0.35$, $\log_{10} [x_2/(1 - x_2)] = \bar{1}.73$. Similarly, if $x_1 = 0.064$, $\log_{10} [x_1/(1 - x_1)] = \bar{2}.84$.

Hence

$$r = \frac{2.3}{5} (\bar{1}.73 - \bar{2}.84)$$

$$= \frac{2.3 \times 0.89}{5}$$

$$= 0.41 \text{ per unit per day}$$

* Independently Kingsolver *et al.* (1959) estimated that 1% rust (i.e., $x = 0.01$) on the modified Cobb scale was equivalent to 10 pustules per culm. This agrees with our interpretation of Asai's figures, and is good evidence for the accuracy of the modified Cobb scale.

c. Use natural, instead of common, logarithms. Read off values of $\log_e [x/(1-x)]$ from Table 2 of the Appendix and use them in Eq. (3.5).

$$r = \frac{1}{t_2 - t_1}\left(\log_e \frac{x_2}{1-x_2} - \log_e \frac{x_1}{1-x_1}\right)$$
$$= \tfrac{1}{5}(-0.62+2.68)$$
$$= 0.41 \text{ per unit per day}$$

3. The proportion x of potato plants infected with virus X was 0.0092 in 1951, 0.0239 in 1952, and 0.0484 in 1953 (Cockerham, 1958). Find r_l between (*a*) 1951 and 1952; (*b*) 1952 and 1953; and (*c*) 1951 and 1953. [Answers: (*a*) 0.95; (*b*) 0.71; and (*c*) 0.83 per unit per year. Note that (*c*) is the average of (*a*) and (*b*).]

4. Wolf (1935) records the following proportions x of plants infected with tobacco mosaic in a field in 1929: 0.015 on June 5; 0.052 on June 17; 0.155 on June 29; and 0.410 on July 9. Find r between (*a*) June 5 and 17; (*b*) June 17 and 29; (*c*) June 29 and July 9; and (*d*) June 5 and July 9. [Answers: (*a*) 0.107; (*b*) 0.101; (*c*) 0.133; and (*d*) 0.112 per unit per day. The figure 0.112 for (*d*) is slightly less than the average, 0.114, for (*a*), (*b*), and (*c*). The reason is that $t_2 - t_1$ varies. It is 12, 12, and 10 days for (*a*), (*b*), and (*c*), respectively. If one weights the figures in proportion to $t_2 - t_1$, the average becomes 0.112, which is correct.]

5. The proportion x of red oak trees with wilt caused by *Ceratocystis fagacearum* rose from 0.055 to 0.37 in 3 years (Boyce, 1957). Find r. (Answer: 0.77 per unit per year)

6. In 7 days the proportion x of unsprayed potato foliage destroyed by *Phytophthora infestans* increased from 0.18 to 0.83 (Hooker, 1956). Hooker used the system of Horsfall and Barratt (1945) for measuring plant disease. Find r. (Answer: 0.44 per unit per day)

7. The proportion x of trees infected with mosaic in a peach orchard was 0.00843 in 1931, 0.894 in 1934, and 1.0 in 1935 (Hutchins *et al.*, 1937). Find r between (*a*) 1931 and 1934; (*b*) 1934 and 1935; and (*c*) 1931 and 1935. [Answers: (*a*) $r = 2.3$ per unit per year; (*b*) and (*c*), no answer is possible when $x = 1$ or 0.]

8. Quite often one finds that r stays nearly constant over the whole observed course of an epidemic. This is another way of saying that the curve for the progress of disease with time is often nearly S-shaped. The reason is obscure. But examples are worth examining, both because S-shaped curves for the progress of disease appear largely in the literature and because the examples illustrate the use of regression coefficients to estimate r or r_l.

Large (1945) observed several epidemics of potato blight caused by *P. infestans* and recorded the proportion x of diseased tissue at different dates. The epidemics were similar. The epidemic at Dartington, Devon, in 1942 is discussed here. On August 11, x was 0.001; on August 18, 0.05; on August 24, 0.43; on August 28, 0.75; and on September 4, 0.98.

In Fig. 3.1 $\log_e [x/(1-x)]$ is plotted against time, and a straight regression line fitted to the points by the standard procedure described in books on statistical methods. The regression coefficient is estimated to be 0.46, with an estimated standard error, based on only three degrees of freedom, of 0.021.

Rewriting Eq. (3.5) in the general form

$$\log_e \frac{x}{1-x} = a + rt$$

shows that the regression coefficient of $\log_e [x/(1-x)]$ on t is r. That is, in this spidemic r had an estimated value of 0.46 per unit per day, with an estimated etandard error of 0.021. This estimate of r gives equal weight to all the observations.

The broken lines in Fig. 3.1 illustrate how the regression line measures r. In the 10 days between August 21 and 31, $\log_e [x/(1-x)]$ increases from -1.96 to 2.64, an increase of 4.6. The increase in 1 day is, therefore, 0.46, which is the regression coefficient of $\log_e [x/(1-x)]$ on time, and also r.

The same data of Large appear again, in slightly modified form, in Fig. 4.1, and yet again, in greatly modified form, in Sections 5.3 and 8.3.

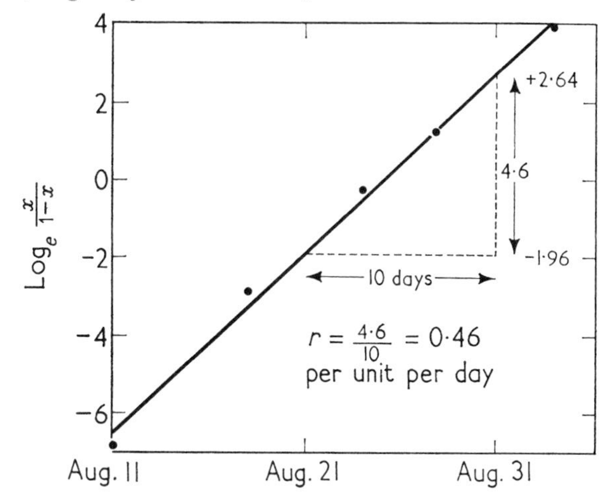

Fig. 3.1. The progress of an epidemic of potato blight caused by *Phytophthora infestans*. The data are those of Large (1945). $\text{Log}_e [x/(1-x)]$ is plotted against time, and the slope of the regression line fitted to the points estimates r.

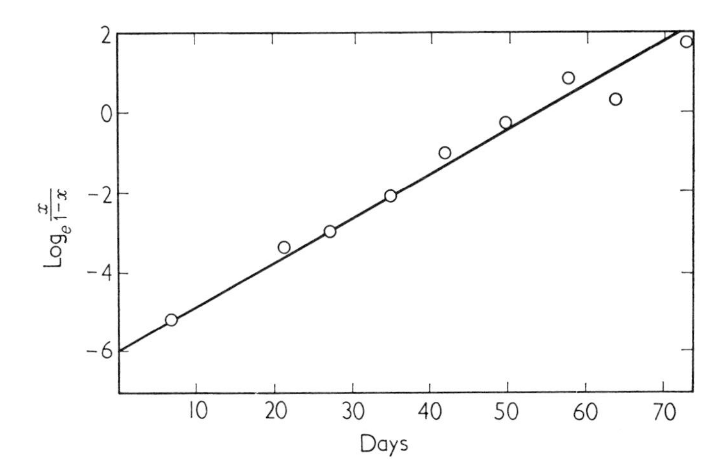

Fig. 3.2. The progress of an epidemic of stripe rust of wheat caused by *Puccinia striiformis*. The data are those of Zadoks (1961).

9. Zadoks (1961) recorded the progress of epidemics of stripe rust of wheat caused by *Puccinia striiformis*. In a detailed experiment in 1960 he measured with a planimeter the area of leaves and the area of lesions on them, and hence

the proportion x of diseased tissue. On April 20, x was 0.0015; on April 27, 0.00 55; on May 11, 0.034; on May 17, 0.050; on May 25, 0.11; on June 1, 0.28; on June 9, 0.43; on June 17, 0.70; on June 23, 0.60; and on July 2, 0.85. In Fig. 3.2, $\log_e [x/(1-x)]$ is plotted against time, and r, the regression coefficient, estimated to be 0.1096 per unit per day, with an estimated standard error of 0.0056. This estimate is based on eight degrees of freedom.

10. Estimates of r obtained from the observations used in the previous examples must be interpreted cautiously, for reasons that will appear in Sections 5.5 and 5.6. Much more important to our narrative are examples in which disease is sampled on two dates, and r estimated between these dates as a regression coefficient without any implication that r stays constant from day to day.

Suppose that one enters a field on date t_1 and makes n_1 independent readings of the proportion x of disease in it. On date t_2 one makes a further n_2 readings. If one uses subscripts 1 and 2 to indicate whether one is referring to dates t_1 or t_2, and writes y for $\log_e [x/(1-x)]$, then we estimate

$$\bar{y}_1 = \frac{1}{n_1} \Sigma(y_1)$$

$$\bar{y}_2 = \frac{1}{n_2} \Sigma(y_2)$$

$$r = \frac{\bar{y}_2 - \bar{y}_1}{t_2 - t_1}$$

The standard error s of this estimate of r is estimated from

$$s^2 = \frac{(n_1+n_2)[\Sigma(y_1-\bar{y}_1)^2 + \Sigma(y_2-\bar{y}_2)^2]}{n_1 n_2 (n_1+n_2-2)\,(t_2-t_1)^2}$$

$$= \frac{(n_1+n_2)[\Sigma(y_1)^2 + \Sigma(y_2)^2 - n_1\bar{y}_1{}^2 - n_2\bar{y}_2{}^2]}{n_1 n_2 (n_1+n_2-2)\,(t_2-t_1)^2}$$

The number of degrees of freedom in this estimate is n_1+n_2-2.

The same method is used to estimate r_l, but with $\log_e x$ instead of $\log_e [x/(1-x)]$.

One may use logarithms to the base 10. Estimates of the regression coefficient and its standard error must then be multiplied by 2.30 to estimate r and its error.

Consider this example. At the first examination, on date t_1, six independent estimates of the proportion x of disease in a field were 0.0095, 0.0085, 0.0092, 0.0094‘ 0.0107, and 0.0078. Ten days later, on date t_2, seven more estimates of x were made They were 0.44, 0.48, 0.28, 0.42, 0.45, 0.63, and 0.40. Hence, $n_1=6$, $n_2=7$, and $t_2-t_1=10$ days. Estimate r and its standard error s.

The data are set out in Table 3.1, from which we get

$$\bar{y}_1 = \frac{-28.116}{6}$$

$$= -4.686$$

$$\bar{y}_2 = \frac{-1.663}{7}$$

$$= -0.2374$$

$$r = \frac{-0.237+4.686}{10}$$

$$= 0.445 \text{ per unit per day}$$

$$s^2 = \frac{(6+7)(131.8101+1.5477-131.7516-0.3946)}{6\times 7(6+7-2)\,10^2}$$

$$= 3.413\times 10^{-4}$$

$$= (0.018)^2$$

TABLE 3.1[a]

DATA FOR ESTIMATING r AND s

Date	x	y	y^2
t_1	0.0095	−4.647	21.5946
t_1	0.0085	−4.759	22.6481
t_1	0.0092	−4.679	21.8930
t_1	0.0094	−4.658	21.6970
t_1	0.0107	−4.527	20.4937
t_1	0.0078	−4.846	23.4837
		−28.116	131.8101
t_2	0.44	−0.241	0.0581
t_2	0.48	−0.080	0.0064
t_2	0.28	−0.944	0.8911
t_2	0.42	−0.323	0.1043
t_2	0.45	−0.201	0.0408
t_2	0.63	+0.532	0.2830
t_2	0.40	−0.405	0.1640
		−1.662	1.5477

[a] See text for details.

CHAPTER 4

How to Plot the Progress of an Epidemic

SUMMARY

When plants infected early in the epidemic are sources of inoculum for later infections, one plots log $[x/(1-x)]$ against time, where x is the proportion of disease. When plants infected early do not contribute to the later infections, or when the source of inoculum is constant, one plots log $[1/(1-x)]$ against time. In both expressions $1-x$ is a correction factor to allow for a decreasing proportion of healthy tissue left for infection. The factor is safe to use, but not wholly adequate.

4.1. The Increase of Disease with Time

For our purpose an epidemic is an increase of disease with time. This definition is admittedly incomplete; epidemics decline as well as rise. But this book is concerned primarily with the control of disease, and control is usually directed at preventing the increase of disease. We can, therefore, afford to take an incomplete view of epidemics.*

The time factor is at the core of our discussions. In the graphs we plot disease against time. The progress of disease with time is probably the first matter to describe in an epidemic.

How should the amount of disease be recorded? Tables give a full record, though the data are not always easy to interpret. Of graphical methods, the simplest is to plot the amount of disease against time on an arithmetical scale. This is done in the top half of Fig. 4.1. Disease

* Many of the diseases against which control measures are directed are always with us. They are, strictly speaking, endemic. Some flare up seasonally or periodically, and these flare-ups are known as recurrent epidemics. Others flare up locally as local epidemics.

progress curves obtained in this way are commonly S-shaped. Purely for the purpose of recording observations the method is adequate only provided that the highest amount of disease is not more than about 50 times as great as the lowest amount.

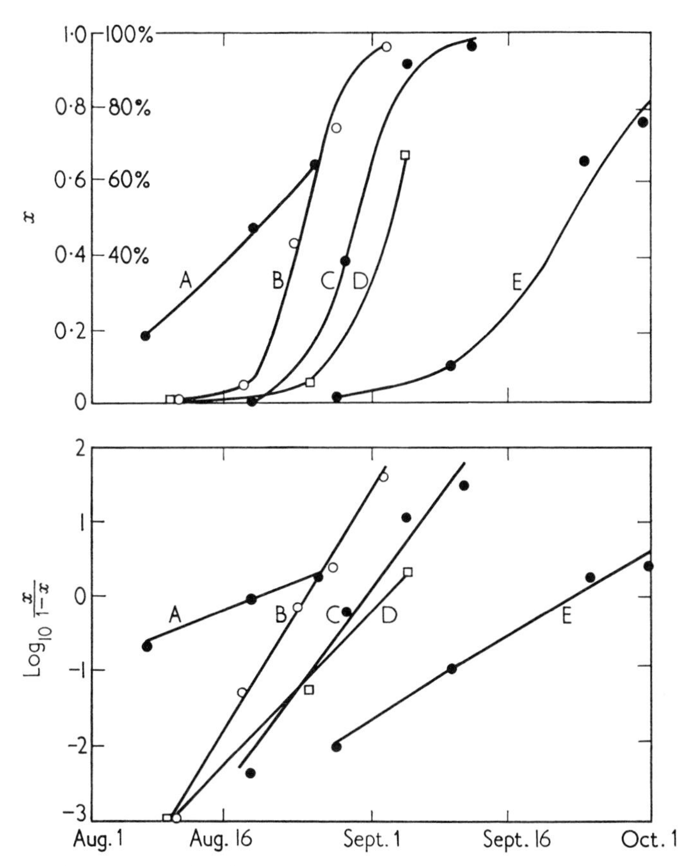

FIG. 4.1. Progress of blight on potatoes caused by *Phytophthora infestans*. The top half shows the increase of x, and the bottom half the increase of $\log_{10}[x/(1-x)]$, with time. Data of Large (1945). A, variety Majestic, Dartington, 1943; B, variety Majestic, Dartington, 1942; C, variety Majestic, Kentisbeare, 1942; D, variety Majestic, Durnsford, 1944; E, variety Arran Consul, Dartington, 1941.

To record a wider range of data graphically, writers have used probits and probability paper, and logarithms, logits (Zadoks, 1961), and logarithmic paper of various sorts.

We plot log $[x/(1-x)]$ or log $[1/(1-x)]$ against time, according to the circumstances.

4.2. Two Ways in Which Disease Can Increase with Time. "Compound Interest Disease" and "Simple Interest Disease"

One can distinguish between two ways in which disease increases. The pathogen may multiply through successive generations in the course of the epidemic. Infection in a wheat field, e.g., may start with only a few pustules of stem rust. These erupt and release uredospores, which in turn form more pustules. So the process can continue until the whole field becomes infected. This process of parent lesions producing daughter lesions, usually in the same crop, is multiplication in a form similar to compound interest. It can occur equally with systemic diseases. If the virus in a cabbage plant with mosaic spreads in successive generations to other cabbage plants and causes mosaic in them, that too is multiplication.

With multiplication of this sort one plots log $[x/(1-x)]$ against time. This follows Eqs. (3.5) and (3.6) in the previous chapter. (See also part *c* of Exercise 2 at the end of Chapter 3.) It implies modified logarithmic increase: it is logarithmic because the increase follows a compound interest pattern, but it is modified to allow for a decreasing proportion, $1-x$, of tissue left for infection.

We shall often use the term, "compound interest disease," when the pathogen multiplies through successive generations in the course of an epidemic.

Note that the multiplication of disease to which we have been referring implies that the pathogen moves from lesion to lesion, or from plant to plant if the disease is systemic. The multiplication of the pathogen within a lesion or the multiplication of a systemic pathogen within a plant is not relevant here.

There is also increase of disease without multiplication in the sense defined by the first sentence of the previous paragraph. For example, the number of wilted plants of cotton growing in soil containing *Fusarium oxysporum* f. *vasinfectum* may progressively increase. During a single season the inoculum present in the soil at the beginning of the season remains the main source of inoculum. The increase in the number of wilted plants as the season advances is not caused primarily by the fungus spreading from one plant to another. Plants that wilt later in the season are likely to be infected from the same source as those that wilt earlier: the soil.

Increase of this sort corresponds roughly to the increase of money at simple interest. Here one plots log $[1/(1-x)]$ against time. This implies that the progress of disease with time is not logarithmic, and allows for a diminishing proportion, $1-x$, of tissue available for infection. (See Section 4.6.)

One also uses $\log [1/(1-x)]$ even with diseases such as rust of wheat, when the source of inoculum is constant for a time. This will be discussed in Section 5.4.

4.3. Increase of Disease by Multiplication. "Compound Interest Disease"

Figures 4.1 and 4.2 show the progress with time of natural epidemics. Figures 4.3 and 4.4 are for epidemics started artificially.

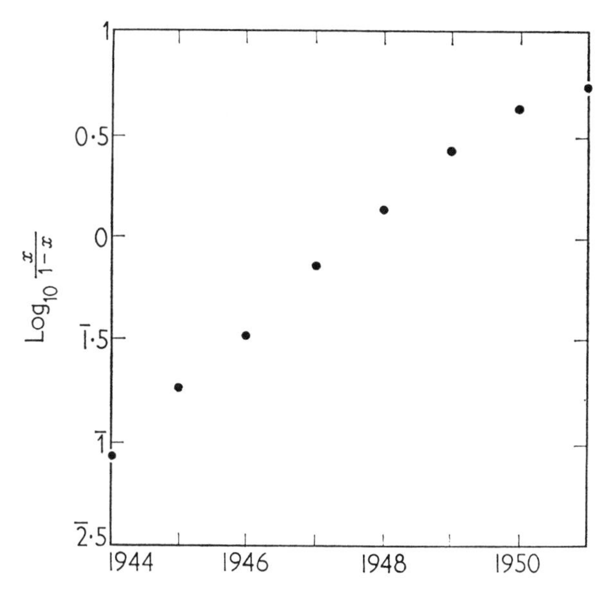

FIG. 4.2. Progress of swollen shoot of cacao. Data of Dale, quoted by Thresh (1958b).

The five sets of data in Fig. 4.1 are from tables given by Large (1945) for the progress of five epidemics of potato blight caused by *Phytophthora infestans*. Disease was estimated by the British Mycological Society's scale for blight (Anonymous, 1947). The top part of Fig. 4.1 plots x and the lower part, $\log_{10} [x/(1-x)]$ at various dates. Straight regression lines have been fitted to the points in the lower part. The fit is reasonably good, which means that the apparent infection rate r stayed reasonably constant over the period of observations. It is, however, not necessary that this should be so for epidemics such as those of potato blight. In

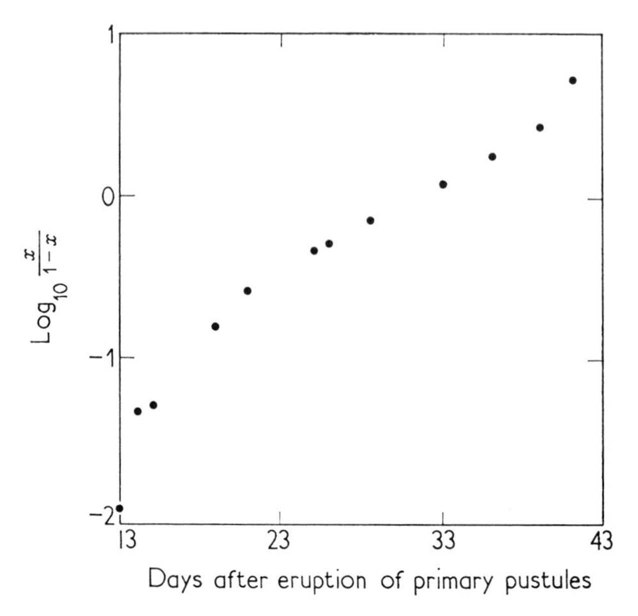

FIG. 4.3. Progress of stem rust in wheat caused by *Puccinia graminis tritici*. Data of Underwood *et al.* (1959).

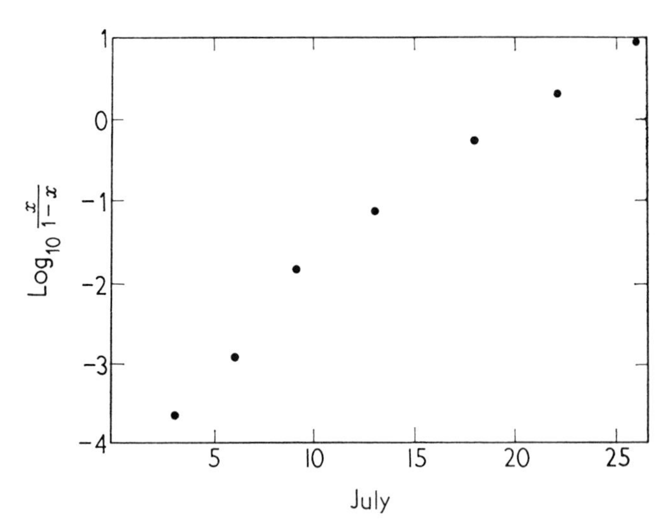

FIG. 4.4. Progress of stem rust in wheat caused by *Puccinia graminis tritici*. Data of Asai (1960).

Figs. 4.2, 4.3, and 4.4 regression lines are not fitted; and it is no part of the argument for the use of log $[x/(1-x)]$ to suggest that regression lines are usually straight.

Figure 4.2 shows the progress of swollen shoot disease of cacao in Trinidad for 8 years. The data are those of Dale (1953), quoted by Thresh (1958b). The disease is systemic, and the proportion x could be obtained by direct counting of infected plants. Figure 4.3 is for an artificially induced epidemic of wheat stem rust caused by *Puccinia graminis tritici*. The data are those of Underwood *et al.* (1959), who estimated disease by means of the modified Cobb scale. Figure 4.4 is also for an artificial epidemic of stem rust in wheat. The data are Asai's (1960), discussed in Exercise 2 at the end of Chapter 3.

These examples are fairly representative. (Other examples are given at the end of this chapter.) Data can be recorded conveniently over a wide range from very low proportions of disease to very high; in Fig. 4.4 the highest proportion of disease x is about 4000 times the lowest. Variations in the apparent infection rate r can be seen at a glance.

4.4. The Increase of Disease without Multiplication. "Simple Interest Disease"

Figure 4.5 records results of Ware and Young (1934) with cotton wilt caused by *Fusarium oxysporum* f. *vasinfectum*. The soil was inoculated with cultures of this fungus on wheat bran several days before planting. Counts were made of wilted plants at intervals during the season. No wilted plants recovered. The data reproduced in Fig. 4.5 are for the check variety, Trice 304, in 1933, the year of most infection.

The top part of Fig. 4.5 records the wilted plants as the proportion x of total plants of the variety. The lower part records $\log_e [1/(1-x)]$. The two parts of Fig. 4.5 are fairly similar, but the line in the lower part is a little steeper.

This way of recording results assumes that there was no multiplication. That is, it assumes that the source of infection of the plants that wilted late in the season was still the inoculum that was cultured on wheat bran or that already existed in the soil before the bran culture was added. It assumes that the plants that wilted in July did not infect the plants that wilted in August.

We use $\log_e [1/(1-x)]$ instead of $\log_{10} [1/(1-x)]$ for two reasons.

First, many pathologists are familiar with the multiple-infection transformation brought to their attention by Gregory (1945, 1948). The derivation of this transformation was statistical (see Chapter 9) and different from the reasoning given in Section 4.6 for $\log_e [1/(1-x)]$. Nevertheless the transformed numbers tabulated by Gregory are simply $100 \log_e [1/(1-x)]$ and are used in much the same way as we have used $\log_e [1/(1-x)]$.

Second, x and $\log_e [1/(1-x)]$ are alike when x is small. For example, if $x = 0.0100$, $\log_e [1/(1-x)] = 0.0100$.

Figure 4.6 brings out the difference between x and $\log_e [1/(1-x)]$. The source is again the data of Ware and Young (1934). The lowest broken line shows the average proportion x of wilted plants for the most

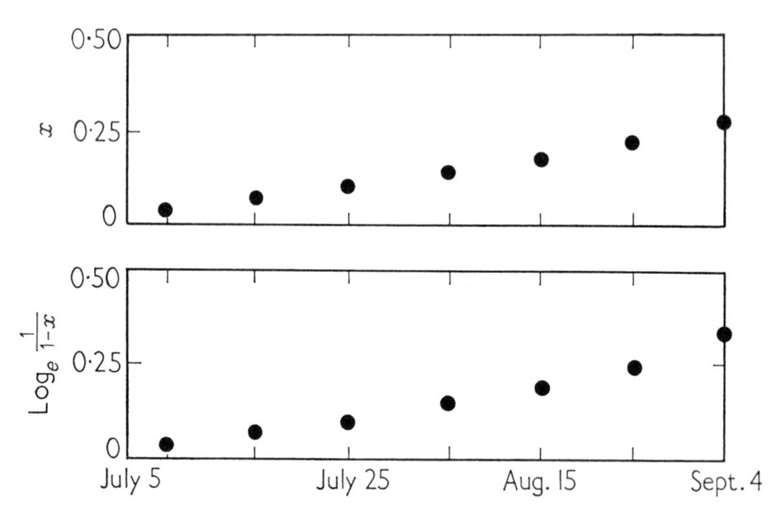

FIG. 4.5. The progress of an epidemic of wilt of cotton caused by *Fusarium oxysporum* f. *vasinfectum*. The upper part plots x against time; the lower part plots $\log_e [1/(1-x)]$ against time. From the data of Ware and Young (1934).

resistant varieties. The middle line is the average for varieties of medium susceptibility, and the top line for very susceptible varieties. The solid lines plot $\log_e [1/(1-x)]$ instead of x. As x increases, the broken and solid lines diverge.

4.5. The Correction Factor $(1-x)$

New infections occur only in tissue not previously infected with the same disease. This is what $1-x$ in $\log [x/(1-x)]$ or $\log [1/(1-x)]$ means.

In Fig. 4.6 compare the bottom pair of curves, for the least susceptible varieties, with the top pair, for the most susceptible varieties. The least susceptible varieties were infected faster after August 5 than before. That is, the curves turn upward on August 5. But the most susceptible varieties were infected more slowly after August 5. The curves turn downward. The change is greater with x than with $\log_e [1/(1-x)]$. Nevertheless, even when judged by $\log_e [1/(1-x)]$, the infection rate of the most susceptible varieties fell after August 5, whereas with the least susceptible varieties it rose; and this change was statistically significant.

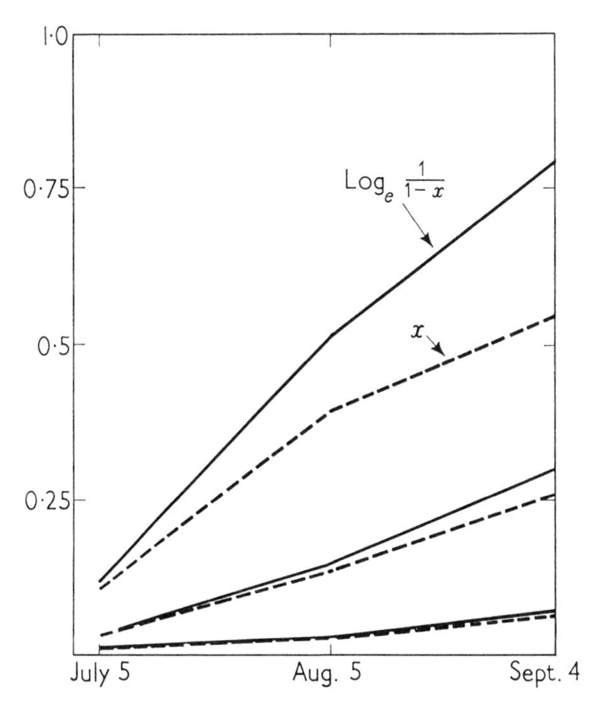

FIG. 4.6. The progress of an epidemic of wilt of cotton caused by *Fusarium oxysporum* f. *vasinfectum*. The changes in infection before and after August 5 are shown as changes in x by broken lines and in $\log_e [1/(1-x)]$ by solid lines. The top, middle, and bottom pairs of lines are for the most susceptible, moderately susceptible, and least susceptible varieties, respectively. From the data of Ware and Young (1934).

This difference between varieties may be real, e.g., varieties may differ in the way plants change their susceptibility as they grow older. But the difference may be only apparent; and the drop in the infection rate of the most susceptible varieties after August 5 may possibly have resulted from undercorrection by the factor $1-x$. This possibility must be examined.

The use of $1-x$ as a correction factor implies that conditions for infection were uniform. For the correction factor to be exact, the uniformity must be absolute. All plants must be equally susceptible and equally exposed to infection. Thus, in the experiments with fusarium wilt of cotton, it is implied that each variety was entirely homozygous with respect to all factors that could affect susceptibility; that the bran cultures of the fungus used for inoculum were uniform and spread uniformly through the whole experimental area; that nematodes were uniformly spread in the soil; and that the soil was chemically and physically uniform. (In the experiments of Ware and Young with which we are dealing, potassium in fertilizers greatly reduced wilting.) All departures from uniformity act in one direction. They all cause the factor $1-x$ to be inadequate and undercorrect for a decreasing proportion of plants left to be infected. Undercorrection explains at least in part why the top curve in Fig. 4.6 bends down on August 5 instead of up.

Use the correction factor—it is better than nothing and always safe because it cannot overcorrect. But be cautious about interpreting results when the proportion of disease is high. We shall say more about this in Chapter 6 when we discuss the interpretation of apparent infection rates.

4.6. The Reason for Using Log $[1/(1-x)]$ When There Is No Multiplication

In the example of fusarium wilt of cotton the basic assumption was that the fungus in the soil, or added in bran cultures to the soil, was the only source of inoculum. One plant did not infect another in the same season. On this assumption the proportion of plants dx that wilt during a short interval of time dt can be expected to be proportional to the amount of inoculum Q and, in uniform conditions, to the proportion $1-x$ of plants still left for infection. Then

$$\frac{dx}{dt} = QR(1-x) \tag{4.1}$$

where R is the infection rate.

This can be changed to

$$\log_e \frac{1}{1-x} = QRt + \text{a constant} \tag{4.2}$$

In Figs. 4.5 and 4.6, $\log_e[1/(1-x)]$ was plotted against time in conformity with this equation. The slope of the curves depends on QR.

With more inoculum Q or a higher infection rate R the curves would rise more steeply.

It is not implied in these equations that Q is constant. Some inoculum can die out during the season or, alternatively, inoculum can increase saprophytically. In the experiments with cotton wilt it is not possible to distinguish between Q and R separately. For example, an increased tempo of wilting could result from either a saprophytic increase of Q or a higher infection rate R. In turn, an increase of R could result from increased susceptibility of the variety of cotton, shortage of potassium in the soil, etc.

In the next chapter (in Section 5.4, which deals with wheat stem rust) we discuss experiments in which it is possible to distinguish between the amount of inoculum and the infection rate.

EXERCISES

1. Bennett and Costa (1949) give figures for the increase of orange trees infected with the tristeza virus. The proportion x was 0.118 on November 28, 1942; 0.195 on October 8, 1943; 0.551 on January 4, 1945; and 0.725 on January 10, 1946. Plot $\log_{10} [x(/1-x)]$ against time. (It would have been better to have had the intervals between examinations in whole numbers of years.)

2. Broadbent (1957) records the increase of cauliflower mosaic. The proportion x of plants infected with mosaic was 0.08 on July 6; 0.142 on August 17; 0.240 on September 7; 0.367 on October 6; and 0.458 on November 6. Plot $\log_{10} [x/(1-x)]$ against time.

3. Beaumont (1954) records the progress of leaf mould caused by *Cladosporium fulvum* on tomato plants in glasshouses. He used a key to assess disease on the foliage. The proportion x of infection on unsprayed plants in one of his trials was 0.005 on July 7; 0.018 on August 4; 0.084 on August 18; 0.190 on September 1; 0.30 on September 21; and 0.57 on October 3. Plot $\log_{10} [x/(1-x)]$ against time. Another set of Beaumont's data are analyzed at the end of the next chapter.

4. Young (1938) found the following proportions x of cotton plants with fusarium wilt in a fertilizer trial. Without fertilizer x was 0.048 on July 5; 0.106 on July 20; 0.135 on August 5; 0.200 on August 20; and 0.343 on September 5. With a 0–0–12 potash fertilizer the corresponding proportions were 0.0095; 0.035; 0.054; 0.063; and 0.077, respectively. Plot $\log_e [1/(1-x)]$ against time, to determine how potash affected the progress of the epidemic.

5. Hewitt *et al.* (1949) found the following proportions x of vines systemically infected with Pierce's disease in a vineyard of Emperor grapes: 0.0015 in 1937, 0.0087 in 1938, 0.022 in 1939, 0.105 in 1940, and 0.25 in 1941. Plot $\log_e [1/(1-x)]$ against time.

One uses $\log [1/(1-x)]$ here instead of $\log [x/(1-x)]$ because Hewitt *et al.* (1949) found that grapevines are attacked mostly by virus from infected alfalfa or other plants. The spread of virus from grapevine to grapevine is relatively unimportant.

6. The proportion x of tomato plants of the variety Burwood Prize systemically infected with spotted wilt (Bald, 1937) was 0.108 on November 6, 0.261 on November 14, 0.481 on November 21, 0.814 on November 28, 0.886 on December 5, and 0.931 on December 12. Plot $\log_e [1/(1-x)]$ against time.

One uses $\log [1/(1-x)]$ instead of $\log [x/(1-x)]$, because the evidence showed that the tomato spotted wilt virus was entering fields from without. Spread of virus within the field from tomato plant to tomato plant could not be detected.

Using $\log [1/(1-x)]$ in this and the previous example does not imply that the source of inoculum remained constant. All it means is that virus did not spread in significant amounts from grapevine to grapevine or from plant to plant in the tomato field. In terms of interest on money, we are dealing here with something similar to simple interest, in the sense that interest does not earn interest. The next two examples will help to explain why $\log [1/(1-x)]$ is used.

7. In his work on tomato spotted wilt Bald (1937) calculated an infection rate by subtracting the natural logarithm (i.e., the logarithm to the base e) of the number of plants remaining healthy at the end of each interval between examinations from the natural logarithm of the number at the beginning of the interval, and dividing by the number of days in the interval. Show that this daily rate is QR in Eq. (4.1) and (4.2).

Consider the last two dates in Exercise 6. On December 5 out of a total of 360 plants in a block, 319 were infected and 41 healthy. On December 12, in the same block, 335 were infected and 25 healthy. The infection rate calculated by Bald was

$$\frac{\log_e 41 - \log_e 25}{12-5} = \frac{2.303\ (\log_{10} 41 - \log_{10} 25)}{7}$$

$$= 0.071 \text{ per day}$$

One can rewrite Eq. (4.2) in the form

$$QR = \frac{1}{t_2-t_1}\left(\log_e \frac{1}{1-x_2} - \log_e \frac{1}{1-x_1}\right)$$

Here t_1 and t_2 are December 5 and 12, respectively; $x_1 = 0.886$, $x_2 = 0.931$ (these values of x being given in Exercise 6); $\log_e [1/(1-x_1)] = 2.17$, $\log_e [1/(1-x_2)] = 2.67$ (from tables for $\log_e [1/(1-x)]$ given in the Appendix). Whence $QR = \frac{1}{7}(2.67-2.17) = 0.071$ per day. This is also the rate calculated by Bald.

The following are the steps:

$$\tfrac{1}{7}\ (\log_e 41 - \log_e 25) = \tfrac{1}{7} \log_e \frac{41}{25}$$

$$= \tfrac{1}{7} \log_e \frac{41/360}{25/360}$$

$$= \tfrac{1}{7} \log_e \frac{1-x_1}{1-x_2}$$

$$= \tfrac{1}{7}\left(\log_e \frac{1}{1-x_2} - \log_e \frac{1}{1-x_1}\right)$$

We choose to write $\log [1/(1-x)]$ instead of $-\log (1-x)$ simply to bring out the contrast with $\log [x/(1-x)]$.

8. In the previous exercise we wrote the infection rate as 0.071 per day, omitting *per unit*. Why?

The infection rate QR is an absolute, not a relative, rate. It is not related to any unit of inoculum.

The factor $1-x$ should be regarded simply as a correction factor. Consider, e.g., the very early stages of an epidemic when x is very small and $1-x$ is practically equal to 1. Equation (4.1) then becomes

$$\frac{dx}{dt} = QR$$

And dx/dt is an absolute rate: it is the increase of x in the interval dt.

Epilog

We can now return to the topic of this chapter: how to plot the progress of an epidemic. In any one graph the slope of the curve should not vary greatly. If it does, points become crowded in one direction or the other. This is equally true for illustrations such as Figs. 4.2, 4.3, and 4.4 in which a curve is not actually fitted to the points. For potato blight, stem rust of wheat, and swollen shoot of cacao, we plotted $\log [x/(1-x)]$ against time, and the slope of the curve measured a relative rate of increase. See Section 3.6 and Exercises 8, 9, and 10 at the end of Chapter 3. Epidemics of these diseases generate their own inoculum as the pathogen spreads within a crop. (*Autocatalytic* is an adjective often used for this sort of increase of disease. We avoid it, for reasons given in the Appendix.) For them the slopes based on relative rates of increase are apt. For diseases such as fusarium wilt of cotton, Pierce's disease of grapevines, and spotted wilt of tomatoes, we plotted $\log [1/(1-x)]$ against time, and the slope of the curve measured an absolute rate of increase. The epidemics of these diseases which we discussed did not generate their own inoculum during the period of recorded observations. (One does not wish to generalize about all epidemics of these diseases.) For them slopes based on absolute rates of increase are apt.

The correction factor $1-x$ is used for all diseases. It stops points from being crowded together in the direction of the y-axis as an epidemic nears the end of its course.

CHAPTER 5

The Basic Infection Rate

SUMMARY

Those who wish now to read about control methods may proceed directly to Chapters 10 to 19, and leave Chapters 5 to 9 for reading later.

This chapter deals with some of the complications introduced by the latent period, p, which is the time newly infected tissue takes to become infectious, and the incubation period, which is the time needed for infection to become visible. The basic infection rate R is estimated per unit of tissue that has already passed the latent period, whereas the apparent rate r is estimated per unit of all infected tissue, including that still in the latent stage. Every factor (such as the susceptibility of the host) that affects the rate of multiplication of disease affects r. This chapter is mostly about the relation of p and R with r or r_l; the part played by removals (the topic of Chapter 8) is ignored here. The models are mainly for continuous infection, but discontinuous infection is also considered, primarily to show that continuous-infection models are probably adequate for our analyses.

One can estimate R directly during the "simple interest" phase of an artificial epidemic. One can also calculate R indirectly from estimates of r or r_l. Values of R obtained in these two ways, from data on wheat stem rust, tally reasonably well. R has a key place in the theory of disease control measures because many control measures act directly on R. For example, if a protectant fungicide halves the proportion of spores that can germinate and start lesions, the fungicide halves R. The extent to which a control measure reduces r or r_l must then be calculated indirectly from the extent to which it reduces R.

5.1. The Basic Infection Rate and the Latent Period

Newly infected tissue takes a period p to become infectious. One should therefore write

$$\frac{dx_t}{dt} = Rx_{t-p}(1-x_t) \tag{5.1}$$

Here x_t and x_{t-p} are x at times t and $t-p$, respectively. At time t, x_t is the total proportion of infected tissue, infectious or not; and x_{t-p} is the proportion infected a period p earlier, i.e., x_{t-p} is the proportion of tissue that has reached or passed the stage of being infectious.

The period p is the latent period. It can be defined by an example. In a summer rust epidemic in wheat p is the time from when uredospores fall on the foliage to when the pustules that form from these spores erupt and release new uredospores. It is the time needed for a generation of the pathogen.

It is unfortunate that latency often has a different meaning in plant pathology. But *latent period* in medical epidemiology has the meaning given here (see Bailey, 1957), and conformity is desirable.

The rate R is the basic infection rate. If one rewrites Eq. (3.2) with subscripts, one gets

$$\frac{dx_t}{dt} = rx_t(1-x_t) \qquad (5.2)$$

A comparison of this with Eq. (5.1) shows that r, the apparent infection rate, is based on x_t, the proportion of infected tissue, whereas R, the basic rate, is based on x_{t-p}, the proportion of tissue that has been infected long enough to be infectious or to have been infectious.*

5.2. The Incubation Period

This is the time needed for symptoms to develop. To use the example of rust in wheat again, it is the time from when spores fall on the foliage to when symptoms can first be seen.

In medical epidemiology the incubation period is important because the start of symptoms is often the signal for isolation or quarantine—Tommy is kept out of school when he is seen to have mumps. In plant pathology the incubation period is important because, apart from other reasons, estimates of the amount of disease are usually based on visible disease. With human diseases the incubation period is usually longer than the latent period. With plant diseases it is usually shorter.

Figure 5.1 illustrates the subject by means of a simple model. It plots an artificial epidemic calculated for $R=1$ per unit per day and $p=6$ days. (The method of calculation is given in the next chapter.) The initial infection, assumed to be started by inoculation with spores, is x_0, taken to be very small so that Fig. 5.1 is all in the logarithmic stage. Infection is regarded as taking place immediately, when $t=0$. (Actually, the infection period, which is part of the latent period, is variable. Variable latent periods are discussed in the next chapter; in this chapter

* In an earlier discussion (van der Plank, 1960) rates were given per cent instead of per unit, and r and R used here correspond to $r/100$ and $r'/100$ used there.

the aim is simplicity, and constant latent periods are assumed.) The incubation period is $a = 4$ days.

The solid line is for all disease, visible and invisible. For the first $p = 6$ days there is no increase, because none of the lesions have formed spores; x remains at the initial level x_0. After $p = 6$ days, the lesions produce spores and x increases. The new lesions formed after the sixth day do not themselves immediately form spores. All spores come from the original lesions, i.e., from the proportion x_0 of the tissue. The increase

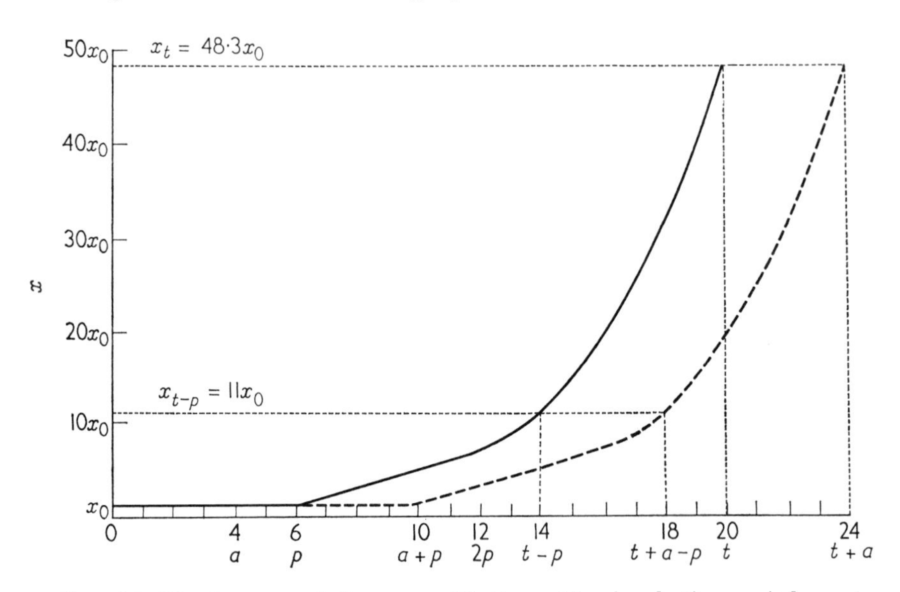

FIG. 5.1. The increase of disease x with time. The incubation period $a = 4$. The latent period $p = 6$. The basic relative infection rate $R = 1$. The solid line is for all disease, visible and invisible. The broken line is for disease that has passed the incubation period and is therefore visible.

between p and $2p$ days, i.e., from the sixth to the twelfth day, is a "simple interest" increase, the "initial capital" being x_0, the amount of disease formed by the initial inoculum. After $2p = 12$ days, the new lesions formed after $p = 6$ days begin to form spores, and the increase swings into a "compound interest" phase.

The broken line is for visible disease. It starts after $a = 4$ days, remains flat until $a + p = 10$ days, goes through a simple interest phase until $a + 2p = 16$ days, and then goes over into a compound interest phase. The two lines are identical, except that the broken line is set $a = 4$ days to the right of the solid line. The time taken for disease to increase from any one level to any other level is the same for both lines.

Therefore estimates of the relative infection rate r would be the same whichever line was used. Consider, e.g., any time, $t=20$; hence $t-p=14$. Then x_t (on the solid line) $=x_{t+a}$ (on the broken line) $=48.3x_0$; and x_{t-p} (on the solid line) $=x_{t+a-p}$ (on the broken line) $=11x_0$. An estimate of r, made over the period p or any other period, and based on visible disease, would hold for all disease, visible and invisible, as well.

The model admittedly gives a simplified view of the matter, because a is taken as constant. Variations in the incubation period affect estimates of infection rates, even if they do not affect the rates themselves. But the effect can be great only if the incubation period varies sharply over short intervals of time. When infection rates are estimated over fairly long intervals, as they usually are in this book, and when scrutiny of the relevant information reveals no large changes in temperatures, etc., one need not fear that estimates of r and R will be greatly distorted.

5.3. Some Biological Meanings

At this stage it is worth pausing to discuss the biological meanings of r, r_l, R, and p.

Many factors affect the rate at which disease increases: temperature, rain, humidity, wind, the resistance or susceptibility of the host, the virulence of the pathogen, the proportion of spores that germinate, the proportion of germinated spores that manage to enter the host and establish infection, the rate of increase in size of the lesion, the speed at which spores are developed and their number, the period over which spores are developed, the abundance, mobility, efficiency, and distribution of vectors, the size of the host plants and their number per acre, the chemical and physical properties of the soil, etc., and also the proportion x of disease.

The last of the list of factors, the proportion x of disease, has a special place. Suppose it were possible during the course of an epidemic to keep every other factor, temperature, humidity, etc., absolutely constant. In spite of this constancy, r would increase as the epidemic proceeds, simply because x increases. (This will be discussed in some detail later.) The logarithmic rate r_l avoids this. For practical purposes it is accurate enough (even if not wholly true) to say that r_l differs from r in that x does not affect it. For this reason r_l is a simpler and more direct measure of the balance of factors in which interest usually lies.

Factors other than x affect r either through R or p or both.* Consider two examples, both concerning blight in potatoes caused by *Phytophthora infestans*. The potato variety King Edward is more susceptible to blight than the variety Champion. Müller (1953) cut pieces of leaf from these varieties, sprayed them with a dilute suspension of zoospores, and kept them under suitable conditions. A few days later the pieces were examined under a microscope and scored as attacked or not attacked. Considerably more King Edward pieces were attacked than Champion pieces. Therefore, one reason why epidemics of blight start earlier in King Edward than in Champion is that spores have a better chance of attacking King Edward. Under the conditions of the experiment, R and hence r are higher in King Edward than in Champion. One can in this example state without doubt that R and not p was involved, because in the experiment there was no possible way in which the latent period of the disease in the two varieties could possibly have mattered. One can contrast this sort of resistance to infection with the so-called and wrongly called "incubation" resistance to blight. If potato plants are sprayed with a suspension of spores and kept in conditions suitable for blight, the lesions on some varieties start forming sporangia freely 4 days after the spraying, whereas on other varieties 5 days or more are needed. (See Table 20.1.) The latter varieties have resistance. Epidemics are less explosive in them, because p is greater and r therefore smaller, other things being equal. In this example one can state without doubt that p is involved, not necessarily alone, because the experiment measures p.

Some factors probably affect both p and R. For example, optimal temperatures for an experiment often go with low values of p and high values of R.

5.4. The Early Stages of Artificially Induced Epidemics of Stem Rust of Wheat

Let us return to Eq. (5.1); it is illustrated by Fig. 5.1. Between times p and $2p$ there is a "simple interest" phase, during which the inoculum stays constant at the initial proportion x_0 produced by artificial inoculation.

Figure 5.2 uses wheat stem rust to illustrate the process. The data are those of Underwood *et al.* (1959), which have already been plotted in Fig. 4.3. Our concern now is only with the data early in the epidemic.

* This statement ignores removals. (Infected tissue is said to be removed when it takes no further part in the epidemic.) It is convenient to discuss removals as a separate topic, which is done in Chapter 8.

The plants (of the variety Knox) were inoculated with race 56 of *Puccinia graminis tritici* when they were 8 to 10 in. tall and in the tillering stage. "Flecking" was first noticed 9 days after inoculation, and 1 or 2 days later the pustules ruptured the host tissues and began to shed uredospores. For the next 13 days the number of pustules were the original erupted primary pustules developed directly by the artificial inoculation. These were estimated at 1.2% of the culm, i.e., $x_0 = 0.012$. Thereafter

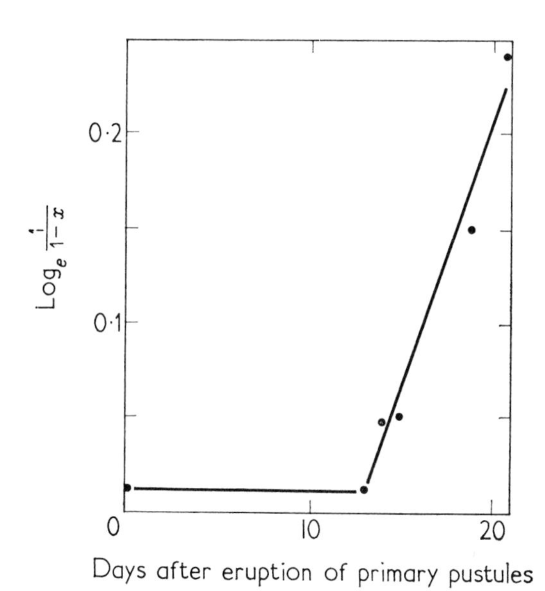

FIG. 5.2. The increase of wheat stem rust caused by *Puccinia graminis tritici* with time. Log$_e$ $[1/(1-x)]$ is plotted against the number of days after the eruption of the primary pustules formed by artificial inoculation. The figure illustrates increase of disease in the "simple interest" stage of an epidemic. Data of Underwood *et al.* (1959).

secondary infection became evident with the appearance of new "flecks" and pustules. Their origin was the uredospores released by the primary pustules. Figure 5.2 follows the course of this secondary infection for 8 days. On the internal evidence, p exceeds 8 days. During the 8 days, then, the increase was at "simple interest," the initial capital of primary pustules being $x_0 = 0.012$.

Because at this phase of the epidemic interest was simple and only later became compound, we plot $\log_e [1/(1-x)]$ against time, instead of $\log_e [x/(1-x)]$.

During this simple interest phase between times p and $2p$, Eq. (5.1) becomes

$$\frac{dx_t}{dt} = Rx_0(1-x_t)$$

or simply

$$\frac{dx}{dt} = Rx_0(1-x) \tag{5.3}$$

From this (by integration)

$$R = \frac{1}{x_0(t_2-t_1)}\left(\log_e \frac{1}{1-x_2} - \log_e \frac{1}{1-x_1}\right) \tag{5.4}$$

where x_1 and x_2 are proportions of disease at times t_1 and t_2. On day 13, which we take to be t_1, $\log_e [1/(1-x_1)] = 0.012$. On day 21, which we take to be t_2, $\log_e [1/(1-x_2)] = 0.242$.

$$R = \frac{0.242-0.012}{0.012(21-13)}$$

$$= 2.4 \text{ per unit per day}$$

Alternatively, if the increase is reasonably uniform, one can fit a straight regression line to the appropriate points in Fig. 5.2. The regression coefficient here is Rx_0, in the equation

$$\log_e \frac{1}{1-x} = a + Rx_0 t$$

which is the general integrated form of Eq. (5.3). By the usual method of estimating regression coefficients,

$$Rx_0 = 0.0269$$

$$R = \frac{0.0269}{0.012}$$

$$= 2.2 \text{ per unit per day}$$

This estimate uses all five relevant points in Fig. 5.2, whereas the previous estimate used only the first and last points.

One can make a similar calculation from the wheat stem rust data of Asai (1960), which have already been presented in Fig. 4.4. From these one estimates R to be 17.5 per unit per day. Details are given in Exercise 1 at the end of this chapter. As this estimate indicates, the epidemic studied by Asai was faster than that studied by Underwood *et al.* (1959).

These estimates of R from observations early in the epidemics are compatible with estimates of r obtained by observations later in the same epidemics. Exercise 2 at the end of this chapter gives details.

Equation (5.3), applied here to artificial epidemics of wheat stem rust, is similar to Eq. (4.1) used in the preceding chapter for an epidemic of fusarium wilt of cotton. The QR in Eq. (4.1) corresponds to Rx_0 in Eq. (5.3). The difference is that Q, the quantity of inoculum in the soil, was not known in appropriate units, whereas x_0 was known in the same units (pustules) as those of disease that developed later. It is, therefore, possible to estimate R from the stem rust data but not from the fusarium wilt data. Otherwise the problems are the same.

5.5. The Relation between r and R

Divide Eq. (5.1) by Eq. (5.2) and rearrange:

$$\frac{R}{r} = \frac{x_t}{x_{t-p}} \tag{5.5}$$

Figure 5.1 was calculated for $R = 1$ per unit per day and $p = 6$ days. If we take $t = 20$, then $x_t = 48.3x_0$ (which is the value of x on day 20), and $x_{t-p} = 11x_0$ (which is the value of x on day 14). Thus, when $t = 20$ in Fig. 5.1, from Eq. (5.5)

$$\begin{aligned} r &= R\frac{x_{t-p}}{x_t} \\ &= \frac{11x_0}{48.3x_0} \\ &= 0.228 \text{ per unit per day} \end{aligned}$$

More often, one wishes to estimate R from r. The simplest examples are where r is constant. An example with r constant follows. An example with r variable is given as Exercise 3 at the end of this chapter.

Figure 4.1 recorded the progress of some epidemics of potato blight observed by Large (1945). Consider the epidemic at Dartington in 1942 (curve B). The curve for $\log_{10}[x/(1-x)]$ plotted against time is practically straight, i.e., r is practically constant. Calculations in the usual way estimate the regression coefficient to be $b = 0.200$. From Eq. (3.6) it can be seen that $r = 2.3b = 0.46$ per unit per day. We assume $p = 4$ days. This is a realistic value for the variety (Majestic) at the temperatures concerned and agrees well enough with the observations of Crozier (1934) and Lapwood (1961b). From Fig. 4.1 one can determine x at an interval of 4 days. For example, on August 17 and August 21, $\log_{10}[x/(1-x)]$ was $\bar{2}.33$ and $\bar{1}.13$, respectively, which corresponds to $x_{t-p}=0.021$ and $x_t=0.12$. (The easiest way to get these figures is to use the tables for $\log_{10}[x/(1-x)]$ in the Appendix.) Hence $R=(0.46\times 0.12)/0.021=2.63$ per unit per day.

5.6. How *R* Changes as an Epidemic Progresses

Let us continue with the example of potato blight represented by curve B in Fig. 4.1; and assume that r stays constant at 0.46 per unit per day, and p at 4 days. In Fig. 5.3 we plot R against x. The result is a

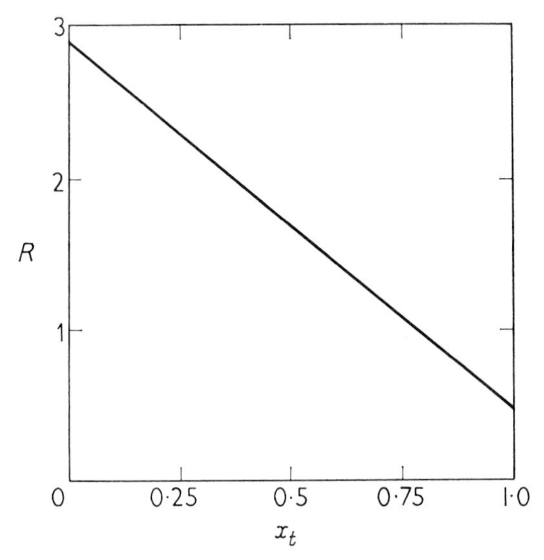

FIG. 5.3. The change of R with x_t, if $r = 0.46$ per unit per day and $p = 4$ days.

straight line from $R=re^{pr}=2.90$, when $x_t=0$, to $R=r=0.46$, when $x_t=1$. That is, with these values of p and r, R will fall steadily as the epidemic progresses, and will end with a value about one-sixth of its starting value.

The hypothesis in the previous paragraph is that r stays constant from beginning to end of the epidemic. In detail such a hypothesis is unreal, because no disease that weather affects will advance steadily from day to day. Nevertheless r often seems to stay fairly steady over long stretches of an epidemic of potato blight. This is the meaning of the straight lines in the bottom half of Fig. 4.1. Other data in the literature often point to same way. One might, for instance, cite the fact that the progress of blight with time is often represented by an S-shaped curve. The exact detail does not matter much here. But it is important to know what a constant value of r means. It does not mean that conditions of infection stay constant. It means just the opposite: that conditions for infection deteriorate steadily as the epidemic runs its course. For r to stay constant, conditions must become progressively less favorable to the fungus. That is, if r is constant, R must decrease.*

Why should R decrease during an epidemic? Why should conditions go against *Phytophthora infestans* as a blight epidemic progresses? We suggest three reasons. First, some parts of a plant are more susceptible to infection, or more vulnerable to infection because of the ecoclimate around them. The fungus often attacks the lower leaves first. If parts of the plant differ in susceptibility or vulnerability and, if, as a result of this, the fungus first attacks some parts preferentially, the parts that the fungus is left to attack later become progressively less susceptible or vulnerable. R falls in consequence. Second, as an epidemic advances and infected foliage withers, the foliage that remains becomes more open to sun and wind. The ecoclimate around the foliage that remains becomes less favorable for infection (Hirst and Stedman, 1960a). As it becomes less favorable, R falls. Third, the middle of an old blight lesion stops forming spores, i.e., it stops being infectious. As an epidemic advances with r constant, the average lesion becomes older, and infected tissue relatively less infectious. As this happens, R falls. The part that loss of infectious tissue plays in an epidemic is mentioned again in Section 8.3 and is discussed quantitatively.

What holds for potato blight seems to hold for other local lesion diseases such as stem rust of wheat. As a rule, R falls as an epidemic progresses. All the evidence in the literature that could be analyzed confirmed this. However, one must expect exceptions sometimes with diseases that the weather affects, and it is not our purpose to assert that R must always fall steadily or follow some particular trend or other. What we specially wish to bring out, and have used potato blight as an illustration to bring out, is the need for caution in interpreting r. The cause of the difficulties of interpretation is p. It is p that causes r and R to differ. It is p that causes the ratio R/r to vary with the proportion x of disease. That is, it is p that brings x in as a factor of r. To get rid of the effect of x one must use the logarithmic infection rate r_l instead of r.

* To be exact one should write: if pr is constant, pR must decrease.

5.7. The Relation between r_l and R

The logarithmic infection rate, it will be remembered from Chapter 3, is the apparent infection rate early in an epidemic when the multiplication of infection is practically unrestricted by lack of healthy susceptible tissue to which the pathogen can spread. It is the rate when the factor $1-x$ is so near to 1 as to be practically negligible.

Writing r_l for r in Eq. (5.5) gives

$$\frac{R}{r_l} = \frac{x_t}{x_{t-p}} \tag{5.6}$$

It is evident that x_{t-p} increases to x_t in p days (or years, if years are the natural units of time) at the rate of r_l per unit per day (or year). Hence, from Eq. (2.1),

$$x_t = x_{t-p}\, e^{pr_l}$$

Rearrange and substitute in Eq. (5.6). This gives

$$\frac{R}{r_l} = e^{pr_l} \tag{5.7}$$

This equation holds only if r_l is constant. On the left-hand side of the equation, r_l is the momentary value at time t; on the right-hand side it is the average value between times $t-p$ and t. The sides are consistent with each other only when r_l is constant. We can interpret this to mean constant from day to day (or year to year), if measurements are taken at intervals of whole days (or years). For an examination of these interpretations, see Sections 5.10 and 5.11.

For further comments on Eq. (5.7) see Sections 6.2 and 8.4.

Equation (5.7) is used in calculations. Consider the example of the error in field experiments with potato blight because of the dispersal of air-borne spores (van der Plank, 1961a). From plots, 20×5 yd., the proportion of spores lost by dispersal in air and wind was estimated to be 0.12. From a square 2-acre field the proportion lost was estimated to be 0.02. If in plots $r_l = 0.46$ per unit per day, what is r_l in a square 2-acre field? Take $p = 4$ days.

Before we attempt the calculation, let us go back to first principles. We are dealing with the logarithmic infection rate, i.e., with the early stages of an epidemic when the factor $1-x$ is negligible So rewrite Eq. (5.1) as

$$\frac{dx_t}{dt} = Rx_{t-p} \tag{5.8}$$

As before, x_{t-p} is the proportion of tissue that has been infected for at least p days; it is the source of spores. Take R_1 to refer to very large fields from which the proportion of spores lost by dispersion is negligible. Accept (in anticipation of Chapter 7) that germinated spores attack independently of one another at the stage of logarithmic increase of disease. In a plot losing 0.12 of its spores, i.e., retaining 0.88 of its spores,

$$\frac{dx_t}{dt} = 0.88R_1 x_{t-p}$$

because the number of spores falling in unit area of foliage is

$$0.88x_{t-p}/x_{t-p}$$

of the number in a large field. Hence, if we write $R_{0.88}$ for the rate in a plot,

$$R_{0.88} : R_1 = 0.88 : 1$$

That is, we assign the effect of the loss of spores directly to R. We shall qualify this generalization when we discuss removals in Chapter 8. Meanwhile accept the generalization and return to the numerical example. In the plot, by Eq. (5.7),

$$R = 0.46e^{4\times 0.46}$$

$$= 2.90 \text{ per unit per day}$$

In a square 2-acre field, losing 0.02 of its spores,

$$R = \frac{2.90 \times 0.98}{0.88}$$

$$= 3.23 \text{ per unit per day}$$

To find r_l in the square 2-acre field, one must therefore solve

$$r_l e^{4r_l} = 3.23$$

The answer found by the method in the next section is 0.48 per unit per day.

5.8. The Products pr_l, pr, and pR

Rearrange Eq. (5.7) and multiply both sides by p to give

$$pR = pr_l e^{pr_l} \tag{5.9}$$

The product pR or pr_l or pr has the property of being independent of the unit of time. We can measure time in milliseconds, days, years, or millennia without affecting the numerical value of the product. For example, if p and R are the values when days are the units, then $p/365$

and $365R$ are the values when years are the units, and the product remains pR. We use the product occasionally, but on the whole rather sparingly except in Chapter 6. There are real units of time—days or years—in epidemiology, and it is unwise to obscure the fact unnecessarily.

One use of the product is in calculations. Consider the problem just mentioned in Section 5.7: $R = 3.23$ per unit per day; $p = 4$ days. What is r_l? $pR = 12.92$ per unit. From Table 5 in the Appendix, pr_l is approximately 1.91 per unit. Hence $r_l = 1.91/4 = 0.48$ per unit per day, correct to two digits.

5.9. The Limit to the Explosiveness of an Epidemic

Just how explosive can an epidemic become? Is there a limit to the rate at which disease can multiply? We can answer that there is a limit and tentatively fix it.

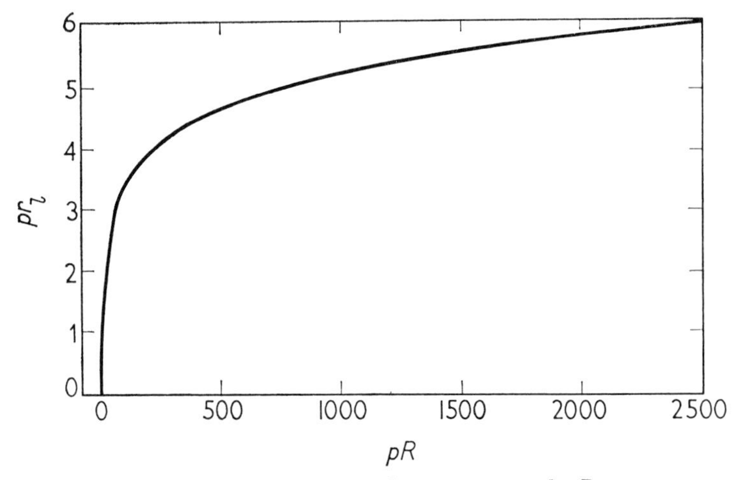

FIG. 5.4. The relation between pr_l and pR.

Figure 5.4 plots pr_l against pR. At first pr_l increases fast as pR increases. But later the increase slows down until, at high values of pR, pr_l is rather insensitive to further change. So far, we have been unable to find a record of an epidemic in which pR exceeds 200 per unit. This corresponds in round figures to $pr_l = 4$ per unit. But to be on the safe side, assume that pR can be as great as 2500 per unit. This corresponds in round figures to $pr_l = 6$ per unit. Tentatively, then, we assume that at its maximum pr_l will be somewhere between 4 and 6 per unit. This means, e.g., that we may tentatively assume that if $p = 10$ days, r_l will not exceed 0.6 per unit per day. For infection to double itself in a day, $r_l = 0.69$ per unit per day (see Exercise 4 at the end of Chapter 2), so p must be less than 9 days.

The relation between pr_l and pR means that there is a governor that puts a limit to the speed of epidemics; and p sets the governor.

5.10. A Discontinuous Infection Model

Equations (5.7) and (5.9) are based on a model in which infection proceeds continuously and at a constant rate day by day, at every moment of the day. To postulate for heuristic purposes a constant rate r from day to day—or from year to year, if years are the appropriate units of time—is not unrealistic. Examples of nearly constant rates have already been noted, for example, in potato blight (Fig. 3.1) or in stripe rust of wheat (Fig. 3.2). But to postulate a constant rate at every moment of the day and night is probably quite unrealistic. For example infection sometimes proceeds only while leaves are covered with dew.

It is, therefore, necessary to inquire how a varying rate during the day would affect the use of Eq. (5.7) or (5.9). We do this by comparing our continuous-infection model with the extreme opposite: a model in which the period of infection is reduced to almost a point—a split second—each day at the same hour every day. In an epidemic of a disease following this model, one could legitimately estimate r experimentally from the estimated increase of x over an interval of time.

Suppose that in this model of discontinuous infection each unit of diseased tissue causes the infection of a units of previously healthy tissue every day after the end of the latent period p, that is, suppose that a is the basic infection-rate for this model. Suppose the epidemic to be in the logarithmic phase. Then, if p and t are integers, and $np \leqslant t < (n+1)p$

$$x_t = x_0 \Big[1 + a\ (t-p+1) + \frac{a^2}{2}(t-2p+1)\,(t-2p+2)$$
$$+ \frac{a^3}{3!}(t-3p+1)\,(t-3p+2)\,(t-3p+3)\ \cdots$$
$$+ \frac{a^n}{n!}(t-np+1)\,(t-np+2)\cdots(t-np+n)\Big] \tag{5.10}$$

(For interest's sake compare this with the corresponding equation—Eq. (6.1)—for continuous infection in the next chapter.)

If, to give a numerical example, $a = 10$ per unit per day and $p = 8$ days, when $t = 32$ days $x = 190{,}551\ x_0$; and when $t = 40$ days, $x = 6{,}051{,}831\ x_0$. Whence, by Eq. (3.3), $r_l = 0{\cdot}4323$ per unit per day between day-32 and day-40. Similarly, if $a = 1$ per unit per day, with $p = 8$ days again, $x = 345\ x_0$ when $t = 32$ days, $x = 1824\, x_0$ when $t = 40$ days, and $r_l = 0.2081$ per unit per day between day-32 and day-40. The reason for choosing high values of t at which to estimate r_l will become obvious in Chapter 6.

Now consider these figures on the opposite assumption, that infection proceeds continuously and at a constant rate. If $r_l = 0.4323$ per unit per day and $p = 8$ days, $R = 13.73$ per unit per day. If $R = 13.73/10$ per unit per day, then, by Eq. (5.7), $r_l = 0.2256$ per unit per day.

Let us interpret these figures biologically. Suppose that a protectant fungicide destroys 9/10 of the spores before they cause infection, i.e., it stops 9/10 of new infections from forming. Suppose that in an unsprayed field $r_l = 0.4323$ per unit per day measured over an interval of a whole number of days, and $p = 8$ days. If the disease is of a sort with infection proceeding continuously and at a constant rate 24 hours a day, the fungicide will reduce r_l to 0.2256 per unit per day. If the disease is of a sort with infection proceeding discontinuously and confined to a split second each day, the fungicide will reduce r_l to 0.2081 per unit per day. The difference between 0.2256 and 0.2081 per unit per day—between 0.23 and 0.21 per unit per day, to shorten results to two digits, as we usually do—is the difference between the extremes of continuous and discontinuous infection.

With diseases like wheat stem rust and potato blight, with which we shall be dealing in later chapters, the germination of spores extends over many hours each day in weather favorable to disease, and the time for germinated spores to establish infection is variable. We, therefore, believe that these diseases are better represented by a continuous infection model. But the substance of the conclusions we reach does not depend on infection being continuous, for it could equally well be reached with a discontinuous-infection model, with the period of infection reduced to a point each day. For this reason, we said in Section 5.7 that for the purpose of using Eq. (5.7) we interpret r_l to be constant if it stays constant from day to day when disease is measured at intervals of a whole number of days.

5.11. Period of Infection is Reduced to a Point Each Year and p is 1 Year

There is an exception. The effect of departure from continuity increases as p is shortened. There are no diseases in which p is known to be as short as 1 day. However, there are diseases for which years are the appropriate units of time and $p = 1$ year. When with such diseases the period of infection is reduced nearly to a point, Eq. (5.7) or (5.9) should not be used in calculations, because the numerical error from discontinuity is too great.

The vectors of leaf roll disease of potatoes, in the environment studied by Broadbent *et al.* (1960), do not stay abundantly in potato fields long enough for plants infected during the season to serve as important sources of infection for other plants during the same season. In this environment, the leaf roll virus is transmitted once a year, the virus being transmitted only during the single annual infestation of fields by vectors. Because the virus cannot be transmitted from plant to plant in the absence of vectors, however abundant it may be in infected plants, p in such an environment is in effect approximately 1 year, and the period of infection each year is short. Infection is discontinuous, and the model described by Eq. (5.10) is reasonably apt.

If $p = 1$ year, and t is measured in years, Eq. (5.10) becomes:

$$x_t = x_0(1+a)^t \tag{5.11}$$

(This is the equation for ordinary discontinuous compound interest on money, discussed in Chapter 2.) because:

$$x_t = x_0\, e^{r_l t} \tag{5.12}$$

$$a = e^{r_l} - 1 \tag{5.13}$$

Consider an example. From figures given for leaf roll disease of potatoes in Exercise 1 at the end of Chapter 3, r_l was 1.95 and 1.01 per unit per year for unsprayed fields and for fields sprayed with DDT, respectively; whence a is 6.0 and 1.7 per unit per year for the unsprayed and the sprayed fields, respectively. (These estimates use Table 4 in the Appendix.) DDT reduced the number of effective transmissions of the virus proportionately from 6.0 to 1.7, i.e., it reduced the number by 72%.

EXERCISES

1. The data of Asai (1960) on wheat stem rust were discussed in earlier chapters. (See Exercise 2 at the end of Chapter 3 and Fig. 4.4.) After artificial infection of wheat (var. Baart) on June 15 with race 56 of *Puccinia graminis tritici* pustules developed and stayed at a constant level of about 0.24 pustules per culm for 10 days until July 3. Thereafter the number of pustules increased. From evidence presented, the latent period p exceeded 8 days, so we can estimate R in a "simple interest" phase at any time between July 3 and July 11. (This was discussed in Section 5.4.) On July 6 and July 9 there were 1.07 and 13.65 pustules per culm. Find R.

When plants are fully infected (i.e., when $x = 1$) there are about 1000 pustules per culm. (See Exercise 2 at the end of Chapter 3.) Thus, $x_0 = 0.00024$;

$x_1 = 0.00107$; $x_2 = 0.01365$; and $t_2 - t_1 = 9–6$ days. At these very low values of x, $\log_e [1/(1 - x)]$ has practically the same value as x itself. Substituting in Eq. (5.4) gives

$$R = \frac{0.01365 - 0.00107}{0.00024(9-6)}$$

$$= 17.5 \text{ per unit per day}$$

Much of this calculation was superfluous. When the proportion of disease is very small, it is unnecessary to convert pustules per culm into proportions. Thus,

$$R = \frac{13.65 - 1.07}{0.24(9-6)}$$

$$= 17.5 \text{ per unit per day}$$

This shows the problem as a true simple-interest problem. It also gives the meaning of R. $R = 17.5$ per unit per day means that each primary pustule formed by the initial artificial inoculation released uredospores that established an average of 17.5 secondary pustules per day during the 3-day period between July 6 and July 9.

2. In Section 5.4 we found that in an epidemic of wheat stem rust studied by Underwood *et al.* (1959) $R = 2.2$ per unit per day. In the previous exercise we found that in the epidemic studied by Asai (1960) $R = 17.5$ per unit per day. Take $p = 10$ days. Assume (in anticipation of Chapter 6) that if conditions stay fairly steady r estimated by Eq. (3.6) can be taken roughly to indicate r_l if x_2 does not exceed about 0.35. Then show that estimates of r are consistent with the estimates of R in these two epidemics.

a. Consider the data of Underwood *et al.* first: $pR = 22$ per unit, whence, from the table in the Appendix, $pr_l = 2.3$ per unit and $r_l = 0.23$ per unit per day.

From the data in their publication, on the fifteenth and twenty-sixth days after the eruption of the primary pustules, x was 0.0505 and 0.35, respectively. That is, $x_1 = 0.0505$, $x_2 = 0.35$, and $t_2 - t_1 = 26 - 15 = 11$ days. Thus, by Eq. (3.6), $r = 0.21$ per unit per day.

These two estimates, 0.23 and 0.21, are close enough to serve as a general check, because the equivalence of r_l and r is only approximate, and because the dates involved in the two estimates were not identical and the weather may have changed.

b. From the data of Asai, $pR = 175$ per unit, hence $pr_l = 3.82$ per unit and $r_l = 0.38$ per unit per day. In Exercise 2 at the end of Chapter 3, r was estimated from Asai's data to be 0.41 per unit per day. The agreement is adequate.

It will be remembered that as an approximation this chapter ignores removals. The agreement in the calculations in this exercise suggests that the approximation is quite good for wheat stem rust. This agrees with the general evidence, given in Section 8.2, that removals often play a small part in epidemics of cereal rusts.

3. In experiments with tomato leaf mold caused by *Cladosporium fulvum*, Beaumont (1954) recorded the following proportions x of disease at Pickering, Yorkshire, in 1951: 0.001 on July 3; 0.012 on July 16; 0.046 on July 30; 0.175 on August 13; 0.40 on August 27, and 0.60 on September 10. Take $p = 10$ days, on the evidence of Bond (1938). Plot R against x_t.

First plot $\log_{10} [x/(1-x)]$ against time, from $t = 0$ for July 3 to $t = 69$ for September 10. This is shown in Fig. 5.5. The curve bends downward, i.e., r decreases as time advances. This occurred constantly in Beaumont's results in various experiments. At any time, r is the slope of the curve, and can be determined by drawing the tangent.

But it is probably quicker in the long run to fit the curve

$$\log_{10} \frac{x}{1-x} = a+bt+ct^2$$

If r decreases with time, c is negative. Write y for $\log_{10} [x/(1-x)]$. Then, by the method of least squares which can be found in appropriate reference works,

$$b = \frac{\Sigma(t^4) . \Sigma(ty) - \Sigma(t^3) . \Sigma(t^2y)}{\Sigma(t^2) . \Sigma(t^4) - [\Sigma(t^3)]^2}$$

$$c = \frac{\Sigma(t^2) . \Sigma(t^2y) - \Sigma(t^3) . \Sigma(ty)}{\Sigma(t^2) . \Sigma(t^4) - [\Sigma(t^3)]^2}$$

$$a = \frac{\Sigma(y) - b\Sigma(t) - c\Sigma(t^2)}{n}$$

$$r = 2.3b + 4.6ct$$

where n is the number of observations (in this example, 6).

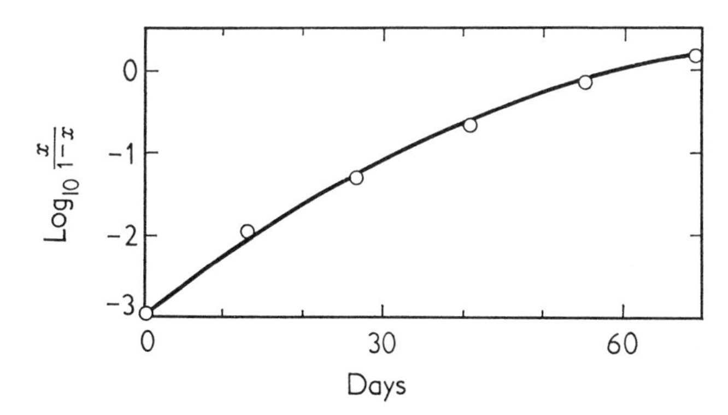

FIG. 5.5. The increase of tomato leaf mold caused by *Cladosporium fulvum*. $\text{Log}_{10} [x/(1-x)]$ is plotted against the number of days after the first observation. Data of Beaumont (1954).

Estimates for Fig. 5.5 are: $a = -2.98$; $b = 0.0751$; and $c = -0.000426$. The earliest possible estimate of R is when $t = p = 10$ days. Then

$$r = 2.3 \times 0.0751 - 4.6 \times 0.000426 \times 10$$

$$= 0.153 \text{ per unit per day}$$

$$\log_{10} \frac{x_t}{1-x_t} = -2.98 + 0.0751 \times 10 - 0.000426 \times 10^2$$

$$= -2.28 \text{ (or } \bar{3}.72)$$

hence

$$x_t = 0.0053$$

For $t-p=0$

$$\log_{10}\frac{x_{t-p}}{1-x_{t-p}} = -2.98 \text{ (or } \bar{3}.02)$$

hence

$$x_{t-p} = 0.00105$$

$$R = \frac{0.153\times 0.0053}{0.00105}$$

$$= 0.77 \text{ per unit per day}$$

When $t=69$ days, $x_t=0.60$, $x_{t-p}=0.48$, $r=0.038$, and $R=0.048$ per unit per day.

As an exercise plot R against x_t from 0.0053 to 0.60. Note how quickly R falls.

CHAPTER 6

The Latent Period

SUMMARY

This chapter and Chapters 7 to 9 may be read after Chapters 10 to 19.

In an epidemic started artificially by a single dose of inoculum, r_l varies with time even if p and R are constant, i.e., even if the interaction of host, pathogen, and environment is constant. This is because there is a latent period. This variation of r_l is fairly rapidly damped out if conditions for infection stay constant. To minimize variation, r_l should be estimated by Eq. (3.3) or (3.4) with $t_2 - t_1$ taken at about $1.2p$.

One must interpret r_l widely. At any instant r_l measures the interaction of host, pathogen, and environment not only at that instant but also before. This is specially relevant for artificial epidemics or natural epidemics after a considerable change of infection rate. With natural epidemics increasing without great change of rate the past history is not very important, and one may with adequate accuracy interpret r_l on present conditions alone.

The natural variation of p from point to point in susceptible tissue reduces the variation of r_l. Using a constant mean value of p in Eq. (5.7) instead of a constant value makes the equation underestimate r_l, but the error is small for most purposes.

As an epidemic proceeds beyond the logarithmic stage, with p and R constant, r increases. This, too, comes about because there is a latent period. The increase does not occur uniformly. It is most marked late in the epidemic when x is greater than 0.5. Earlier in the epidemic a large increase is delayed. Because of this, if one estimates r by Eq. (3.5) or (3.6) between times t_1 and t_2 and if x_2 is less than about 0.35, one may assume accurately enough that r and r_l are numerically almost equal. That is, one may use r to estimate r_l.

An epidemic may conveniently be divided into three stages. The first is a logarithmic stage. Increase of the pathogen is unhindered by overlapping of lesions or multiple infections by a systemic pathogen. The logarithmic infection rate reflects the interaction of host, pathogen, and environment (before and at the time of measurement) and is independent of x. The second stage extends until x is about 0.35. Lesions overlap, and multiple infection occurs; but host, pathogen, and environment still affect r in much the same way as they affect r_l. In the third stage, especially after x is greater than 0.5, one enters a jungle of empiricism into which little light has entered; and r estimated here cannot easily be compared with r estimated earlier or with r_l.

6.1. Logarithmic Increase of Infection with p and R Constant

In this chapter we do not analyze many data that have appeared in the literature of plant diseases. Instead, we examine a few hypothetical models to see what the consequences are of different sets of conditions.

For the simplest model consider an epidemic started artificially by inoculation on a single date. The proportion x of infection caused by this artificial inoculation is supposed to be very low, so that the epidemic can be supposed to stay in the logarithmic phase, with no competition between lesions for space. That is, we suppose that the proportion of infection stays low enough to obviate the need for the correction factor $1-x$. In this simple model infection following inoculation is supposed to occur all at one time. Neither p nor R varies throughout the whole period of analysis. On these assumptions disease increases according to the Eq. (5.8):

$$\frac{dx_t}{dt} = Rx_{t-p}$$

where, as before, x_t and x_{t-p} are the values of x at times t and $t-p$, respectively.

The simple model represented by this equation is nearer to, say, the cereal rusts than to *Phytophthora* blight of potato. We shall discuss some complications in a later chapter.

The early phase of increase by this equation has already been shown graphically in Fig. 5.1.

The initial proportion x_0 stays unchanged for the period p. Hence, for $0<t<p$,

$$x_t = x_0$$

$$r_l = \frac{dx_t}{x_t\,dt}$$

$$= 0$$

This is followed by a "simple-interest" phase, when $p<t<2p$,

$$x_{t-p} = x_0$$

$$\frac{dx_t}{dt} = Rx_0$$

$$x_t = x_0 + Rx_0(t-p)$$

$$r_l = \frac{R}{1+R(t-p)}$$

Thereafter, for $2p < t < 3p$,

$$\frac{dx_t}{dt} = Rx_0 + R^2x_0(t-2p)$$

$$x_t = x_0\left[1 + R(t-p) + \frac{R^2}{2}(t-2p)^2\right]$$

$$r_l = R\frac{1 + R(t-2p)}{1 + R(t-p) + \frac{R^2}{2}(t-2p)^2}$$

For $np < t < (n+1)p$, where $n = 1, 2, 3 \ldots$,

$$x_t = x_0\left[1 + R(t-p) + \ldots \frac{R^n(t-np)^n}{n!}\right] \tag{6.1}$$

$$r_l = R\frac{1 + R(t-2p) + \ldots \frac{R^{n-1}(t-np)^{n-1}}{(n-1)!}}{1 + R(t-p) + \ldots \frac{R^n(t-np)^n}{n!}}$$

$$= R\frac{\sum_{s=1}^{n} \frac{R^{s-1}}{(s-1)!}(t-sp)^{s-1}}{\sum_{s=0}^{n} \frac{R^s}{s!}(t-sp)^s} \tag{6.2}$$

in other notation.

The numerator of the fraction in Eq. (6.2) is x_{t-p}/x_0 and the denominator x_t/x_0.

Figure 6.1 shows the increase of x with time, calculated by Eq. (6.1) for $pR = 60$. For convenience time is recorded with p as the unit. The equation of the line fitted to the points is*

$$x_t = \frac{x_0}{1 + pr_l}e^{r_l t}$$

where r_l has the value calculated from Eq. (5.9). That is

$$pr_l = pRe^{-pr_l}$$

hence

$$pr_l = 3.0$$

* The writer is much indebted to A. P. Burger, Director of the National Research Institute for Mathematical Sciences, Pretoria, for proving that the initial value should be $x_0/(1+pr_l)$ for the line to fit the points when t/p is large.

After about $4p$ the line begins to fit the points fairly well.

Figure 6.2 for $pR = 6$, is similar. Here $pr_l = 1.432$.

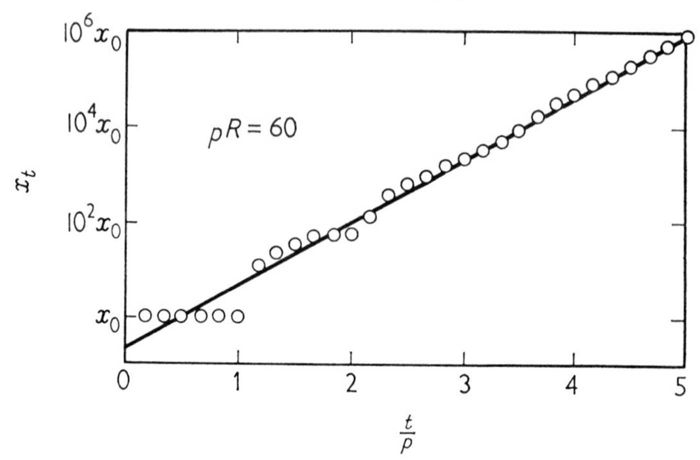

FIG. 6.1. The increase of x with time, when $pR = 60$. The line fitted to the points is for the increase of x when $pr_l = 3.0$.

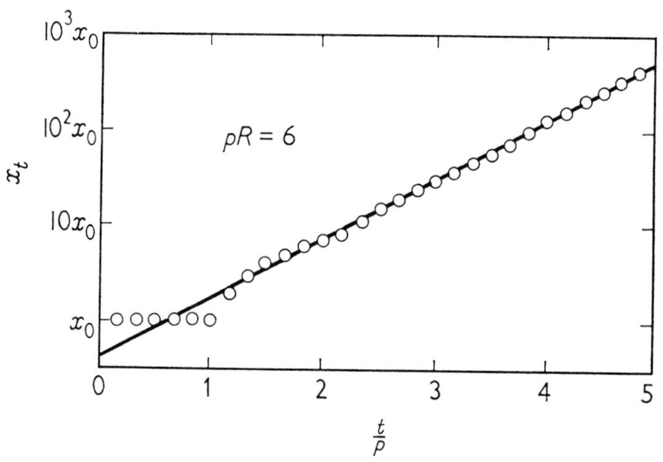

FIG. 6.2. The increase of x with time, when $pR = 6$. The line fitted to the points is for the increase of x when $pr_l = 1.432$.

6.2. The Variation of r_l with Time, with p and R Constant

The initial variation of r_l, clearly shown in Figs. 6.1 and 6.2, is better studied by using Eq. (6.2). When $t = p$, r_l reaches an absolute maximum R. Then it falls to an absolute minimum $R/(1+pR)$. Thereafter r_l

varies through a succession of maxima and minima, the variation decreasing with time, until finally r_l settles down at the value given by Eq. (5.7).

The variation is greatest when pR is great.

Figure 6.3 shows how r_l varies with time when $pR = 6$. For convenience pr_l is shown instead of r_l.

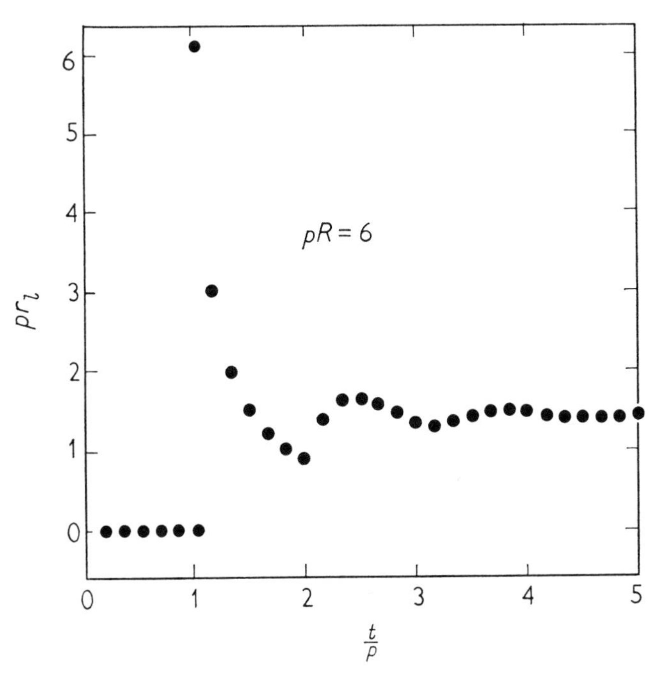

FIG. 6.3. The variation of pr_l with time, when $pR = 6$. The values of pr_l are instantaneous values.

These results are for the model of an artificial epidemic in which p and R are constant and the epidemic starts from a single inoculation. Can we infer from this model that in a natural epidemic r_l varies greatly, even when the interaction of host, pathogen, and environment is constant? We have assumed in earlier chapters that r_l is an index that measures this interaction and gives the measure in a single composite figure. (See Sections 3.5 and 5.3.) Is this assumption wrong? Variation of r_l for any reason other than variation in the interaction of host, pathogen, and environment necessarily reduces the importance of r_l as an index.

The answer is that r_l can vary even when the interaction of host, pathogen, and environment is constant. But it varies much less than our model suggests, for several reasons. In any case it is necessary only to widen the interpretation of r_l for it to become a perfectly valid index of host–pathogen–environment interactions.

6.3. The Variation of r_l as an Average Value over an Interval of Time

Equation (6.2) gives r_l at an instant of time. For example, when $t=p$, $r_l=R$, but only for an instant: immediately afterward, r_l declines.

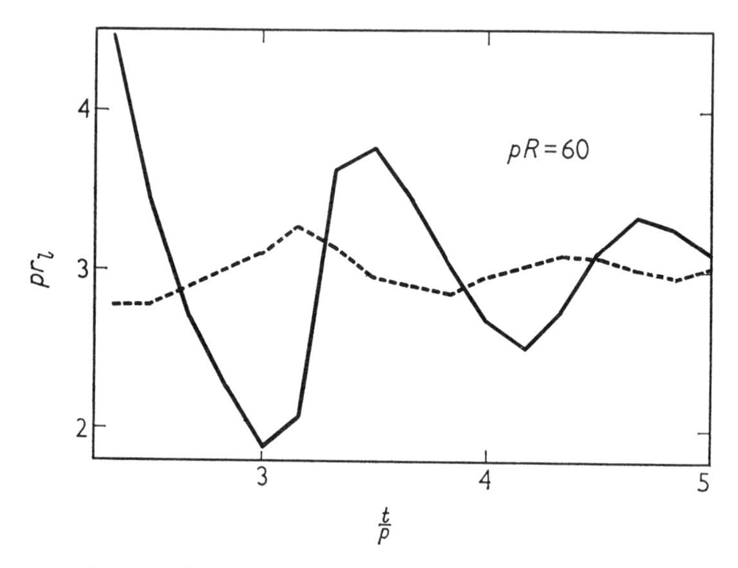

FIG. 6.4. A comparison of instantaneous values (solid line) of r_l with values averaged over an interval of $7p/6$ (broken line). (See text for details.)

With experimental observations one estimates the average value of r_l between times t_1 and t_2 by Eq. (3.3) or (3.4). The average value varies less than the instantaneous value.

To minimize variation it is best to make the interval between t_1 and t_2 approximately $1.2p$. (The factor 1.2 is a compromise; the best factor depends on pR and t/p.) If this is done most of the variation of r_l is damped out when $t_1 > 2p$, i.e., when t_1 is taken after the simple interest phase of the epidemic. But even if t_1 is taken earlier, so that most of the simple interest phase falls between t_1 and t_2, the variation is reduced greatly. These remarks about t_1 refer to total disease, visible or invisible. Because one is concerned experimentally only with visible disease, one must allow for the incubation period. One counts not from the date of inoculation but from the date on which the lesions caused by inoculation first become visible.

Figure 6.4 is for $pR=60$. This is a rather high value—higher than in most epidemics. The variation of r_l is, therefore, higher than usual, and the figure presents a fairly extreme case. Points are plotted both for instantaneous values of r_l and for average values between t_1 and t_2 when $t_2-t_1=7p/6$. These average values are plotted against t_2. Figure 6.4 shows that variation of average values of r_l is considerably less than for instantaneous values.

Because variations persist beyond the logarithmic stage into later stages of an epidemic, it is a wise precaution to use an interval between t_1 and t_2 of about $1.2p$ for estimating r as well as r_l. There are, of course, occasions when this precaution cannot be taken. In Exercise 2 at the end of Chapter 5 the interval t_2-t_1 used to analyze the data of Underwood *et al.* (1959) was about $1.1p$, which is satisfactory. But, to analyze Asai's data in Exercise 2 of Chapter 3, circumstances allowed an interval of only $0.5p$.

6.4. The Effect of a Prolonged Period of Inoculation on the Variation of r_l

In our model the epidemic started from a single dose of inoculum. This is not the way of natural epidemics. Inoculum comes in over a stretch of time, and this inevitably reduces variation of r_l.

Ascospores of *Venturia inaequalis*, the cause of apple scab, come in spring from dead leaves that have lain for the winter on the orchard's floor. With suitable weather they reach and infect the spring growth of the trees over a stretch of many weeks.

The sporangia of *Phytophthora infestans* that start epidemics of blight in potato fields come from infected shoots out of blighted tubers culled when stores are opened after winter and dumped in piles; or they come from blighted tubers planted as seed; or they blow in from infected fields in the neighborhood. Whatever the source, they come over a stretch of time.

The uredospores of *Puccinia graminis* that cause stem rust in wheat blow from field to field or from State to State, from south to north along the great rust track of North America. They blow, not just for an hour, but whenever climate and wind direction are right.

A single concentrated dose of inoculum is needed to cause the sharp variation of r_l in our model. In natural epidemics doses of inoculum stretched over a long interval inevitably reduce the variation.

6.5. The Effect of Variation of the Latent Period p on the Variation of r_l

All living processes vary. We can picture p at different points of a plant having a constant mean value and being distributed with constant variance. But we cannot picture p being constant from point to point. This is a biological impossibility.

Other things being equal, variation of p reduces variation of r_l with time.

Zadoks (1961) found p to vary randomly in experiments with wheat stripe rust caused by *Puccinia striiformis*, and Lapwood (1961b) with potato blight caused by *Phytophthora infestans*. Lapwood inoculated detached potato leaflets with drops of a spore suspension taken up in pieces of filter paper. He cut discs from these leaflets, incubated them at 15°C. and at high humidity, and examined them daily to see which had produced sporangia. With the variety Majestic, e.g., no disc had yet formed sporangia 3 days after inoculation, but 58% of the discs had formed sporangia after 4 days, 87% after 5 days, and 100% after 6 days. That is, p was between 3 and 4 days in 58% of the discs, between 4 and 5 days in 29%, and between 5 and 6 days in 13%. With the less susceptible variety Ontario, p was between 3 and 4 days in 13% of the discs, between 4 and 5 days in 63%, between 5 and 6 days in 17%, and between 6 and 7 days in 8%. These figures give a rough idea of the variance one can expect. Little is known about the distribution, except that it is often discontinuous in natural epidemics because sporangia often form only at night and early in the morning when the relative humidity is high.

One can determine the effect of variation of p from models. An example of the sort of calculation involved is given as an exercise at the end of this chapter. In Fig. 6.5 the solid line is for a constant value, $p=6$. The broken line is for p varying as follows: $p=4$ in $\frac{1}{5}$ of the tissue; $p=5$ in $\frac{1}{5}$; $p=6$ in $\frac{1}{5}$, $p=7$ in $\frac{1}{5}$; and $p=8$ in $\frac{1}{5}$. Here $\bar{p}=6$, where $\bar{p}$ is the mean value. It is assumed that the inoculum is common to all tissues at all times. The values plotted in Fig. 6.5 are instantaneous values; $R=1$. The figure shows clearly how variation of p reduces the variation of r_l.

In another comparison r_l was estimated on the same assumptions as in the previous paragraph (that $p=4$ in $\frac{1}{5}$ of the tissue, $p=5$ in $\frac{1}{5}$ of the tissue, and so on, and that inoculum is common to all tissues).

These estimates were compared with those for a constant latent period, with $p=6$. With $R=10$, the variance of r_l with p variable was less than 1/29 of the variance with p constant, this comparison being made with nine estimates of r_l from $t=7$ to $t=15$, inclusive, at equal intervals.

Variation of R has not the same effect. A constant mean value acts just as a constant value so far as variation of r_l is concerned, provided that inoculum is common to all tissues at all times.

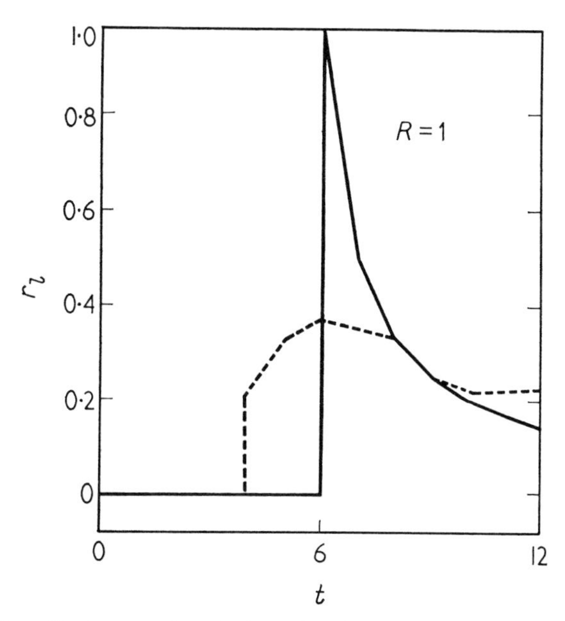

FIG. 6.5. The effect on r_l of variation of p. For the solid line p has a constant value of 6; for the broken line it has a constant mean value of 6. (For other details, see text.)

6.6. A Wider Interpretation of r_l

If in the course of an epidemic r varies, it continues to vary, to an ever decreasing degree, even after the source of variation is removed—this results directly from the latent period p inevitably associated with disease. Because of p, the rate of infection at any instant depends on the interaction of host, pathogen, and environment not only at that instant but also earlier. Today's rate depends not only on today's rain, but also yesterday's and, to a smaller extent, last week's; and today's rain will affect the rate next week. This will happen in addition to any effect that persisting high atmospheric humidity or soil moisture may have.

In Sections 3.5 and 5.3 we pictured r_l as summarizing in one single figure all the factors of the host, pathogen, and environment that affect the infection rate. That picture is true. But it is true only in the wider sense in that it summarizes the factors not only at the time when r_l is measured but also before. There is continuity in an epidemic. Because of p one cannot isolate the present from the past.

Artificial inoculation changes conditions abruptly. Before inoculation host and pathogen are apart; after inoculation they are together. The rate r_l of an artificial epidemic, especially in its early stages, reflects not only conditions of the present but also changes in the past, and it necessarily varies. To assess quantitatively how it varies we must know the time since inoculation, $\bar{p}R$, the variance of p, the length of the interval from t_1 to t_2, and other factors. These are details. The essence is that r_l varies after inoculation because inoculation brings about a change.

In natural epidemics changes are normally not so abrupt as those brought about by artificial inoculation. In an epidemic that has been proceeding at a fairly steady rate, one can as a fair approximation assume that the present value of r_l reflects present conditions.

6.7. The Error from Using a Constant Mean Value $\bar{p}$ in Eq. (5.7) Instead of a Constant Value p

To return to the topics of Sections 5.7 (the relation between r_l and R) and 6.5 (the variation of p), what error does one incur in Eq. (5.7),

$$r_l = Re^{-pr_l}$$

if one substitutes a constant mean value $\bar{p}$ for p? Answers of a sort are given in Table 6.1. The entries for constant p were calculated by

TABLE 6.1

VALUES OF r_l FOR DIFFERENT VALUES OF R, WITH p CONSTANT AND p VARIABLE

	$R = 0.1$	$R = 1.0$	$R = 10$
p Constant			
$p = 6$	0.0669	0.239	0.50
p Variable			
Case A	0.0670	0.241	0.50
Case B	0.0672	0.245	0.53

Case A: $p = 5$ in $\frac{1}{4}$ of the tissue; $p = 6$ in $\frac{1}{2}$; and $p = 7$ in $\frac{1}{4}$. $\bar{p} = 6$.
Case B: $p = 4$ in $\frac{1}{5}$ of the tissue; $p = 5$ in $\frac{1}{5}$; $p = 6$ in $\frac{1}{5}$; $p = 7$ in $\frac{1}{5}$; $p = 8$ in $\frac{1}{5}$. $\bar{p} = 6$.

Eq. (5.7.). Those for variable p were calculated by a modification of Eq. (6.1) that can easily be deduced from the answer given for Exercise 1 at the end of this chapter. Estimates are given for two cases. Case B probably represents greater variation than the average. Even so, the error of r_l calculated from a known value of R is not very great, even at the highest entered value of R. Until experimental observations are accurate enough to justify greater precision in calculation, one can probably use mean values of p in the equation without fearing too large an error.

The error of R calculated from a known value of r_l would be greater. But, with few exceptions, the end of our calculations is r_l, not R.

It is assumed throughout that inoculum is common to all tissues at all times.

6.8. Increase of Infection beyond the Logarithmic Phase with p and R Constant

So far in this chapter only logarithmic increase of infection has been considered. We must now consider increase of infection over the whole range $0 < x < 1$, when x_t increases according to Eq. (5.1).

$$\frac{dx_t}{dt} = Rx_{t-p}(1-x_t)$$

and p and R are constant.

Figures 6.6, 6.7, and 6.8 show the increase of x_t with time. Data have been given by van der Plank (1961c). They were originally computed by J. D. Neethling of the National Research Institute for Mathematical Sciences, Pretoria, who solved Eq. (5.1) by solving the integral equation

$$x_t = 1-(1-x_0)\exp\left[-\int_0^t pRx(t-p)\,dt\right]$$

using Simpson's rule. For convenience $\log_e[x_t/(1-x_t)]$ is plotted against t/p.

Figure 6.6 is for $pR = 266.4$ and $x_0 = 10^{-9}$; Fig. 6.7 for $pR = 60$ and $x_0 = 10^{-9}$; and Fig. 6.8 for $pR = 5.544$ and $x_0 = 10^{-4}$.

The equation of the line drawn in Fig. 6.6. is

$$\log_e \frac{x_t}{1-x_t} = \frac{4{\cdot}158t}{p} + \text{a constant}$$

The value 4.158 is that of pr_l calculated by Eq. (5.9) for $pR = 266.4$. Similarly, for the line in Fig. 6.7 $pr_l = 3.0$; and for the line in Fig. 6.8, $pr_l = 1.386$. These straight lines show how $\log_e[x_t/(1-x_t)]$ would have increased if r had stayed numerically the same as r_l. In other words, they show how disease would have increased if the relative infection rate was numerically the logarithmic rate.

At first, early in the epidemic, r stays close to r_l. Only later, when $\log[x_t/(1-x_t)]>0$, i.e., when $x_t>0.5$, does it increase markedly.

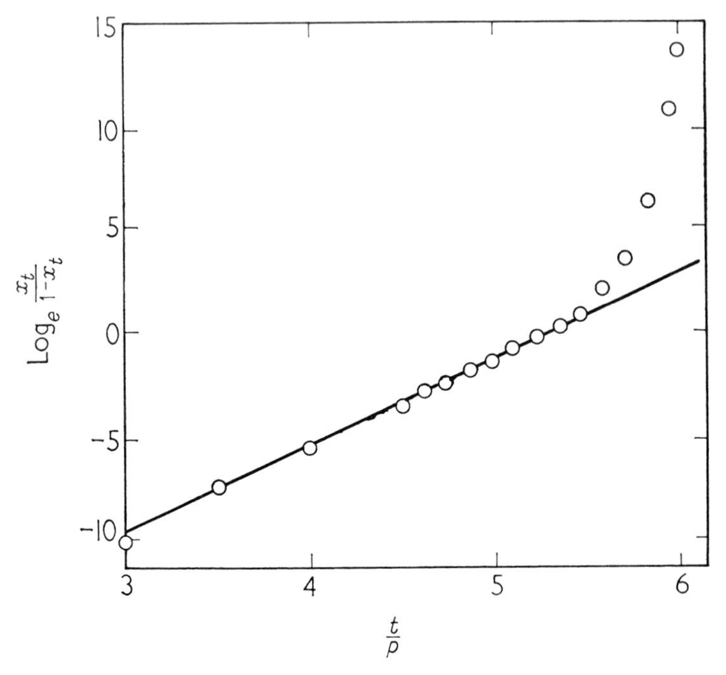

FIG. 6.6. The increase of $\log_e [x_t/(1-x_t)]$ with time according to Eq. (5.1). $pR=266.4$; and $x_0=10^{-9}$. (For other details, see text.)

Note the contrast. When p and r are constant, R decreases, and the decrease with increasing x is uniform throughout the epidemic from start to finish. (See Section 5.6 and Fig. 5.3.) When p and R are constant, r increases. But the increase is not uniform. Instead, a marked increase is delayed until the epidemic has progressed far.

6.9. The Estimation of r_l after the Logarithmic Stage of an Epidemic

The delay before r increases is of capital importance. The logarithmic stage of an epidemic is often an inconvenient time for estimating infection rates. The amount of disease is then often too small for convenient measurement. When the epidemic is explosive the estimation of r_l between times t_1 and t_2 is especially difficult. Even if t_1 is chosen to fall within the logarithmic stage, t_2 is likely to fall outside it. But the results in the previous paragraph show that this does not necessarily matter. Within limits one may estimate r between t_1 and t_2 outside the logarithmic stage and infer that this would have been the estimate of r_l under the same conditions.

What are these limits? To get a clearer answer the data used in Figs. 6.6, 6.7, and 6.8 have been reproduced in Fig. 6.9. The top, middle,

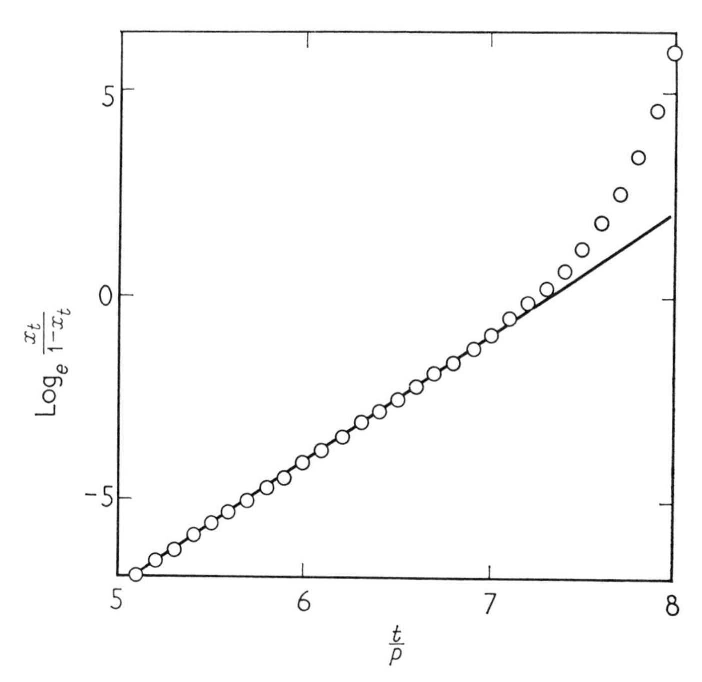

FIG. 6.7. The increase of $\log_e [x_t/(1-x_t)]$ with time according to Eq. (5.1.). $pR = 60$; and $x_0 = 10^{-9}$. (See text for other details.)

and bottom sets of data are for $pR = 266.4$, 60, and 5.544, respectively. The values of x_0 are those recorded in the previous section. Equation (3.5) was used to calculate r between t_1 and t_2, this interval being $1.17p$, $1.2p$, and $1.33p$ for the three sets, respectively. For convenience results have been recorded as pr and are plotted against x_2, the value of x at time t_2. The horizontal lines show the corresponding values pr equal to pr_l calculated by Eq. (5.8).

One cannot state the limits precisely because for this a knowledge of the precision of the experimental estimates is needed. But, to generalize, the difference between r_l and r is negligible until $x_2 > 0.15$, and not very important until $x_2 > 0.35$.

The three sets of data reproduced in Fig. 6.9 correspond with fairly fast epidemics in which $pR > 5$. This would cover epidemics of diseases such as wheat stem rust and potato blight. A full investigation of slower epidemics has not yet been undertaken.

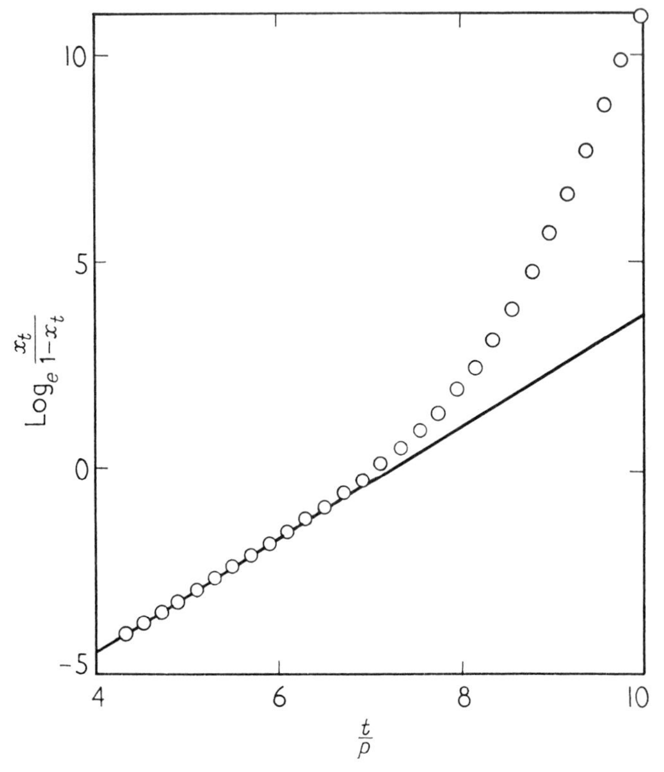

FIG. 6.8. The increase of $\log_e [x_t/(1-x_t)]$ with time according to Eq. (5.1). $pR = 5.544$; and $x_0 = 10^{-4}$. (See text for other details.)

6.10. Three Arbitrary Stages in an Epidemic

One may think of an epidemic as divided into three stages: a logarithmic stage, a stage in which x does not exceed about 0.35, and a final stage with high proportions of disease.

The logarithmic stage is the simplest. The infection rate r_l is independent of x (see Section 7.2). The pathogen population can be considered to increase unhindered by the overlapping of lesions or multiple infections if the disease is systemic. Consequently this is the stage at which one can most easily understand and analyze the factors that influence infection.

The second stage retains some of the simple features of the logarithmic phase. But one now must use a correction factor $1-x$ to allow for overlapping of lesions and multiple infections. The latent period p and the overlapping of lesions or multiple infections combine to increase r above r_l, but, as shown in the previous section, the increase in the second stage

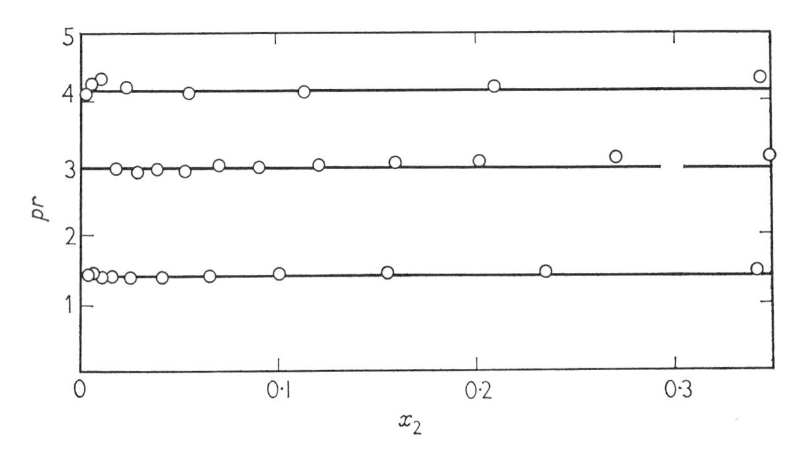

FIG. 6.9. The change of pr with x at time t_2. The top, middle, and bottom sets of data are for $pR = 266.4$, 60, and 5.544, respectively. (See text for other details.)

is not large. On the other hand the correction factor $1-x$ is inadequate except with hypothetical perfectly uniform conditions for infection. (See Section 4.5.) This inadequacy leads to undercorrection, which is probably not normally very large during the second stage. The two errors of using r as a measure of r_l are in opposite directions and balance one another to some extent. If the interaction of host, pathogen, and environment stayed nearly constant r would stay nearly constant until x reached 0.35 or even increased a little beyond.

The third stage of an epidemic is altogether different, especially when x exceeds 0.5. If the interaction of host, pathogen, and environment stayed constant, r would increase, and it would increase steeply if pR were great. At this stage, too, the inadequacy of $1-x$ as a correction factor becomes more pronounced. If r stays nearly constant throughout an epidemic as in Figs. 3.1 and 3.2 (i.e., if the curve for the increase of disease with time is nearly sigmoid), this means that conditions have varied, not stayed constant. New factors become important in the third stage that were not important before. If they balance one another in a way that keeps r constant or nearly constant, we have no knowledge yet that enables us to draw up a balance sheet. One must just accept an empirical result. Only if pR is very small and not much greater than pr_l can one expect r to stay nearly constant throughout the three stages of an epidemic as it proceeds with host, pathogen, and environmental factors constant.

EXERCISE

An epidemic proceeds according to the equation

$$\frac{dx_t}{dt} = Rx_{t-p}$$

R is constant. In one-quarter of the tissue, $p = 5$; in one-half of the tissue $p = 6$; and in one-quarter of the tissue, $p = 7$. Inoculum is common to all tissues at all times. Show that when $t = 32$

$$\begin{aligned}
\frac{x}{x_0} = {} & 1 + R(t-6) \\
& + \frac{R^2}{2!\,4^2}[(t-10)^2 + 4(t-11)^2 + 6(t-12)^2 + 4(t-13)^2 + (t-14)^2] \\
& + \frac{R^3}{3!\,4^3}[(t-15)^3 + 6(t-16)^3 + 15(t-17)^3 + 20(t-18)^3 + 15(t-19)^3 \\
& \qquad + 6(t-20)^3 + (t-21)^3] \\
& + \frac{R^4}{4!\,4^4}[(t-20)^4 + 8(t-21)^4 + 28(t-22)^4 + 56(t-23)^4 + 70(t-24)^4 \\
& \qquad + 56(t-25)^4 + 28(t-26)^4 + 8(t-27)^4 + (t-28)^4] \\
& + \frac{R^5}{5!\,4^5}[(t-25)^5 + 10(t-26)^5 + 45(t-27)^5 + 120(t-28)^5 + 210(t-29)^5 \\
& \qquad + 252(t-30)^5 + 210(t-31)^5] \\
& + \frac{R^6}{6!\,4^6}[(t-30)^6 + 12(t-31)^6]
\end{aligned}$$

CHAPTER 7

Average Values of Infection Rates. Increase of Populations of Lesions and of Foci. Independent Action of Propagules

SUMMARY

This chapter and Chapters 8 and 9 may be read after Chapters 10 to 19.

The chapter brings together some miscellaneous topics.

More general proof is given that the values of r_l and r are averages of all values between t_1 and t_2 irrespective of what changes occur between these times.

The growth of populations of lesions or foci (i.e., local concentrations of disease) can be logarithmic even though the growth of individual lesions and foci is not. To apply the arbitrary and approximate rule that disease increases logarithmically until $x = 0.05$, one uses the value of x in the field as a whole even though x within foci is much higher. Foci necessarily occur in all fields in which disease is multiplying.

For the equations of earlier chapters to apply when disease increases beyond the logarithmic stage, propagules must act independently of one another. The experimental evidence supports this postulate of independent action, at least when the concentration of propagules does not exceed that found in natural epidemics.

7.1. The Relative Infection Rate as an Average

If the relative infection rate varies from hour to hour or from day to day, the average value for the whole period between times t_1 and t_2 is the average of all instantaneous values between these times. This follows from the equations already given. But for clarity let us review the matter shortly.

Consider first logarithmic rates. At any instant the relative infection rate is

$$\frac{1}{dt}\frac{dx}{x} = \frac{d}{dt}\log_e x$$

hence it follows that the increase of $\log_e x$ between t_1 and t_2 is

$$\log_e x_2 - \log_e x_1 = r_l(t_2 - t_1)$$

which is Eq. (3.3); and this equation gives r_l as the average value of the relative infection rate over the period between t_1 and t_2, whatever may be the nature of the changes in the logarithmic infection rate over this period.

Similarly, after the logarithmic phase of an epidemic, the relative infection rate at any instant is

$$\frac{1}{dt}\frac{dx}{x(1-x)} = \frac{d}{dt}\log_e \frac{x}{1-x}$$

Hence r calculated as

$$r = \frac{1}{t_2 - t_1}\left(\log_e \frac{x_2}{1-x_2} - \log_e \frac{x_1}{1-x_1}\right)$$

which is Eq. (3.5), correctly gives the average value of the relative infection rate over the period between t_1 and t_2, however much the relative rate may have changed in that period.

7.2. The Logarithmic Infection Rate

Picture a large field of potatoes. The field is very lightly infected with blight, and the few lesions are widely separated from one another. By hypothesis the lesions do not interfere with one another, i.e., they do not overlap. Picture further that spores form in these lesions, move away, and start new lesions. By hypothesis these new lesions too do not overlap either with one another or with the old lesions. On this hypothesis one deduces that the number of new lesions is proportional to the number of old. If the spores from 1000 old lesions start N new lesions, the spores from 2000 old lesions will start $2N$ new lesions, random variation apart. This is where we began in Chapter 2. The absolute infection rate dx/dt is proportional to x; the increase is logarithmic. The relative infection rate dx/xdt is independent of x. For example, Eq. (6.2) which relates r_l to p, R, and t mentions neither x nor x_0.

It is as useful a definition of the logarithmic relative infection rate r_l as any to say that it is independent of x. And a logarithmic infection rate is a direct deduction from a hypothesis that lesions do not overlap or interfere with one another.

7.3. The Growth of an Individual Lesion in Relation to the Growth of a Population of Lesions

Picture a single lesion of potato blight on a leaf. The lesion grows as the fungus extends outward from the lesion's perimeter into the healthy leaf. It grows radially. The relative growth rate of the lesion decreases as the lesion grows, because the ratio of the lesion's perimeter to its area decreases as it grows. The relative rate is not independent of the lesion's area, and is therefore not logarithmic.

The way an individual grows and develops does not decide the way a population grows—a child goes on teething whether his country's population is rising or falling. So also the form of growth of a lesion, which is not logarithmic, does not decide the form of increase of a population of lesions, which is logarithmic as long as the lesions are too few to overlap. The growth of a lesion and increase of a population are connected only in the rate. Favorable temperature which makes a lesion grow fast is also likely to make a population of lesions grow fast.

There is another way of looking at it. The logarithmic infection rate, which is the rate of increase of the population, is reflected in the distribution of lesions in age groups. The faster the rate, the higher is the proportion of young lesions. A constant rate means a constant distribution of age groups (i.e., a stable age structure) and a constant average age of the lesions. This is not easy to calculate directly. But some figures will illustrate the point. If $r_l = 0.5$ per unit per day, which means fast increase, 39.4% of the infected tissue has been infected less than 1 day, 23.9% between 1 and 2 days, 14.4% between 2 and 3 days, 8.8% between 3 and 4 days, 5.3% between 4 and 5 days, 3.2% between 5 and 6 days, and 5.0% more than 6 days. While r_l stays constant, these proportions do not change. If $r_l = 0.05$ per unit per day, which means slow increase, 4.9% of the infected tissue has been infected for less than 1 day and 74% for more than 6 days.

7.4. The Growth of Foci

A focus has been defined (Anonymous, 1953) as the site of local concentration of diseased plants or disease lesions, either about a primary source of infection or coinciding with an area originally favorable to establishment, and tending to influence the pattern of further transmission, of disease.

Foci vary in form. Sometimes they are concentrated and sharply defined. Within the foci the proportion of diseased tissue is great, and the boundary between the foci and the rest of the field is clear and sharp. Other foci are less conspicuous and merge into the rest of the field.

Zadoks (1961) described focus development of stripe rust of wheat caused by *Puccinia striiformis* around the isolated infected leaves in which the fungus normally survives the winter. The first sign of a focus is three to five infected leaves close together around the leaf that was the initial source of infection. Sometimes the leaves around this leaf, though not themselves forming spores, are covered with spores; and Zadoks suggested that dispersal of spores within foci at this stage is by leaves rubbing together in the wind rather than through the air. Later the leaf that was the source of overwintering fungus dies, and its shrivelled remains can usually be seen in the center of the small focus. When the focus involves up to ten leaves and extends about 10 cm. in the row, lesions appear on the erect leaves and offer their spores to the wind. When the focus extends over about half a meter in the original row, the two adjoining rows are also infected, and the soil between the rows is yellow with spores. Up to this stage the focus is inconspicuous from a distance, and infected leaves are obscured by overhanging young green leaves. But by the time the focus covers an area of a square meter, it can clearly be seen from a distance. Wind disperses spores freely at this stage; and if the weather is favorable new daughter foci—secondary foci—develop elsewhere in the field. These develop in much the same way as the parent focus, and in their turn form spores which disperse to start tertiary foci. So the process goes on, weather permitting, until the whole field is infected. If weather conditions allow disease to multiply fast, the field in the later stages of an epidemic is severely and, in outward appearance, fairly uniformly infected with only inconspicuous focal infection.

Hirst and Stedman (1960b) have described focal development of potato blight caused by *Phytophthora infestans* around infected tubers planted as seed. The fungus enters shoots growing from these tubers, forms spores on them, and then spreads to other shoots and plants, or under certain conditions of soil moisture, passes directly through the soil to infect leaves near it. At first new lesions appear only in plants near the initial source of fungus. Further spread intensifies and expands the focus until it contains one or two hundred plants. Then, if conditions favor distant dispersal, spread occurs beyond the focus.

This pattern, however, is not general for potato blight. Often the fungus enters a field, not in infected seed tubers, but only when sporangia blow in from other fields. This occurs during the phase of distant dispersal. When this happens, foci are not evident. Disease seems uniformly spread in the field, although in fact infection almost certainly develops in a large number of small foci close enough together to give the general impression of uniformity.

Posnette (1943), Thresh (1958a, b), and others have described the spread of swollen shoot of cacao. This virus disease is transmitted by

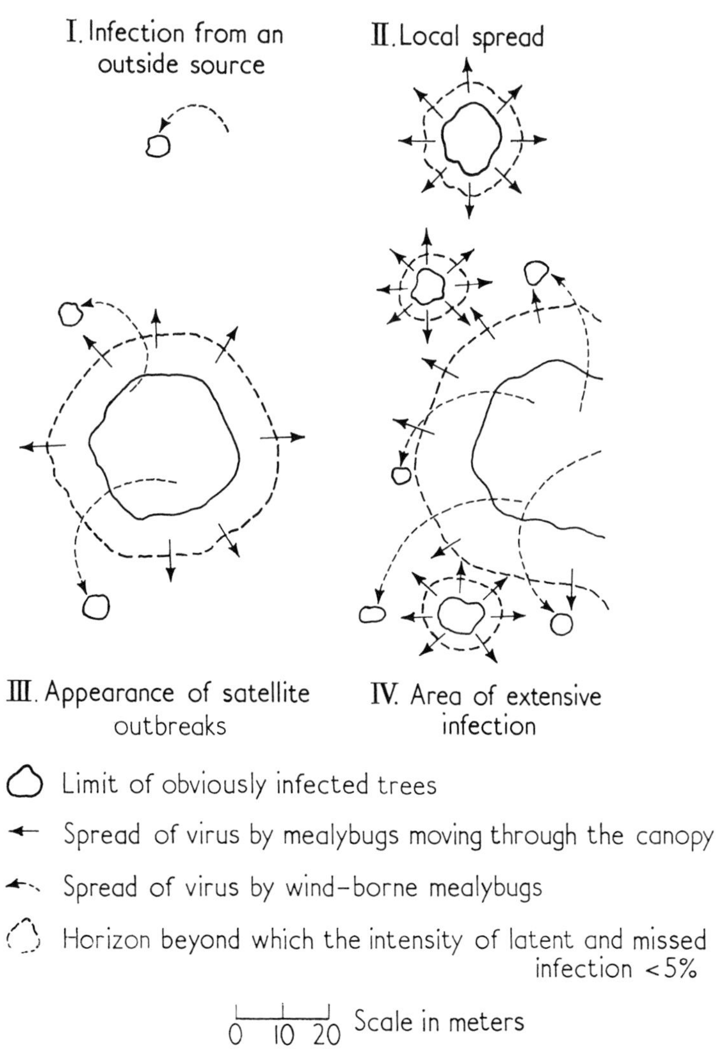

FIG. 7.1. The pattern of spread of cacao swollen shoot disease. After Thresh (1958b).

mealy bugs, especially *Pseudococcus njalensis*. A focus is developed around an infected tree. Much of the spread is from tree to tree in contact. This spread is assumed to be caused by crawling mealy bugs, and causes a compact and well-defined focus. New secondary foci start at some distance away. These are assumed to be started by mealy bugs, particularly first instar nymphs, blown out of the parent focus by wind. The new foci in turn increase in size, and start newer foci. And so the process continues. It is illustrated in Fig. 7.1.

Tobacco mosaic virus is spread mainly by men working in the fields. Much of this spread is in foci one row wide (Wolf, 1935), because plants are handled row by row up to the time of topping. Then the grower usually tops two rows at a time, and the foci become two rows wide.

7.5. The Spread of Pathogens in Relation to the Focal Pattern

The spread of pathogens over distances is the topic of a later chapter. But we cannot ignore spread entirely until then, and a few remarks are pertinent here.

When a pathogen has spread from a point (say, a lesion) and we draw a line from that point we find that daughter lesions are more abundant on the line near the point than further away. Spread of infection from a point source is along a gradient; it falls off along a line with distance from the source. This seems to be an invariable rule.

Foci, according to definition, are local concentrations of disease. It follows that where disease spreads and multiplies, there foci must arise. There can be no multiplication without gradients arising, and no gradients without foci arising. The focus is as much a natural unit of plant disease as the lesion itself.

Foci appear when disease multiplies. To adopt the reasoning in Chapter 4 one expects foci whenever it is logical to plot $\log [x/(1-x)]$ against time but not necessarily when it is logical to plot $[1/(1-x)]$ against time. (Foci can occur even when one uses $\log [1/(1-x)]$. For example, they can occur with fusarium wilt of cotton if the *Fusarium* or nematodes are unevenly distributed in the soil. But here they occur because of unevenness, not because of multiplication.)

Foci need not be conspicuous. A large number of small foci might give the appearance of uniform distribution of lesions. But with multiplication foci are necessary, conspicuous or inconspicuous, in one form or another.

The gradient influences the form. An interesting part is played by the inverse-square gradient: a gradient along a line in which the amount of disease varies inversely with the square of the distance from the point that is the source of inoculum. With gradients less steep than inverse-square gradients there will be foci without clear margins. Near the source of inoculum there will be more disease, say, per square meter than elsewhere, but the disease will extend away from the center, in ever decreasing amounts per square meter but, nevertheless, indefinitely. With gradients steeper than inverse-square gradients the margins are clear. The steeper the gradient, the more easily one can say, this is a focus, and this is healthy encircling field.

A spore-forming fungus extends a lesion by mycelium invading the healthy tissue around it. But it starts new lesions by the release of spores. A focus extends its limits continuously into the encircling field and grows bigger; but it also extends discontinuously to start new secondary foci beyond its own limits. The two processes, growing bigger and starting new foci, may involve different methods of dispersal. A focus grows bigger by consolidation and mopping up. The gradient can be steep, and steep gradients are often observed, but a steep gradient is not apt for discontinuous spread to start new foci.

In discussing Zadoks' work on stripe rust of wheat we noticed that *Puccinia striiformis* seems to spread both by leaves rubbing together and through air. It is possible that rain drops also disperse it, as Hirst (1961) has found them to disperse spores of *Puccinia graminis* apparently by collision with the spore-bearing surface. Leaf rubbing can only increase a focus in size. Under what conditions other forms of dispersal could start new foci is unknown.

The sporangia of *Phytophthora infestans* are borne by air. They are also dispersed by water (Hirst and Stedman, 1960a). Sporangia are formed only at high relative humidity, and infection occurs only with dew or rain. Water dispersal and dispersal in air of low turbulence during wet weather or on dewy nights probably increase foci in size. But *P. infestans* also spreads far (see Chapter 22). It is significant that the greatest concentrations of sporangia in the air above potato fields are in bright sunny spells before noon (Hirst, 1953, 1958), when conditions would seem adverse for infection but favorable for sporangia to move. Possibly *P. infestans*, a water lover, makes the traveler's sacrifice of comfort for the sake of getting around.

Workers seem to agree that mealy bugs crawling through the canopy enlarge existing foci of cacao swollen shoot disease and that new foci start where wind has blown viruliferous bugs.

The vagaries of man, the chief vector of tobacco mosaic virus, are enough to diversify the pattern of spread of mosaic in tobacco fields.

All pathogens can enlarge an existing focus; to stop being able to do this is to stop being pathogenic. To cause widespread epidemics they must be able to form new foci and for this they, perhaps, use more than one way of spreading.

7.6. The Growth of an Individual Focus in Relation to the Growth of a Population of Foci

In Section 7.2 it was deduced that an epidemic is in the logarithmic stage when lesions are too few to overlap. To adapt this to systemic disease, an epidemic is in the logarithmic stage when multiple infections do not occur. Earlier, in Section 3.4, we arbitrarily defined the logarithmic stage as ending when x exceeds 0.05. That is, we assume overlapping and multiple infection to be negligible until x exceeds 0.05. For the purpose of discussion, accept this arbitrary definition.

At what place do we determine when x exceeds 0.05? In the field as a whole? If $x = 0.05$ in the field as a whole, then it is much greater in the foci themselves, and within the foci there will be much overlapping and multiple infection. In the foci only? If $x = 0.05$ in the foci, it will be much less in the field as a whole.

The answer is, we determine x in the field as a whole. What then about overlapping of lesions or multiple infection with systemic disease within the foci? Does this not destroy the whole concept of the logarithmic stage as the stage at which lesions (or systemically infected plants) do not interfere with one another, and the relative infection rate is independent of the number of lesions? Within the foci, lesions obviously interfere with one another, and the relative infection rate in the focus is not independent of the number of lesions in the focus.

To understand the anomaly one must regard foci, not lesions, as the units of disease. Each individual lesion grows in a way that is not logarithmic, but the population of lesions can nevertheless increase logarithmically. This was the theme of Section 7.3. So, too, each individual focus may grow in a way that is not logarithmic, but the population of foci may nevertheless increase logarithmically.

If lesions do not overlap, the epidemic is in the logarithmic stage; if foci do not overlap, the epidemic is in the logarithmic stage. The relative infection rate in an epidemic is reflected in the distribution of lesions in age groups; the relative infection rate in an epidemic is reflected in the distribution of foci in age groups.

Changing from lesions to foci brings in only one important difference. One needs to deal with populations of foci instead of populations of lesions. One must estimate disease over an area large enough to include a whole population of foci.

As foci enlarge infection spreads largely from the perimeter where diseased or healthy plants or leaves meet. Spread in this way means that the rate of growth of the focus slows down as the focus enlarges, other things being equal. Thresh (1958b) measured the change with cacao swollen shoot disease. The smallest foci in his experiments contained an average of 3.9 visibly infected trees. Surrounding them were 3.2 trees with disease still in the incubation stage. This ratio of new to old infections is $3.2/3.9 = 0.82$. The largest foci contained an average of 142 visibly infected trees. Surrounding them were 51 trees with disease still in the incubation stage. The ratio here is 0.36. The relative growth rate of individual foci fell off considerably as they enlarged.

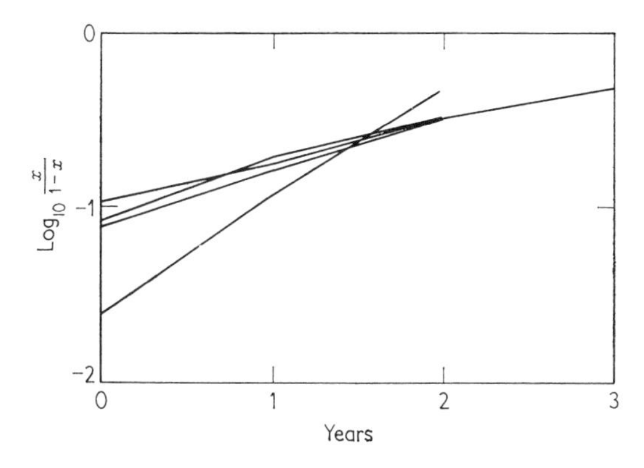

FIG. 7.2. The progress of swollen shoot disease of cacao in four plantations. From highest to lowest disease at time 0, they were in Ghana, Trinidad, Nigeria, and Ghana, respectively. From tables published by Thresh (1958b).

If during the course of an epidemic the relative infection rate of a population of foci is independent of the relative growth rate of individual foci, the population is large enough for r_l and r to be given the meaning we have given them in earlier chapters.

Figure 4.1 illustrates five epidemics of potato blight in which r maintained its values as the epidemics proceeded. Figure 3.2 illustrates an epidemic of stripe rust of wheat in which this also happened. Zadoks (1961) found this to be the rule for stripe rust epidemics. This proves nothing. But it at least shows that r does not reflect the growth rate of individual lesions or foci.

Cacao swollen shoot disease gives better evidence. Climate changes less from year to year than from day to day. Yearly relative infection rates of systemic disease in a perennial crop are likely to vary less for climatic reasons than daily rates in an epidemic of a seasonal disease such as potato blight. Also, because a systemic pathogen extends throughout the tree, foci are especially clear, with few healthy trees within them. This makes the test more stringent; if infection rates of a population of foci reflect the growth rates of individual foci this should be clearest with systemic disease, and especially systemic disease in a perennial crop. Figure 7.2 plots $\log_{10}[x/(1-x)]$ against time for four separate epidemics of cacao swollen shoot disease. These are epidemics in which at least three yearly readings were taken while x was less than 0.35. This limit was set because within this limit r in an adequate population of foci is likely to stay constant and equal to r_l if environmental

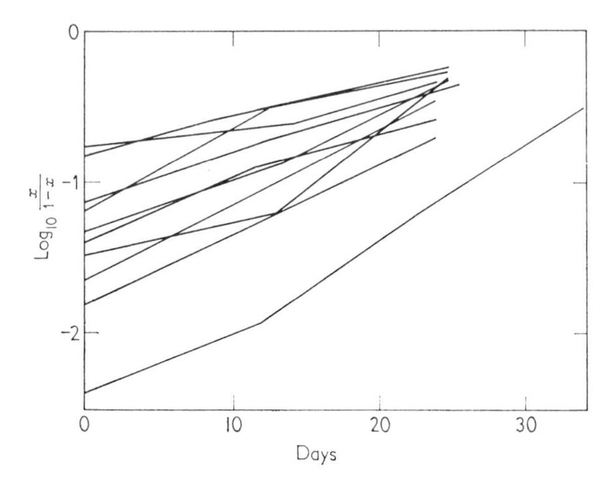

FIG. 7.3. The progress of tobacco mosaic in ten fields of tobacco. Data of Wolf (1935).

conditions stay constant (see Section 6.9). The first year of observation is recorded as 0 on the time axis. Reading $\log_{10}[x/(1-x)]$ from top to bottom in year 0, the epidemics were recorded by (*a*) Posnette and Todd in Ghana in 1661 trees, (*b*) Dale in Trinidad in 576 trees, (*c*) Thresh in Nigeria in 827 trees, and (*d*) Posnette and Todd in Ghana in 1139 trees. All the information is given by Thresh (1958b). In three of the epidemics the curve is almost straight; r stayed almost constant. In the fourth [epidemic (*b*) recorded in Trinidad] r falls slightly. This could be owing to natural variation or to the inadequacy of $1-x$ as a correction factor. (See Section 4.5.) But this was also the epidemic in the smallest number, 576, of trees. This number is perhaps too small for an adequate population of foci. The matter cannot now profitably be pursued; but one can note that in the other three epidemics, each in more than 800 trees, r reflected nothing of the growth rate of individual foci. With 800 trees, it seems, the population of foci is large enough to obscure the behavior of each individual focus.

Tobacco mosaic is another apt disease for illustration. The relative growth rate of foci that spread mainly up and down a row in one dimension should fall quickly and differ greatly from a logarithmic rate. Figures 7.3 and 7.4 plot $\log_{10} [x/(1-x)]$ against time for a number of epidemics. They analyze the data of Wolf (1935) for different tobacco fields. Figure 7.3 uses Wolf's Tables 10 and 12; Fig. 7.4 his Table 11. The date of the first examination of each field is recorded as time 0, and entries continue for as long as x was less than 0.35. Fields in which

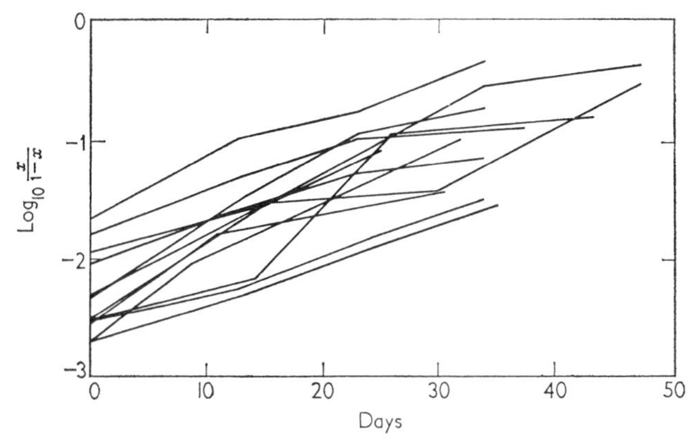

FIG. 7.4. The progress of tobacco mosaic in twelve fields of tobacco. Data of Wolf (1935).

less than three examinations were made while x was less than 0.35 are necessarily excluded. The data in mass show no general tendency for r to fall with time and reflect focal growth rates. Only two of the fields had less than 1000 plants, eleven had between 1000 and 2000, seven between 2000 and 3000, and two had more than 3000. No difference was noticed between these groups. It seems that they all gave an adequate population of foci.

7.7. The Mass Increase of Foci

This chapter has proceeded from the growth of a single lesion to the increase of disease in a population of foci in a field or plantation. We must now go on to the massed increase of disease in whole countries (we shall deal in this section with three islands) across all boundaries of fields or plantations.

We follow the same argument as before. Disease varies from field to field and from plantation to plantation. It varies with altitude, aspect, soil type, and technique of cultivation. But as long as there are enough fields (to mention only fields) to make a population of fields, disease will increase as in a simple population of lesions. It will increase logarithmically if x is small enough (i.e., less than 0.05 by our arbitrary rule) for the country as a whole despite local heavy incidence. For this we assume only that disease occurs throughout the whole country. This makes islands useful areas for study.

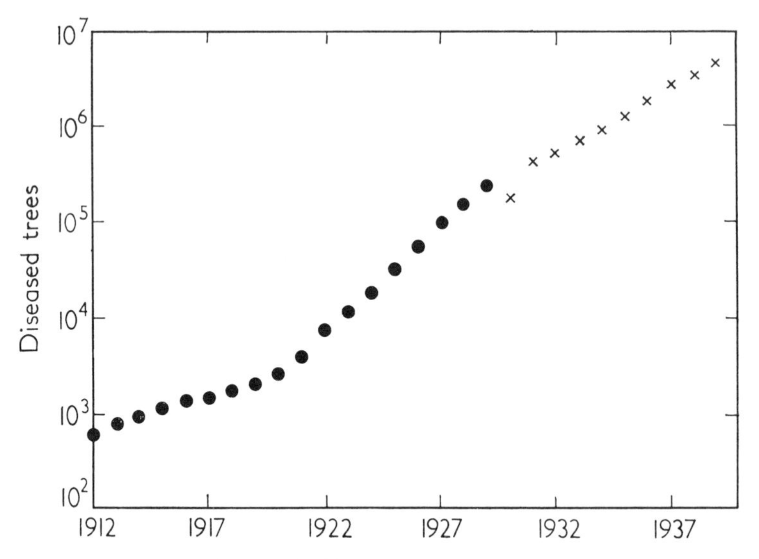

FIG. 7.5. The cumulative total of banana trees in Jamaica affected with Panama disease caused by *Fusarium oxysporum* f. *cubense* from 1912 to 1939. Data of Watts Padwick (1956). From 1930 on, the records exclude the parish of Portland.

Panama disease of bananas is caused by a soil-borne fungus *Fusarium oxysporum* f. *cubense*. In Jamaica it was first found in 1911. Soon thereafter legislation was passed requiring infected banana plants to be removed and the soil they were in to be treated with a fungicide. Inspectors were appointed to carry out the job. They recorded the number of plants destroyed from 1912, when there were 625 infected trees, to 1939, by which time over 4 million plants had been destroyed. Watts Padwick (1956) extracted the information from the annual reports of the Jamaican Department of Agriculture, and from his table Fig. 7.5 has been constructed. It shows the cumulative total of diseased trees year by year. There is a break from 1930 on, when the records excluded the parish of Portland. For calculation from 1930 on, we have assumed that 58% of the previous infections were in this parish, on the basis of the 1929 inspections. This point is not very important in the long run. Results in Fig. 7.5 are plotted on a logarithmic scale. It was not possible to estimate $1-x$ and apply a correction factor. But to judge from the fact that in 1937 27 million stems of bananas were exported from Jamaica, $1-x$ must have been near enough to unity to be unimportant except possibly in the last few years.

From 1920 to 1939, the number of infected plants rose from 2000 to 4 million, and the relative infection rate stayed fairly steady. From 1920 to 1929—which, so far as we know, was an interval of uniform records—$r_l = 0.51$ per unit per year. Before 1920, the rate was slower. In this year the regulations were changed. Until 1920 all apparently healthy plants within a distance of one chain (about 20 meters) from a diseased plant had to be destroyed along with the diseased plant itself (Wardlaw, 1961). This meant that many apparently healthy plants had to be destroyed. Then the regulations were relaxed in order to reduce the number of apparently healthy plants that had to be destroyed each time. The stricter regulations of the earlier period are reflected in a slower infection rate. From 1915 to 1919, inclusive r_l was only 0.14 per unit per year. (We start at 1915, by which time more than 1000 plants had already been infected—which seems to be enough for the massed data we need.)

If one allows for the change in 1920, the evidence is for a substantially steady logarithmic increase of Panama disease, despite the fact that disease varied from plantation to plantation and parish to parish.

Sudden death disease of the clove tree, caused by fungus *Valsa eugeniae*, is the name given to a wilt disease that causes seemingly healthy clove trees to die within a few days. It appeared in epidemic form in the islands of Zanzibar and Pemba, mainly but not entirely in mature trees. The government established clove nurseries and gave away free young trees to farmers to replace losses from sudden death. New plantations were rare so it can be assumed that the nursery trees were used for replacement. From the relative number of nursery trees issued year by year, Nutman and Sheffield (1949) were able to estimate the annual death rate. They also estimated that in 1946 half of the mature trees in Zanzibar had been killed by swollen shoot disease.

By taking $x = 0.5$, i.e., $\log [x/(1-x)] = 0$, in 1946, Fig. 7.6 has been drawn from Nutman and Sheffield's data for Zanzibar. More than half a million trees were killed in Zanzibar, so Fig. 7.6 reproduces massed data. Plotting $\log [x/(1-x)]$ against time shows r to have stayed fairly constant at roughly 0.42 per unit per year.

In the 2.5 million trees on Pemba, total losses were smaller, and the epidemic was in the logarithmic stage. Nutman *et al.* (1951) found the annual rate of increase to be 14%. Hence $r_l = 0.13$ per unit per year ($e^{0.13} = 1.14$).

It is confirmation of the theme of much of this chapter that although the infection rate of the population of trees was logarithmic, individual foci on Pemba enlarged radially and their frequency distribution in size

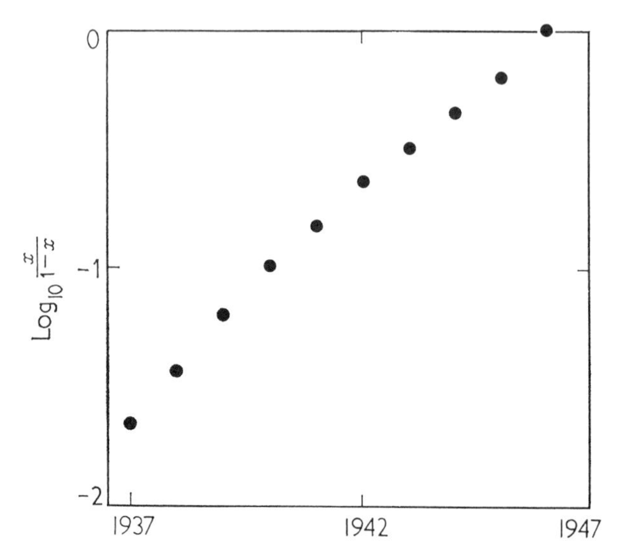

FIG. 7.6. The course of an epidemic of sudden death disease of clove trees caused by *Valsa eugeniae*, from 1937 to 1946. Data of Nutman and Sheffield (1949) for Zanzibar.

groups stayed constant, the average size through the years staying at 25 trees per focus (Nutman *et al.*, 1951). That is, the foci had the stable age structure expected from a constant, logarithmic, relative infection rate.

7.8. Massed Foci of Potato Blight. The Epidemic *S*

There is a similar increase of potato blight in massed fields. So long as there are enough fields to make a population of fields, disease increases as in a simple population of lesions. The epidemic tends to build up in a rough *S*, though in detail its progress inevitably depends, among other things, on changes of weather from day to day.

Figure 7.7 reproduces data on 117 fields in the sand areas of the Netherlands in 1953 (Anonymous, 1954). The fields were not treated with a fungicide. The Netherlands system of blight assessment was interpreted by Cox and Large's (1960) conversion curve. Data for the three varieties, Bintje, Eigenheimer, and Voran, are analyzed separately.

For each variety the blight progress curve is roughly S-shaped. Another view of the epidemic S is given by Fig. 22.1, which shows the cumulative total of fields in which blight had been recorded.

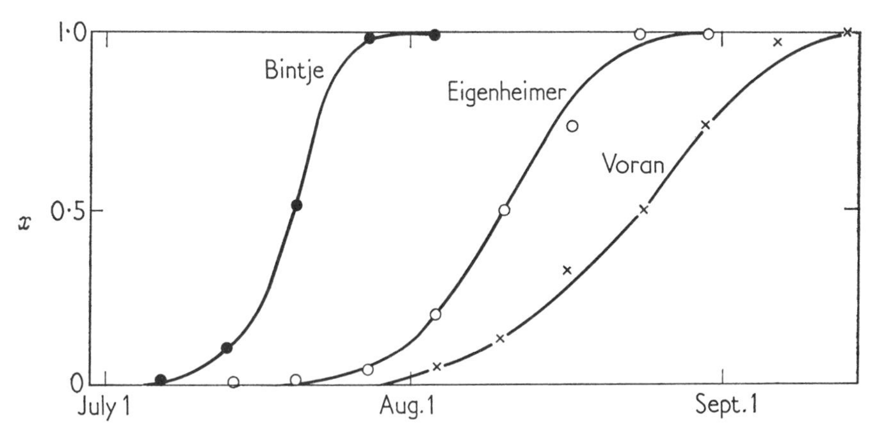

FIG. 7.7. The progress of blight in 117 fields of three potato varieties. The data are for the sand area of the Netherlands in 1953 (Anonymous, 1954).

7.9. The Independent Action of Propagules. The General Problem

We are concerned here with the relation between the amount of inoculum and the number of lesions it produces (or, if the disease is systemic, with the number of plants it infects).

The relation does not affect our equations for the logarithmic stage of an epidemic. Provided that there is no overlapping of lesions (or multiple infection, if the disease is systemic), an epidemic will proceed logarithmically, irrespective of whether propagules act independently of one another or in concert with one another, synergically. Even with synergic action between a number of propagules, r_l will remain independent of x. Synergism, or lack of it, would, indeed, be one of the factors, like atmospheric temperature and humidity, that influence the numerical value of r_l; beyond that its effect would not go.

But our equations after the logarithmic stage assume the independent action of propagules. Synergic action would make r increase progressively as x increases beyond the logarithmic stage. No evidence for a tendency for such an increase has been noticed; to this extent the general evidence from epidemics is against a hypothesis of synergic action. But a more detailed study of the matter is needed.

It is common experience that to be sure of causing infection an experimenter should use more than one spore or propagule. In experiments soon to be described in some detail Petersen (1959) found that it often took about 100 viable uredospores of *Puccinia graminis tritici* to produce one pustule of wheat stem rust, and Knutson and Eide (1961) found that about 1 to 4% of the sporangia of *Phytophthora infestans* that germinate cause blight infection on potato leaves.

The hypothesis of independent action assumes that all these spores act independently of one another. In any group of spores, the proportion capable of initiating infection may be small: this proportion will be determined by many of the various factors of host, pathogen, and environment that influence R. But those that initiate infection do so without help (or hindrance) from others.

The hypothesis of synergic action has been put forward largely to explain why so many spores are needed if the experimenter wishes to be sure of starting an infection. As sometimes propounded the hypothesis carries the rider that one single spore on its own cannot possibly infect. If this rider were true, one could immediately exclude any suggestion of synergism from the attack of *P. infestans*, many powdery mildews, and rust fungi because routine single spore inoculations with these fungi are now an experimental commonplace.

But the rider is unwarranted. It does not logically follow the hypothesis. If spores in large numbers help one another to infect—i.e., if there is synergic action—it does not necessarily follow that a single spore without help cannot possibly infect.

Single spore infections are known to occur when uredospores of *Puccinia graminis tritici* attack wheat. Yet, contrary to the rider, the results of Petersen (1959) show that uredospores of *P. graminis tritici* act synergically when they are present in large numbers. Uredospores were allowed to fall on wheat plants which were then kept for 24 hours in an incubation chamber and a further 4 days in a glasshouse. Leaves were then stained by the method developed by Bell (1951) and examined for infection points. Figure 7.8, drawn from Petersen's data, shows how the number of uredospores affected the number of points of infection. The first nine points, up to 2810 spores and 330 infection points per cm.^{2}, have been treated as a group. The regression coefficient of the number of infection points on the number of uredospores is 0.108. The regression line passes practically through the origin, which is evidence that the spores acted independently of one another. The last six points, up to 5640 spores and 1520 infection points per cm.², have also been treated as a group. The regression coefficient is 0.490. This is significantly

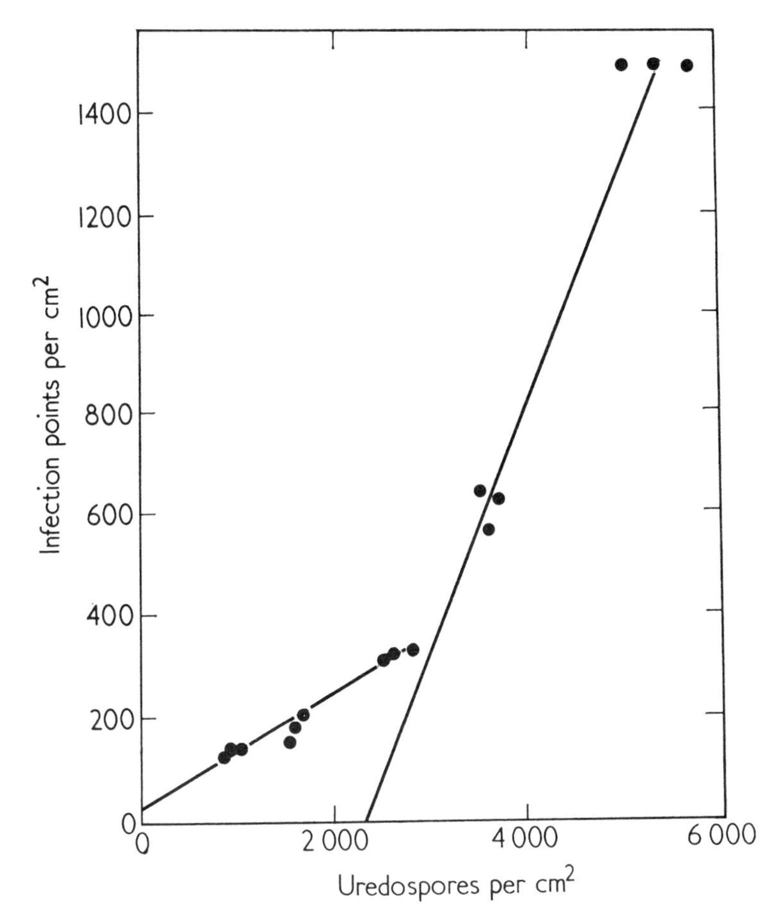

FIG. 7.8. The relation between the number of infection points and the number of uredospores of *Puccinia graminis* per cm.² Data of Petersen (1959).

higher statistically than the previous figure. At the higher concentrations spores infected more easily. There were more infection points per 100 spores. Spores helped each other to infect; they infected synergically. The regression line cuts the x-axis to the right of the origin, which is also evidence of synergic action.

These results were with infection points. Results with pustules themselves were not quite so clear, because at higher concentrations of spores there was great competition for space, with more infection points per pustule. With about 900 uredospores per cm.^{2}, there were approximately 2.7 pustules per 100 infection points. With about 5300 spores per cm.^{2}, there were only 1.4 pustules per 100 infection points. Nevertheless even pustule numbers gave evidence of synergic action. With 850 uredospores per cm.^{2}, 100 spores produced an average of 0.43 pustules; with 6200 uredospores per cm.^{2}, 0.64 pustules. This difference is statistically significant.

This synergic action can perhaps be traced to the volatile product released by uredospores which stimulates spore germination (Allen, 1957). Petersen found clear evidence that high concentration of spores stimulated germination. With 920 uredospores per cm.^{2}, 28% of the spores germinated and formed appressoria; with 5300 uredospores per cm.^{2}, 52% germinated and formed appressoria.

One can probably usefully drop the term, synergic action. It is better to talk of mutual interference. Interference can be in either direction. Spores (or other propagules) may help one another to infect, or they may hinder one another. Uredospores of *Puccinia graminis* produce a substance that promotes germination (Allen, 1957), but they also produce a substance that inhibits germination (Allen, 1955). In principle, at any rate, spores in mass can either help or hinder one another by their own products.

7.10. The Independent Action of Propagules. *Puccinia graminis*, *Phytophthora infestans*, and Some Other Pathogens

Puccinia graminis and *Phytophthora infestans* carry much of the narrative in the chapters on disease control, and must be examined in some detail.

Petersen's data which have just been discussed show that uredospores of *Puccinia graminis tritici* act independently of each other up to a concentration of 2600 spores per cm.^{2} which experimentally produced 18 pustules per cm.^{2}. This number of pustules is roughly equivalent to 100% infection on the modified Cobb scale commonly used in the United States for estimating the percentage of rust infection. That is, spores act independently of one another through the whole course of a natural epidemic. The data of Rowell and Olien (1957) also show independent action up to the highest incidence of disease studied by them: about 20 pustules per leaf.

When the quantities of spores are consistent with what one might find in natural epidemics, there is no evidence of mutual interference. When the quantities are too high to be relevant to natural epidemics, there is evidence of mutual interference. In scanning the literature for evidence about interference, one must ensure that evidence irrelevant to epidemiology is excluded.

The results of Knutson and Eide (1961) show that spores of *Phytophthora infestans* act independently up to concentrations that produce 250 lesions per potato plant. The potato plants were small (only 12–15 in. high); and 250 lesions per plant represents severe infection, as much as can occur in the field. For potato blight, just as for stem rust, we can assume that spores act independently, which is an implied postulate in our equations.

In a later chapter information is given on wilt of tomatoes caused by *Fusarium oxysporum* f. *lycopersici*, and smut of wheat caused by *Tilletia* spp. (see Figs. 13.1, 13.2, and 13.3). With both these organisms, too, the evidence suggests that propagules act independently of one another over the concentrations likely to be found in natural epidemics.

In the literature there are examples of the lack of a simple relation between the number of spores and the incidence of disease, but we need not necessarily infer from them that spores interact with one another. For example, Colhoun (1961) studied the incidence of club root of cabbages caused by *Plasmodiophora brassicae*. To quote just two figures from his tables, 42% of the plants became infected in soil containing one spore per gram and 51% in soil containing 10 spores per gram. Incidence of disease did not keep pace with increased concentration of inoculum. Club root is sensitive to environmental conditions and to the nutrition of the host plants. That the plants were not all equally susceptible or vulnerable to infection seems more likely than that spores began to interact with one another at concentrations of the order of one spore per gram of soil. To put the same idea differently, we cannot from Colhoun's data relate incidence of club root to the number of spores simply; we suggest that this is largely because the correction factor $1 - x$ is quite inadequate for the club root data, whereas the factor seems reasonably adequate for the data on bunt of wheat and fusarium wilt of tomatoes presented in Figs. 13.1, 13.2, and 13.3.

CHAPTER 8

Corrected Infection Rates

SUMMARY

This and Chapter 9 may be read after Chapters 10 to 19.

Corrections to infection rates are needed for one reason or another.

As plants grow, new healthy tissue tends to dilute the concentration of infected tissue if the pathogen causes local lesions. Appropriate corrections are sometimes needed.

Infected tissue is removed from taking further part in an epidemic when it grows old and sterile, as it often does at the center of a lesion or focus. It is, also, at least partly, removed when infected plants at the center of a focus become spatially isolated from healthy plants and thus are less able to pass on infection.

The chapter studies in some detail what happens when tissue is infectious for a limited period after the latent period, and then becomes sterile and removed from the epidemic. A basic infection rate corrected for removals can be calculated as well as a threshold value of this rate below which an epidemic cannot start.

8.1. Correction of r_l and r for the Growth of the Host Plants

A correction is needed because of the growth of the host plants. To estimate r_l and r we use the proportion x of infected tissue. The growth of the host plants—the increase in the mass of susceptible tissue—dilutes the proportion of disease. Indeed it is possible for x to decrease even while disease is increasing, if the host grows faster than the pathogen.

Consider a pathogen that causes local lesions. Let y be the mass of susceptible host tissue. At any instant the relative growth rate of this host tissue is

$$\frac{dy}{ydt} = \frac{d}{dt} \log_e y$$

At any instant in the logarithmic stage of the epidemic the relative increase in the number of lesions is

$$\frac{d}{dt}\log_e x + \frac{d}{dt}\log_e y$$

Hence the relative infection rate averaged between times t_1 and t_2 is

$$\rho_l = \frac{1}{t_2 - t_1}\log_e \frac{x_2 y_2}{x_1 y_1}$$

where the subscripts 1 and 2 have the same meanings as before.

For convenience we write

$$\rho_l = \frac{1}{t_2 - t_1}\log_e \frac{m x_2}{x_1} \tag{8.1}$$

where

$$m = \frac{y_2}{y_1}$$

If the mass of susceptible tissue doubles between t_1 and t_2, then $m = 2$.

Similarly, at any instant after the logarithmic stage the relative increase in the number of lesions, corrected by $1-x$, is

$$\frac{d}{dt}\log_e \frac{x}{1-x} + \frac{d}{dt}\log_e y$$

Hence the relative infection rate corrected for the growth of the host and averaged between t_1 and t_2 is

$$\rho = \frac{1}{t_2 - t_1}\log_e \frac{m x_2(1-x_1)}{x_1(1-x_2)} \tag{8.2}$$

It is not ordinarily necessary to correct for the growth of the host plants if the disease is systemic, because the growth of the host does not necessarily affect x.

Zadoks (1961) has studied the growth of a field of wheat and the increase in area of foliage infected by *Puccinia striiformis*, the cause of stripe rust. His results are illustrated in Fig. 8.1. Some detailed figures were kindly supplied by Dr. Zadoks and serve as an example. On April 20 the total area of foliage (healthy and diseased) was 1.89×10^8 cm.^{2} per hectare, and the proportion of foliage diseased was $x = 0.0015$. On June 1 the corresponding figures were 3.98×10^8 cm.^{2} per hectare and $x = 0.275$. Estimate r and ρ.

$$r = \frac{2.30}{42} \log_{10} \frac{0.275 \times 0.9985}{0.0015 \times 0.725}$$

$$= 0.132 \text{ per unit per day}$$

$$\rho = \frac{2.30}{42} \log_{10} \frac{3.98 \times 0.275 \times 0.9985}{1.89 \times 0.0015 \times 0.725}$$

$$= 0.149 \text{ per unit per day}$$

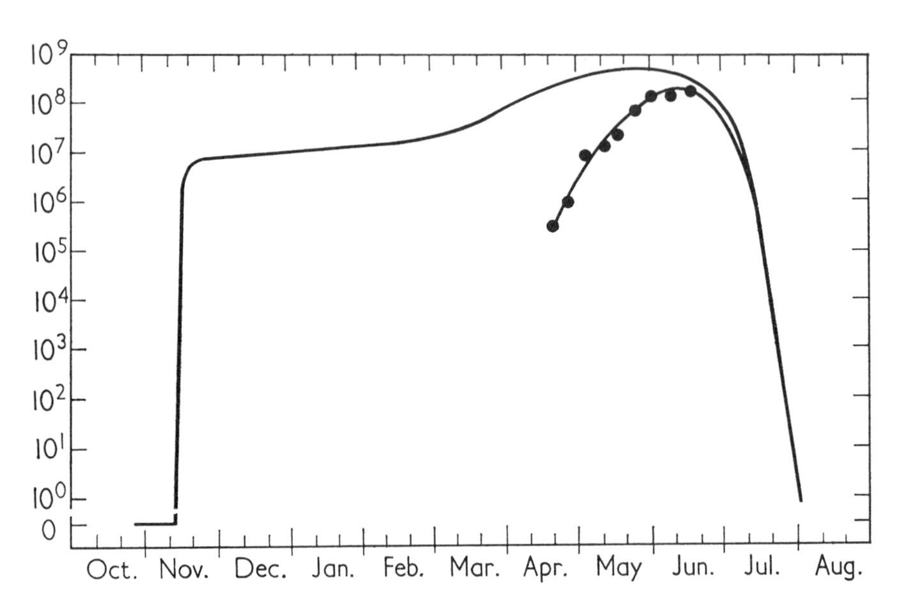

FIG. 8.1. The growth of a field of wheat and the increase of stripe rust caused by *Puccinia striiformis*. The upper line shows the development of foliage in a wheat field from the date of seeding in October until maturity at the end of July. The figures are in square centimeters per hectare, and refer to all foliage, healthy and diseased. The lower line shows the increase in the area of infected foliage, also in square centimeters per hectare, from April when disease was first recorded. After Zadoks (1961).

The correction is usually relatively small with diseases such as potato blight, both because r itself is usually high and because blight usually attacks when the plants are nearly fully grown. But with diseases such as those of fast-growing young spring foliage of fruit trees infection rates must be corrected.

8.2. Correction for Removals

Infected tissue may be removed from active participation in an epidemic in various ways.

As a lesion of potato blight grows through a leaf, there is an outer zone of mycelium working its way into healthy tissue. Behind it is a zone in which sporangia are formed if the humidity is high enough. Inside this sporing zone is a sterile zone, containing perhaps an occasional undispersed sporangium but producing no new sporangia even under optimal conditions of temperature and humidity. This sterile zone is removed from the epidemic. It neither produces inoculum nor is susceptible to reinfection by *Phytophthora infestans*.

A parallel form of removal occurs sometimes with foci instead of lesions. The focus like the lesion sometimes has an outer zone in the latent stage of infection, within this an infectious zone, and within this a sterile zone removed from the epidemic. This happens with swollen shoot disease of cacao; trees in the center of a large focus are killed by the virus, and carry neither virus nor vector.

Another form of removal is mainly spatial. Infected plants in the center of a large focus of systemic disease are normally in contact only with other infected plants. The larger the focus, the further they are away from healthy plants and the more innocuous they become, even if they do not die. They become removed, partially at any rate, from the epidemic.

One can visualize a rough and very irregular order of increasing importance of removals.

When a stock of "seed" potatoes is planted year after year, viruses M, S, and X can increase virtually without removal. Foci are broken up and destroyed every year when the crop is harvested, stored in bulk, and the tubers planted the following season in random order. These viruses increase in varieties in which they produce only mild symptoms without conspicuous stunting or harm. They are spread by contact and are therefore not dependent on visits of vectors. There is in fact nothing to remove infectious material on an appreciable scale except in those countries where man interferes by using serological tests and removing infected plants.

With tobacco mosaic, removals are more considerable, since foci develop along the rows and diseased plants become increasingly distant from healthy plants.

With cereal rusts, removals are probably not great when inoculum enters a field fairly uniformly, as with uniform artificial inoculation or when spores blow in from distant fields. Foci are then inconspicuous. Removal by the death of host tissue is small. Some cereal rust fungi live in balance with the susceptible host. Allen (1926) records that with 16-day-old infections of *Puccinia recondita* in wheat only 1% of the host cells, at most, are dead. Chester (1946) states that each pustule of wheat leaf rust caused by this fungus produces some 2000 spores a day for about 2 weeks. In corn infected with *Puccinia polysora*, each pustule produces from 1500 to 2000 spores a day for 18 to 20 days (Anonymous, 1958; Cammack, 1961). On large, rank leaves of oat plants, pustules of *Puccinia coronata* have been kept for a month, all the time producing fresh crops of spores (Durrell and Parker, 1920). Asai (1960) records that in the early stages of an epidemic of stem rust in wheat caused by *Puccinia graminis* the number of pustules and spore production remained stable for at least 10 days.

For the sake of calculation accept that a pustule of wheat stem rust takes 10 days to develop from the time of initial infection to the stage of spore production (i.e., $p = 10$ days) and thereafter produces spores for at least a further 10 days. Assume also from Asai's data that $r_l = 0.41$ per unit per day. (See Exercise 2 at the end of Chapter 3.) Then in the logarithmic stage of an epidemic 98.4% of the infected tissue is in the latent stage and has not yet produced spores; 1.6% of the infected tissue has been forming spores for up to 10 days; and only 0.027% of the infected tissue has been forming spores for more than 10 days or has stopped forming spores. That is, less than 0.027% of the infected tissue has been removed because pustules have aged.

Potato blight caused by *Phytophthora infestans* is further along the series of increasing removals. As we have already noticed, the fungus advances through a leaf leaving behind the sporing zone a zone of sterile tissue. The process has been studied in detail by Lapwood (1961a, b, c). The width of the sporing zone or ring between newly infected tissue on the outside and the sterile center on the inside is an important factor in the resistance or susceptibility of potato varieties. As such, it will be discussed in a later chapter. For present purposes we note that the sporing ring is approximately a single day's growth of a lesion in susceptible varieties in conditions favorable for spore formation. It is less than a day's growth in more resistant varieties or when the relative humidity of the air is not continuously high. By using droplets of water with spores to infect four susceptible varieties, Lapwood (1961b) found that the average diameter of the area bounded by and including the sporing ring was 8.3 mm. on the fourth day after inoculation, and 13.9 mm. on the fifth day. On the fifth day the diameter of the sterile area within the sporing ring was 7.6 mm. The sterile area on the fifth day had almost caught up the area bounded by and including the sporing ring of the fourth day. In other experiments Lapwood marked the outer boundary of the sporing ring with pin pricks. Under conditions of continuous high humidity sporing was continuous between consecutive days. The pin pricks that marked the outer boundary of the sporing ring on the one day marked the inner boundary of the ring on the next day. Tissue was infectious for only 1 day, whereas with the cereal rusts it can be infectious for 10 to 30 days.

Accept that tissue with blight is infectious for 1 day. Accept also that the latent period $p=4$ days and that $r_l=0.46$ per unit per day. These are all figures obtained experimentally for the potato variety, Majestic. On them, 84.2% of infected tissue is in the latent stage and has not started to form sporangia; 5.9% of the infected tissue forms sporangia; and 10.0% of the infected tissue is past forming sporangia and is sterile. In the language of this chapter, 10% of the infected tissue has been removed. These figures are for the logarithmic stage of an epidemic.

Finally, we come to diseases such as loose smut of wheat caused by *Ustilago nuda*. The fungus enters the florets, then the seeds. When infected seed is planted, the fungus moves systemically through the growing part to the grain. This disintegrates and releases spores to infect the new season's florets and to complete the cycle. Infectious tissue passes on infection and then dies. Removals are heavy. The infection period is reduced practically to a point in time. Infection proceeds year after year in a series of generations each 1 year long.

8.3. Relation between the Corrected Basic Infection Rate R_c and r

Consider the effect of removals quantitatively. Potato blight caused by *Phytophthora infestans* is a good example. As we have seen, a blight lesion on a leaf consists of an outer zone not yet forming sporangia. The tissue in this zone has been infected for less than p days. Inside this is the sporing zone or ring. The tissue in this zone has been infected for more than p days but less than $i+p$ days, where i is the period over which tissue stays infectious. Inside this again, at the center of the lesion, is tissue that has stopped sporing. This sterile tissue has been infected for more than $i+p$ days.

The appropriate equations now are:

$$\frac{dx_t}{dt}=rx_t(1-x_t)$$

$$\frac{dx_t}{dt} = R_c(x_{t-p} - x_{t-i-p})(1 - x_t) \tag{8.3}$$

Here R_c is the basic infection rate corrected for removals.

By division and rearrangement

$$R_c = r \frac{x_t}{x_{t-p} - x_{t-i-p}} \tag{8.4}$$

Let us apply Eq. (8.4) to the data of Large (1945) on potato blight. These same data were given in Figs. 3.1 and 4.1 (curve B) and analyzed in Sections 5.5 and 5.6 to show the relation between r and R. The potato variety was Majestic. For this variety we take $p=4$ days (see

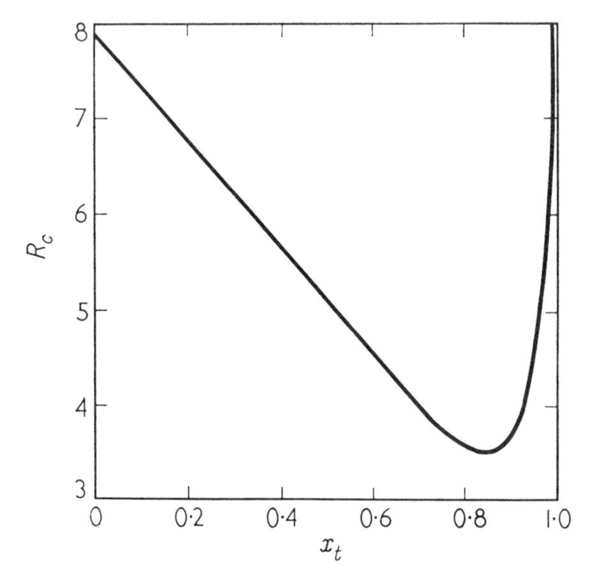

FIG. 8.2. The variation of R_c with x_t, when $r = 0.46$ per unit per day, $i = 1$ day, and $p = 4$ days.

Sections 5.5. and 5.6) and $i=1$ day (see Section 8.2). From Large's data, $r=0.46$ per unit per day. (See Exercise 8 at the end of Chapter 3 and Section 5.5.)

Figure 8.2 applies these data to Eq. (8.4). Compare it with Fig. 5.3 which uses the same data presented in the same way except that removals are not allowed for. In Fig. 5.3, without allowance for removals, R falls steadily as x_t increases over the whole course of the epidemic.

In Fig. 8.2, with removals allowed for, R_c falls steadily until x_t exceeds 0.8. In the process, R_c falls to somewhat less than half its initial value, from 7.86 to about 3.5 per unit per day. In Fig. 5.3, R fell from 2.90 to 0.46 per unit per day, a fall to somewhat less than one-sixth of its initial value. Removals are clearly a factor affecting the decrease of R during an epidemic. But they are clearly not the only factor. Two others were suggested in Section 5.6. *Phytophthora infestans* selectively attacks the most susceptible or vulnerable tissue (e.g., the bottom leaves) first. As the attack proceeds, the foliage opens and the ecoclimate becomes less favorable for infection.

8.4. Balance in Epidemics

There are no experimental data that adequately cover the last stages of a blight epidemic when x exceeds 0.8 or 0.9. At this stage Fig. 8.2 shows R_c to increase sharply if r stays constant. But no factors are known that would give R_c a final boost; and it is more reasonable to interpret the curve in the opposite way and take r to fall sharply if R_c stays constant or continues to fall. The removal of infectious tissue slows down an epidemic much more near its end than at its start.

If blight strikes a field of mature potato plants that have stopped growing, x will continue to increase even if r falls, provided that $r > 0$. But with disease in, say, natural vegetation that is growing, a balance is normally struck between new infections, on the one hand, and removals and new growth, on the other. The more the removals and the quicker the growth, the less is the value of x at the balance. Natural epidemics seldom run to completion. The blight of chestnuts caused by *Endothia parasitica* is one of the few that virtually did.

The point of balance varies as conditions change from time to time. No epidemic has been studied enough to justify an attempt at quantitative calculations.

8.5. Relation between R_c and r

Consider now only the logarithmic stage of an epidemic. In Eq. (8.4) one substitutes

$$x_{t-p} = x_{t-i-p}\, e^{ir_l}$$

$$x_t = x_{t-i-p}\, e^{(i+p)r_l}$$

These equations are easily seen from Fig. 8.3. Hence

$$R_c = r_l \frac{e^{(i+p)r_l}}{e^{ir_l}-1} \tag{8.5}$$

Equation (8.5) holds only if r_l is constant after time $t-i-p$. It corrects Eq. (5.7) by allowing for removals.

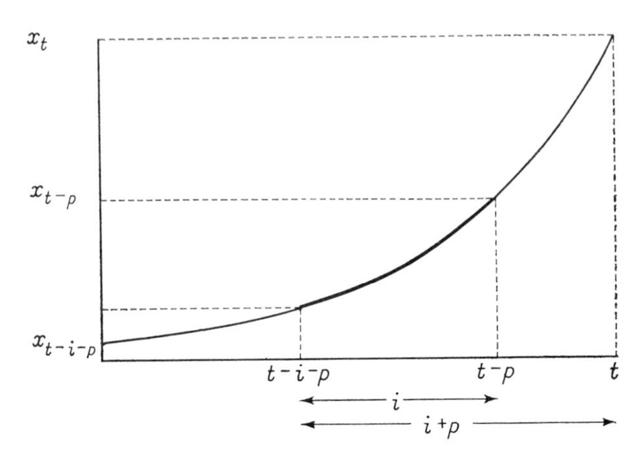

FIG. 8.3. The increase of disease x with time t at the infection rate r_l. Take x_{t-i-p} at time $t-i-p$ as a base. By the compound interest equation, (2.1), this increases in time i to $x_{t-i-p}\,e^{ir_l}$, which is x_{t-p}; and it increases in time $i+p$ to $x_{t-i-p}\,e^{(i+p)r_l}$ which is x_t. The thick part of the curve represents infectious disease.

8.6. A Threshold Theorem

From Eq. (8.5) it follows that $r_l > 0$ only if $iR_c > 1$. That is, no epidemic can start unless conditions of host, pathogen, environment, and fungicide interact to make $iR_c > 1$. If $i = 1$ day, R_c must exceed 1 per unit per day for an epidemic to start. If $i = 0.5$ year, R_c must exceed 2 per unit per year.

If i varies from point to point in the susceptible tissue, one may use the mean value of i to estimate the threshold value of R_c. That is, the variance of i is irrelevant to this particular problem, except, of course, insofar as it affects the accuracy of estimates of the mean.

The assumption here is that inoculum is shared equally by all points. When inoculum is not shared equally, and some parts of the plants are more susceptible or vulnerable than others (e.g., when the lower leaves are attacked first), the threshold value must be calculated for these parts.

8.7. The Threshold Theorem and Control of Disease by Fungicides

To illustrate what the threshold theorem means consider some spraying experiments against potato blight caused by *Phytophthora infestans*.

Figure 8.4 is drawn from results in Maine reported by Brandes *et al.*

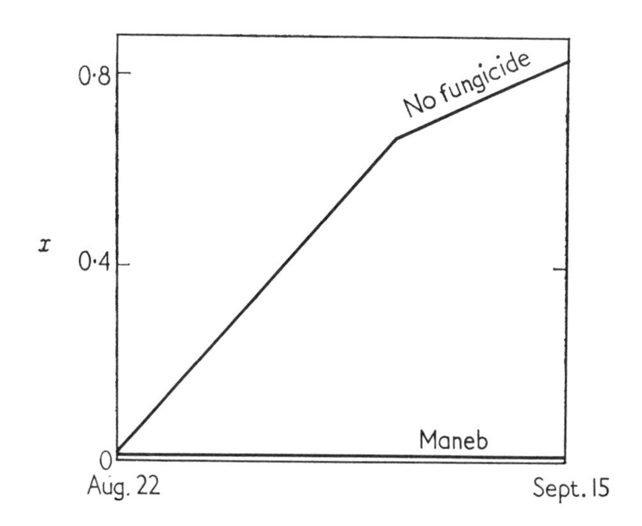

FIG. 8.4. The effect of treatments with maneb on the progress of an epidemic of blight caused by *Phytophthora infestans* in the potato variety Katahdin. Data of Brandes *et al.* (1959).

(1959). Maneb was applied at various strengths and intervals. In unsprayed plots blight was first noticed on August 22. It increased steadily to 85% on September 15 when readings were discontinued. All the maneb treatments except the weakest (with 0.5 lb. maneb in 100 gal. of water, applied every 10 days) kept blight entirely under control, and on September 15 there was less than 1% of blight. The experimental plots were only four rows wide, so this control by maneb was achieved despite an evidently heavy supply of inoculum from the unsprayed plots in the experimental field. Nevertheless maneb kept $iR_c < 1$ for the duration of the experiment.

In this experiment treatment started early, and blight in the maneb treated plots never got a start. But fungicides can also stop a blight epidemic once it has started. Figure 8.5 records some results by Hooker

(1956) in Iowa. Spraying began on July 18 and continued until September 14 at weekly intervals. There is no record of the amount of blight on July 18, but it must have been considerable, because a week later 4–5% of the foliage was blighted.

Blight was not stopped by the first spraying, but continued to develop for about 2 weeks, although at a slower rate in the sprayed plots. Then the better fungicides—maneb, zineb, and Bordeaux mixture—stopped

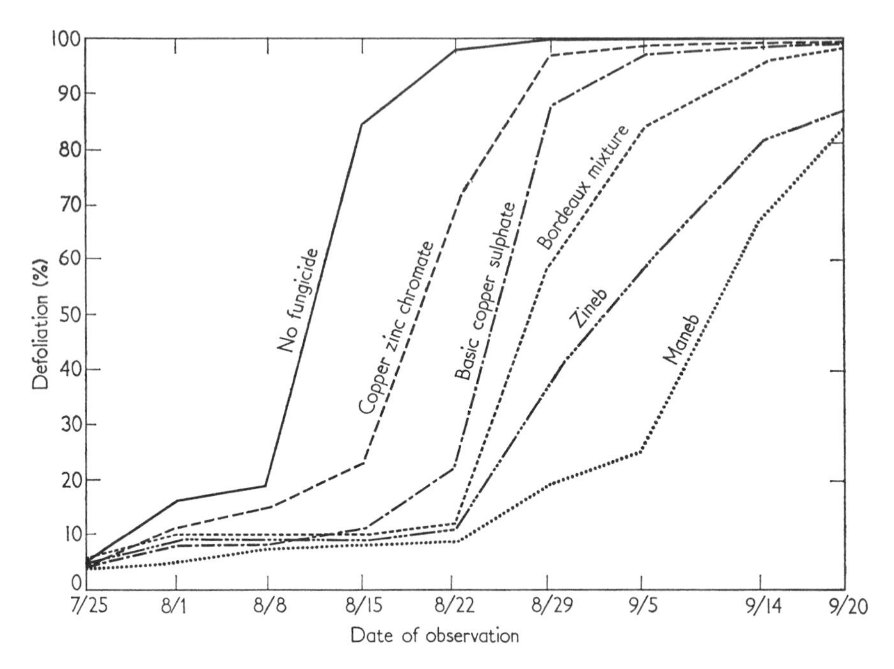

FIG. 8.5. The progress of an epidemic of blight caused by *Phytophthora infestans* in the potato variety Irish Cobbler. Details in text. Data of Hooker (1956). By kind permission of Dr. W. J. Hooker, Iowa State University, Ames, Iowa.

further blight development for 2 weeks. Finally at the end of August, blight got away despite weekly sprayings, and increased steadily in September.

Consider these events in more detail.

The first stage, immediately after spraying began, is a transition stage from unsprayed to sprayed fields. In an unsprayed field the proportion of infected tissue that is infectious is characteristic of an unhindered epidemic. For example, it was estimated in Section 8.2 that 6% of the infected tissue is infectious if $p = 4$ days, $i = 1$ day, and $r_l = 0.46$ per unit per day. The first sprays to be applied reduce this proportion and bring it gradually to a value characteristic of sprayed fields. This takes time. It took roughly 2 weeks in Hooker's experiment.

In the next stage, the better fungicides gave complete control. They stopped blight from increasing further. That is, they kept $iR_c < 1$.

In the final stage, the fungicides lost control, and iR_c increased above the threshold value. The phenomenon is common; and other examples are given in Chapter 21.

Consider quantitatively the stage of complete control of blight. The proportion x of infected tissue was slightly above the limit of the logarithmic stage of an epidemic (i.e., $x > 0.05$). There is therefore a small error in applying Eq. (8.5) to the results, but this is unimportant for the present purpose of illustration. We take $p = 4$ days, $i = 1$ day, and $r_l = 0.44$ per unit per day in the unsprayed plots, this being the value of r calculated from Hooker's results in Exercise 5 at the end of Chapter 3. Hence, by Eq. (8.5), $R_c = 7.2$ or, in a round figure, 7 per unit per day in the unsprayed plots. Because $i = 1$ day, R_c needed only to be reduced to 1 per unit per day to bring iR_c to the threshold value of 1 per unit and stop the epidemic entirely. That is, to maintain complete control of blight the fungicide had only to reduce the proportion of spores able to germinate and start a lesion to one-seventh of the proportion on unsprayed foliage. The same reduction would have prevented an epidemic from starting in the first place. In other words, in the conditions of Hooker's experiments *Phytophthora infestans* could not have started an epidemic, however slow, in foliage protected by a fungicidal deposit that killed 6 germinating spores out of 7. With 6 spores killed out of 7, removals would be relatively so fast that infectious tissue would not accumulate enough to allow an epidemic to exist. Reducing the proportion of spores able to germinate and start a lesion to one-seventh of what it would have been on unsprayed leaves represents only moderate killing. But it is enough if host, pathogen, and environment interact to keep i short.

For comparison, consider how much killing (i.e., fungicidal action in the narrow, literal sense of the term) would be needed to stop *Puccinia graminis* from starting an epidemic of wheat stem rust. To keep to a comparison with potato blight, we again take $r_l = 0.44$ per unit per day; that is, the epidemics of potato blight and wheat stem rust are taken to mount at the same rate in unsprayed fields. (This value of r_l is near the value 0.41 found for wheat stem rust by Asai and discussed in earlier chapters.) But for wheat stem rust $p = 10$ days and $i = 10$ days. Hence by Eq. (8.5), $R_c = 36$ per unit per day and $iR_c = 360$ per unit. To stop an epidemic of wheat stem rust from starting, a fungicidal deposit on the plants would have to let not more than 1 germinating spore in 360 survive. This compares with 1 in 7 for potato blight. One may vary the terms of the comparison considerably without essentially modifying the conclusion that a fungicide must be vastly more efficient to control wheat stem rust than to control potato blight. There are possibly several reasons why wheat stem rust has defied chemical control whereas potato blight is amenable to it. This is one of them.

The matter will be taken up in a later chapter when we discuss the fungicide square: host, pathogen, environment, and fungicide. There is more to fungicidal action in the broad sense—the control of fungus plant diseases by chemicals—than just killing fungi.

8.8. The Threshold in Epidemics of Two Systemic Diseases

When a lesion of *Phytophthora* blight grows through a potato leaf, the fertile sporing ring leaves a zone of sterile tissue behind it in the center of the lesion. A similar sequence is seen with some systemic diseases.

Peach trees infected with rosette virus pass through a latent period and then an infectious period. After a short infectious period they die, the disease in peaches being quickly lethal. That is, i is short. Peach growers make it even shorter by roguing out infected trees. Also, the disease spreads slowly. That is, R_c is small. One cannot evaluate i and R_c on the data available. But clearly the threshold value is seldom exceeded, because the disease is easily controlled if infected wild hosts are kept away from the orchard's edge. In Peach County, Georgia, KenKnight (1961) estimated the annual incidence to be only 1 in 100,000 peach trees. But occasionally the threshold value is exceeded, and orchards are devastated. Possibly in such instances R_c is raised by a greater abundance or activity of vectors. (The vector has not yet been determined.)

When the pathogen kills the host plant and dies with it, as peach rosette virus does, a slight change of conditions may be all that is needed to cause a disease normally rare and relatively innocuous to become devastatingly destructive.

Swollen shoot disease of cacao illustrates this too. The focus with swollen shoot disease is much like the lesion with *Phytophthora* blight of potato. There is an outer zone of trees in the latent stage of infection; inside this is a zone of infectious trees; and inside this again, if the focus is large, a zone of dead or dying trees carrying only a low population of vectors and separated widely from healthy trees. (See Section 7.4.) In West Africa the virus is indigenous, and cacao met the virus there after it crossed the Atlantic Ocean. Widespread epidemics did not appear until later. In broad outline the history seems to be fairly clear in the Eastern Province of Ghana. Cocoa production in Ghana increased rapidly this century. In 1898 only 185 tons of raw cocoa were exported. The quinquennial average from 1906 to 1910 was 15,000 tons exported; from 1916 to 1920 it was 106,000 tons; and from 1926 to 1930 it was 219,000 tons. It rose further to a peak of 281,000 tons in 1939. In 1936, Steven (1936) reported swollen shoot disease for the first time. But it was clear that the disease had been known in isolated outbreaks many years earlier. In the 1930's the disease became epidemic. Isolated outbreaks started to fuse, although as late as 1938 it was still possible to see a multitude of discrete outbreaks (Posnette, personal communication). At some time, possibly in the 1920's, the epidemic threshold was crossed, probably because of the growth of the industry.

The mealy bug vectors move around slowly, by crawling or being blown about. Their spread from tree to tree is greatly helped when there is a continuous interlacing canopy. This depends on tree density (Cornwell, 1958) and tree age. Strickland (1951) obtained a positive correlation between girth of the trees and the percentage infested with mealy bugs and found that large trees had more mealy bugs than small ones. Trees in young plantations have little chance of intercepting mealy bugs blown in from outside, and when outbreaks occur in young trees they spread slowly (Benstead, 1951, 1953). The 1200-fold increase in raw cocoa exports in 30 years represents a great increase in the number and age of trees, and explains why the threshold was crossed. We can be reasonably certain of this, even if there are not enough data to allow us to calculate a threshold value.

8.9. Looking Back

Let us quickly look back at two matters in previous chapters to see how they are affected by removals.

Figure 8.6 shows the relation between pr_l and pR calculated by Eq. (5.9), and between pr_l and pR_c by Eq. (8.5) with $i = p/4$. This value of i is suggested by *Phytophthora* blight of potatoes. There is a difference between the relations, shown by the inset, when values are very low. But at high values of pR or pR_c, the relations follow much the same trend. Removals do not alter the previous conclusion (see Section 5.9) that p sets an upper limit to r_l.

In Section 6.10 we divided an epidemic into three stages: an early logarithmic stage; an intermediate stage, with values of x up to about 0.35, in which r approximately retains the value of r_l if conditions stay constant; and a final empirical stage about which little is understood. This was for epidemics without removals. With removals the early stage remains logarithmic; all that is needed is to change the appropriate equations for quantitative calculations, e.g., instead of Eq. (5.7) use Eq. (8.5). The second stage is extended; in constant conditions the rise in the value of r is delayed, and r can ordinarily be used longer to estimate r_l. However, for the sake of caution, we retain $x = 0.35$ as the useful upper limit for estimating r_l as r. Factors in the third and final stage

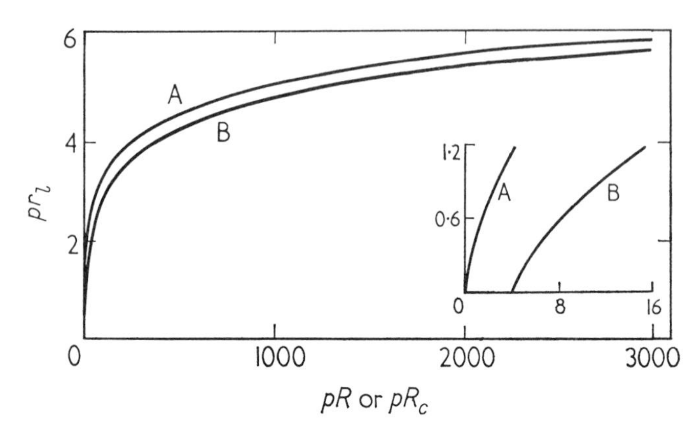

FIG. 8.6. The relation of pr_l to pR in curve A and pR_c in curve B. For curve B, $i = p/4$.

remain as obscure with removals as without. As an epidemic slows down for lack of susceptible tissue for the pathogen to attack, removals become increasingly more important, and may bring the epidemic to a stop. This was discussed in Section 8.4.

EXERCISES AND EXAMPLES

1. In Section 5.7 we considered the problem of the error in field experiments with potato blight caused by the dispersal of spores. From plots, 20×5 yd., the proportion of spores lost by dispersal was estimated to be 0.12. From a square 2-acre field the proportion lost was estimated to be 0.02. If $r_l = 0.46$ per unit per day in the plots, what is it in the field? In Section 5.7 we did not allow for removals in the calculation; we must now do so. Take $i = 1$ day and $p = 3.75$ days. (We use $p = 3.75$ days because this is the figure in the original publication. The findings of Lapwood (1961b) have appeared since, and indicate $p = 4$ days to be a better figure. The difference is insubstantial for present purposes.)

In the plots, from Eq. (8.5)

$$R_c = 7.00 \text{ per unit per day.}$$

In the field, following the reason in Section 5.7

$$R_c = \frac{7.00 \times 0.98}{0.88}$$

$$= 7.80 \text{ per unit per day}$$

Hence, from Eq. (8.5),

$$r_l = 0.486 \text{ per unit per day.}$$

This compares with the value 0.477 obtained without correction for removals.

2. In a field experiment with fungicides against potato blight, the unsprayed plots had 59% blight whereas the sprayed plots had 0.1%. It was estimated by van der Plank (1961a) that the sprayed plots were then receiving 20.7 spores of *Phytophthora infestans* from the unsprayed plots for every 1 spore they produced themselves. In the sprayed plots $r_l = 0.34$ per unit per day. What would r_l have been in the sprayed plots if they had not been bombarded with spores from unsprayed control plots? That is, what would r_l have been in sprayed fields isolated from outside sources of infection? Ignore as relatively trivial the loss of home grown spores from sprayed plots and fields. Take $i = 1$ day and $p = 4$ days.

In the sprayed plots bombarded with spores, by Eq. (8.5)

$$R_c = 4.60 \text{ per unit per day}$$

This rate was achieved by $20.7 + 1$ spores for every 1 spore an isolated sprayed field would have received. In an isolated sprayed field

$$R_c = \frac{4.60}{21.7}$$

$$= 0.212 \text{ per unit per day}$$

Hence $iR_c = 0.212$ per unit. This is below the threshold value for an epidemic. An epidemic would not yet have started in an isolated sprayed field. On the evidence of these figures, the relatively fast infection rate of $r_l = 0.34$ per unit per day in the sprayed plots is the result of contamination by spores from the unsprayed plots. The sprayed plots do not represent sprayed fields because even in intensively cultivated regions sprayed fields are better isolated from contamination than these sprayed plots were. To that extent there was a serious error of experimental design which had nothing to do with the calculation of ordinary statistical tests of significance.

3. Colwell (1956) determined the course of an epidemic of oat stem rust caused by *Puccinia graminis avenae*. The oats were a heterozygous mixture in which one-quarter of the plants were susceptible to the prevalent races of the fungus, and three-quarters were immune. On September 14 and 24 the proportion x of diseased tissue was 0.01 and 0.10, respectively. Estimate r_l in a population homozygously susceptible to the prevalent races. Take $i = 10$ days and $p = 10$ days, and assume that spores from the susceptible plants fell uniformly on all plants.

To calculate r in the mixture one must change the usual correction factor from $1 - x$ to $0.25 - x$. Instead of Eq. (3.6) use

$$r = \frac{2.3}{t_2 - t_1} \log_{10} \frac{x_2(0.25 - x_1)}{x_1(0.25 - x_2)}$$

$$= \frac{2.3}{10} \log_{10} \frac{0.1 \times 0.24}{0.01 \times 0.15}$$

$$= 0.28 \text{ per unit per day}$$

We take this to represent r. This stretches our limit slightly. It was agreed in Chapter 6 that r could be used to estimate r_l if x_2 did not exceed 0.35. Here $x_2 = 0.1$, but because only one-quarter of the plants were susceptible this is equivalent to a limit of 0.4. However, the results in Chapter 6 show that this limit does not bring in a great error, and we shall use it.

With $i = 10$ days, $p = 10$ days, and $r_l = 0.28$ per unit per day, Eq. (8.5) gives $R_c = 4.9$ per unit per day. In the varietal mixture of oats, three-quarters of the spores contributed nothing to the epidemic because they fell on immune plants. In a mixture with all plants susceptible, $R_c = 4.9 \times 4 = 19.6$ per unit per day. Hence by Eq. (8.5), $r_l = 0.39$ per unit per day.

Reducing the proportion of susceptible plants to one-quarter reduces r_l from 0.39 to 0.28 per unit per day. The assumption is that spores fell equally on all plants. But spores tend to fall on the plant they come from. For this reason the value 0.39 is probably too high. However, the calculation has heuristic value. At least it shows that diluting susceptible plants with immune plants in a varietal mixture does not proportionately reduce the logarithmic infection rate.

4. Lapwood (1961b, c) has shown that the width of the sporing ring of *Phytophthora infestans* on potato leaves is reduced by dry conditions or by using more resistant varieties. We have been using $i = 1$ day for the variety Majestic in moist conditions. In drier conditions or on more resistant varieties than Majestic, $i < 1$ day. Redraw Fig. 8.2, using the same values of p and r but with $i = 0.75$ day.

CHAPTER 9

Stochastic Methods in Epidemiology

SUMMARY

When a pathogen multiplies by spreading from lesion to lesion and plant to plant, and the progress of disease with time is measured, it is better to transform proportions of disease to log $[x/(1-x)]$ than to equivalent angles or probits. The main purpose of this chapter is to discuss methods that estimate the probability that a plant will become infected or that a random point in susceptible tissue will be infected. The two problems discussed are the multiple infection of plants with systemic disease and the overlapping of local lesions. Applied to large numbers of plants, the probability methods give the same answers as deterministic methods. This book is concerned with large numbers of lesions on large numbers of plants, for reasons given in Chapter 1, which justifies using deterministic methods.

9.1. Transforming Proportions of Disease

One estimates the progress of disease in a field by estimating the pro portion (or percentage) of disease in samples of the field. If the disease is systemic, the proportion is estimated by direct counting. Infected plants A found in a total of plants B estimates the proportion $x=A/B$. If the pathogen causes local lesions, one may estimate the proportion x directly, as Zadoks (1961) did with stripe rust of wheat caused by *Puccinia striiformis*. In a detailed experiment he measured the area of infected foliage with a planimeter. Commonly, one uses an established scale. Cereal rusts are often measured in the United States by the modified Cobb scale. Diagrams show the abundance of pustules at various levels of disease (e.g., 5, 10, 25, 40, 65, and 100%), and disease in a sample is assessed by comparison. The British Mycological Society key (Anonymous, 1947) for assessing *Phytophthora* blight in potatoes fixes various percentages as the result of long experience and satisfactory results in practice. For example, about 50 spots per plant, or up to 1 leaflet in 10 attacked, is taken as 5% blight. For many diseases the system of Horsfall and Barratt (1945) is used. It applies to disease assessment the Weber-Fechner law that the human eye distinguishes according to the logarithm of the light intensity.

Whatever the system, one normally ends with a percentage or proportion. For the purpose of statistical analysis plant pathologists commonly transform a proportion into an equivalent angle or into a probit.*

Transforming into equivalent angles postulates that disease is binomially distributed in a homogeneous field. The postulate is usually

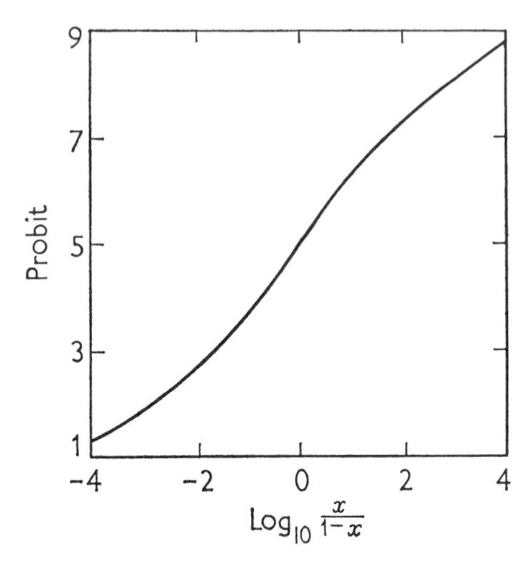

FIG. 9.1. The relation between the probit of x and $\log_{10}[x/(1-x)]$.

incorrect. The variance is often substantially greater than for a binomial distribution when the disease is systemic and vastly greater when the pathogen causes local lesions.

When disease multiplies, it multiplies in foci. (See Section 7.5.) Pathogens spread in a gradient; more spread a short distance along a line from the source than far. The result is foci of disease. And foci are the antithesis of the homogeneity needed for transformation into equivalent angles.

Probits (Bliss, 1934; Finney, 1952) are a device to straighten a sigmoid (S-shaped) curve. So too is $\log[x/(1-x)]$. But *sigmoid* is a loose adjective, and the sigmoid curve that probits straighten is not exactly the sigmoid curve that $\log[x/(1-x)]$ straightens. The difference is brought out in Fig. 9.1. Instead of being straight, the line in this figure is curved.

* The first transformation is to the angle whose sine is $x^{\frac{1}{2}}$, where x is the proportion of disease. The probit of x is the abscissa corresponding to a probability x in a normal distribution with mean 5 and variance 1.

To the type of sigmoid curve that we have been concerned with—a sigmoid disease progress curve—probits are not relevant, but $\log[x/(1-x)]$ is.

$\text{Log}[x/(1-x)]$ has been used before as a transformation, especially for quantities that increase autocatalytically or logistically. Berkson (1944) called $\log_e[x/(1-x)]$ the logit of x, but Finney (1952) later changed the definition. The use of *autocatalytic*, *logistic*, and *logit* for disease problems will be discussed in the Appendix.

Transformation to $\log[x/(1-x)]$ is indicated when one is dealing with a pathogen that spreads from lesion to lesion or plant to plant (i.e., a pathogen that "multiplies" in the sense given in Chapter 4) and when the independent variable is time. General rules for choosing transformations in epidemiology are given in the Appendix.

9.2. Sampling Errors of Estimates of Infection Rates

The logarithmic and the apparent infection rates, r_l and r, are the linear regression coefficients of $\log_e x$ and $\log_e[x/(1-x)]$, respectively, on time. The error variance of estimates of these rates is estimated as for other regression coefficients. The matter was discussed in Exercises 8, 9, and 10 at the end of Chapter 3.

Similarly Rx_0 is the regression coefficient of $\log_e[1/(1-x)]$ on time. (See Section 5.4.)

9.3. Deterministic and Probability Methods in Epidemiology

This book relies on deterministic methods. The use of these methods implies that new infections are determined not by chance but by the amounts of inoculum and susceptible tissue, and by the infection rate. In probability methods one assumes an element of chance in the formation of new lesions. In a deterministic model one thinks in terms of the proportion of tissue that is infected; in a probability model one thinks in terms of the probability that a point chosen randomly in susceptible tissue will be infected.

Consider two probability problems as examples.

9.4. Multiple Infections with Systemic Disease

Gregory (1945) analyzed records in the literature of the incidence of disease at varying distances from the source of inoculum. Many of the records stated the percentage of plants that were diseased. What he wanted was the average number of lesions per 100 plants. To change from one form of recording to the other he used a method first applied in entomology by Thompson (1924). An example will illustrate the problem. Suppose that with a leaf spot disease, or any other local lesion disease, it is recorded that 45% of the plants are infected. This gives definite information about the other 55% of the plants—they had no lesions at all. But about the 45% the information is not precise; some can be expected to have 1 lesion, some 2, some 3 lesions, and so on. On the assumption that the lesions are randomly distributed and that both lesions and host plants are numerous, one can estimate the probable total number of lesions per 100 plants.

Gregory (1948) discussed the matter again, and published a table of the multiple-infection transformation, which transforms percentages into infections. The discussion is mainly of historic interest here and not generally applicable to our problems, because in the system of epidemiology used in this book we do not consider a plant with one local lesion to be fully infected. But one can use the method for systemic infections.

Consider a systemic virus disease spread by an insect vector. Suppose that there are A plants and that the virus is transmitted to them B times. By "transmitted" we mean that the virus is injected in such a way that it leads to the infection of healthy plants and would lead to the infection of diseased plants if they had not already been infected. For B/A write m. If the transmissions are randomly distributed and if A and B are large, one estimates from the Poisson distribution that the probability of a plant being healthy is

$$q = e^{-m} \tag{9.1}$$

Hence

$$m = -\log_e q$$

When A and B are large the probability q that a plant is healthy is also the expected proportion of healthy plants, which in the symbols used in earlier chapters can be written as $1-x$. Therefore

$$m = -\log_e (1-x)$$

$$= \log_e \frac{1}{1-x}$$

In meaning, m is the average number of times the virus was "transmitted" to a plant. When therefore one plots $\log_e [1/(1-x)]$ against time, as in Fig. 4.5, one is, in effect, plotting m, the average number of "transmissions," against time. By plotting in this way one allows for multiple infections, i.e., for more than one transmission per plant. In earlier chapters we used $1-x$ as a correction factor, on the argument that we are concerned only with the infection of tissue that is not already infected, i.e., on the argument that we exclude or ignore the possibility of multiple infection. Now we reach the same result by assuming that multiple infections do occur and by allowing for them. The paths are different, but the destination is the same.

The essential assumptions are that the pathogen's attack should be random and that the environment and the resistance of the host plants should be uniform. Departures from these assumptions inevitably cause multiple infections to be underestimated. The underestimation is the same as the undercorrection by the factor $1-x$ discussed in Section 4.5.

If one compares Gregory's table for the multiple-infection transformation with the table for $\log_e[1/(1-x)]$ in the Appendix, it will be seen that the transformed numbers are $100 \log_e[1/(1-x)]$. The factor 100 appears because Gregory deals with the percentage of disease and we with the proportion.

9.5. The Overlapping of Local Lesions

One can investigate the problem with local lesions simply as follows. Consider a leaf with leaf spot disease. Let a and b be the areas of the leaf spot and leaf, respectively. When there is 1 spot on the leaf, the possibility that any random point on the leaf will be in healthy tissue is $q=1-(a/b)$; when there are 2 spots on the leaf, $q=[1-(a/b)]^2$; and so on.

Consider now a large number n of leaves. An acre of potatoes has about 10^7 leaves. So if one were discussing disease in an acre of potatoes, n would be of the order of 10^7. When n leaves have kn leaf spots, the probability that any random point in the leaves will be in healthy tissue is

$$q=\left(1-\frac{a}{bn}\right)^{kn}$$

$$=e^{-ak/b} \tag{9.2}$$

When n is large, it does not matter if a and b vary. They can be regarded as means.

From Eq. (9.2), writing $1-x$ for q, as in the previous section, one gets

$$\frac{ak}{b}=\log_e \frac{1}{1-x}$$

One interprets ak/b as the area of diseased tissue, relative to total leaf area, that one would expect if there was no overlapping of lesions.

More simply, as long as $ak/b < 1$, ak/b is the proportion of diseased tissue that one would expect if lesions did not overlap. It means for local lesions what m means for systemic infections; that is, Eq. (9.2) is for local lesions what Eq. (9.1) is for systemic infections.

When with local-lesion diseases, one plots $\log_e[1/(1-x)]$ against time, as in Fig. 5.2, one is, in effect plotting ak/b and thus allowing for overlapping. The assumption, as before, is that disease is distributed randomly over tissue that is all equally susceptible and equally exposed to infection.

9.6. The Influence of Numbers

In Sections 9.4 and 9.5 it was assumed that the number of plants, leaves, and systemic infections was large. This assumption simplifies the mathematics. It also makes probability methods give the same answer as deterministic methods.

When one deals with small numbers the answers are not necessarily the same, and the answer by the probability method must be preferred. To discuss the control of plant disease we choose deterministic methods because they are simpler. But the consequence of this choice is that results are meant not for just a few lesions on just a few ornamental plants in a garden, but for large fields, orchards, plantations, and forests. If disease is very unevenly distributed, the number of plants must be enough to include many foci of disease. (See Chapter 7.) The choice of deterministic methods is justified because the book is concerned primarily with the general strategy against disease: with how to control thousands and millions of lesions on thousands and millions of plants.

9.7. Comparisons with Medical Epidemiology

The history of the epidemiology of diseases of man is instructive. The earlier work was deterministic, stochastic analysis followed. It consolidated what had been found deterministically, and it extended the analysis to problems for which deterministic methods are inept.

Insofar as this type of history can be applied to the epidemiology of plant diseases, a deterministic start would seem proper, but there are at least two obstacles in the way of following up with stochastic methods:

a. Stochastic models in medical epidemiology usually assume that persons can be counted and divided into four classes: the susceptible, the infected, the infectious, and the removed (by isolation, immunity, or death). This classification is useful for systemic diseases, but it is difficult to construct a useful stochastic model for the local lesion diseases which are so important in plant pathology. This difficulty remains even if one undertakes to count the lesions and not just assess the proportion of infection.

b. There is less incentive to use stochastic methods to study epidemics of plant disease. In medicine stochastic methods are needed particularly to analyze data from small households. A comparable need does not seem relatively quite so urgent in plant pathology, because the observer is normally able to collect evidence from comparatively large samples.

CHAPTER 10

A Guide to the Chapters on Control of Disease

10.1. The General Proposition

This chapter starts the part of the book that deals with the control of disease.

The general proposition is simple. At any time during the course of an epidemic the amount of disease is determined by how much inoculum there was at the start and how fast disease has developed since.

For illustration we shall oversimplify the relation and use the equation for compound interest

$$x = x_0 e^{rt} \tag{2.1}$$

The proportion of disease x at any time is determined by the initial inoculum x_0, the average infection rate r, and the time t during which infection has occurred. The equation applies only to the logarithmic phase of an epidemic, and we have blurred the distinction between r and r_l. But for simplicity of illustration let that pass.*

Some methods of control reduce t; they are mainly the planting of early maturing varieties, or early sowing to escape disease. Another method, using tolerant varieties, reduces neither x_0, r, nor t; for example, orange scions are grown on tolerant rootstocks when there is a danger of infection by citrus tristeza virus. The orange trees become infected, but without being obviously harmed.

But most methods of control reduce x_0 or r or both.

* Also for simplicity of illustration this chapter is written as if all diseases were "compound interest diseases." To adapt the sections that follow to "simple interest diseases," replace r by R, and where necessary replace x_0 by Q (see Section 13.9).

10.2. Control Measures That Reduce x_0: Sanitation, Vertical Resistance, and Chemical Eradication

Phytophthora infestans survives the winter in infected potato tubers. Infected tubers get planted as seed which may give rise to infected plants; or they are discarded on cull piles where they later produce infected shoots which release sporangia that blow into neighboring fields. Reducing x_0 by planting healthy seed or destroying infected cull piles is sanitation.

So too with bunt of wheat x_0 is determined primarily by the amount of *Tilletia caries* and *Tilletia foetida* on seed grains and in the soil. Weather and other conditions determine how much of this infects the growing seedlings. Reducing x_0 by using clean seed or chemically treated seed or by crop rotation is sanitation.

Sanitation is the topic of the next three chapters.

The species *P. infestans* comprises many races. The foliage of certain potato varieties is resistant to some races but susceptible to others. If 99% of the sporangia reaching a field of potatoes are of races to which the variety is resistant, the epidemic must build up from the remaining 1%. The effect of resistance is the same as that of sanitation. If by sanitation—by using healthy seed potatoes or destroying infected cull piles—99% of the initial inoculum is destroyed, the epidemic must build up from the remaining 1%. The two methods—planting varieties resistant to some races of the pathogen but not to others, and sanitation—are identical in effect. Both, in our numerical examples, reduce x_0 to 1/100 of what it otherwise would have been.

So too some varieties of wheat are resistant to some races of bunt and susceptible to others. Sowing a variety resistant to 99% of the bunt spores on the seed and in the soil is identical in effect with sanitation that destroys 99% of the spores on the seed and in the soil. Resistance and sanitation in these examples each reduce x_0 to 1/100 of what it otherwise would have been.

Resistance to some races of a pathogen but not to others will be called vertical or perpendicular resistance. Vertical resistance is essentially sanitation in another form.

Vertical resistance has been called racial resistance against potato blight (Niederhauser *et al.*, 1954), specific resistance against wheat stem rust (Stakman and Christensen, 1960), and other names. It is a form of resistance much used in plant breeding.

When the vertical resistance is high, the pathogen cannot reproduce on the host. If we spray spores of a race of *P. infestans* onto leaves of a potato variety with vertical resistance to the race, small necrotic spots develop where the fungus penetrates. No spores are produced in these spots, and the race cannot reproduce itself. On the other hand, if onto leaves of the same potato variety we spray spores of a race against which the variety has no vertical resistance, reproduction of the fungus is apparently normal. The vertical resistance reduces x_0 by confining inoculum only to races that can attack the variety. But vertical resistance does not reduce r,* because reproduction is normal in races not governed by the resistance.

The initial inoculum x_0 can also be reduced chemically. Seed can be treated with chemicals, and seedbeds can be fumigated. These treatments can conveniently be considered as sanitation. Chemical eradicants can be used on foliage, in the same way as organic mercury compounds are used against apple scab. This too is conveniently considered as sanitation. It is also discussed in Section 21.9.

10.3. Control Measures That Reduce r: Horizontal Resistance and Protectant Fungicides

In contrast with vertical resistance, horizontal resistance is spread against all races. Horizontal resistance has been called field resistance against potato blight, generalized resistance against wheat stem rust, and other names.

Horizontal resistance reduces r. Potato plants with horizontal resistance against blight become infected, but the epidemic progresses more slowly than in the susceptible varieties. The reduction of r can be traced to various factors: fewer spores can initiate lesions; fewer spores are produced; infected tissue takes longer before it starts to sporulate; and others that will be discussed in Chapter 14.

Although horizontal resistance primarily affects r, it also sometimes affects x_0 indirectly. For example, if the foliage of a potato variety has horizontal resistance, it will be less infected. Infected foliage is the source of inoculum to infect tubers. Healthier foliage means healthier tubers; and healthier tubers mean healthier seed and reduced x_0 the next season.

Horizontal resistance can also affect x_0 directly if one considers the initial inoculum to be the plants initially infected, instead of spores,

* This statement is for vertical resistance high enough to stop the pathogen from reproducing on the host. Resistance high enough to stop, or nearly stop, reproduction is what plant breeders aim at when they choose vertical resistance. Our generalization seems good enough for most practical purposes.

overwintering mycelium, and the like. Numerical examples about this are given for wheat stem rust in Section 20.3.

But, although exceptions do occur, by far the most usual effect of horizontal resistance is on r.

The difference between vertical and horizontal resistance is discussed in Chapter 14 in relation to potato blight. Potato blight is specially apt for the discussion, because the literature is relatively large and distinctions between the two forms of resistance relatively clear. The discussion is continued in Chapters 15 to 20, inclusive, largely in relation to wheat stem rust and other diseases.

Protection with fungicides also reduces r. A fungicide layer on the surface of a leaf reduces the proportion of spores forming germ tubes that successfully initiate lesions. By doing this it reduces r.

Fungicidal action is discussed in Chapter 21.

For Chapter 20—about the factors involved in horizontal resistance—and Chapter 21 we must return to the topics of Chapters 5 to 8.

There are other ways of reducing r. Cultural methods and control of the environment are among them. Many are applied automatically; wheat is not grown commercially in summer in warm, wet areas where severe rust epidemics can be expected every year. But, although these methods are well known and widely used, there are few data about them that can be used for epidemiological analysis; and no special analysis is attempted.

CHAPTER 11

Sanitation with Special Reference to Potato Blight

SUMMARY

Sanitation is the process of reducing the inoculum from which an epidemic starts. A campaign in Aroostook County, Maine, to destroy potato cull piles from which *Phytophthora infestans* comes showed how complex the process is when disease starts from small amounts of inoculum and then multiplies until all the foliage is destroyed.

11.1. A Definition of Sanitation

We define sanitation as a process that reduces, or completely eliminates, or completely excludes the initial inoculum from which epidemics start.

Some pathogens are still localized. *Hemileia vastatrix*, the cause of coffee rust, has not yet reached the Western Hemisphere. Keeping it out by every possible device is sanitation.

Xanthomonas citri, the cause of citrus canker, was successfully eradicated in Florida and South Africa. This was sanitation.

Seed is commonly infected or contaminated with pathogens. Wheat seed carries *Tilletia caries* and *Tilletia foetida* that cause bunt. Tomato seed carries *Corynebacterium michiganense,* the cause of canker. Bean seed carries, among other pathogens, the bacteria that cause blight and *Colletotrichum lindemuthianum* that causes anthracnose. Planting healthy seed or seed that has been treated chemically or by heat or other physical agents is sanitation. Sowing certified seed which has been inspected for disease is sanitation. So too is planting seed obtained from areas where some particular pathogen is rare, as, for instance, planting seed of beans grown west of the Rocky Mountains.

Using seedlings from healthy seedbeds and propagation material from healthy nurseries is sanitation. Health certificates based on nursery inspection and quarantine further sanitation.

Seed potatoes carry many viruses, bacteria, and fungi. The grower of table potatoes who plants certified seed practices sanitation. So too does the seed grower who isolates his seed fields from diseased potato crops and rogues out virus-infected plants as they are seen.

Many pathogens survive in the soil. Crop rotation that reduces their abundance is sanitation. So too is soil fumigation and treatment with steam.

Sclerotium rolfsii is dangerous mainly in the upper few inches of soil. Plowing sclerotia in deep is sanitation.

Venturia inaequalis, the cause of apple scab, infects leaves in summer and later falls to the ground with the leaves. In spring it fructifies on the fallen leaves and releases ascospores which reach and infect the new flushes of growth. Spraying apple orchards with phenyl mercuric chloride late in summer just before the leaves fall kills the fungus in the leaves. This reduces the amount that enters and survives the winter. It is sanitation. In temperate climates with unfrozen soils earthworms bury many infected leaves on the orchard's floor. Keeping a high earthworm population is sanitation, as is any other method of disposing of the leaves.

Vertical resistance as sanitation will be considered separately.

It may be claimed that this definition of sanitation is too wide and loose. For example, scab in apples can be reduced if one waits in spring until infection has occurred and immediately afterwards applies phenyl mercuric chloride as an eradicant fungicide. Is this sanitation? Does spraying young growth with phenyl mercuric chloride in spring differ essentially from spraying old leaves just before the fall? Or is it a fungicide treatment not different essentially from spraying with lime sulfur? The answer is, it does not matter. One can consider spraying with phenyl mercuric chloride in spring as sanitation, because it reduces the amount of inoculum from which the summer scab epidemic can develop; or one can consider it as an eradicant fungicide treatment and apply to eradicant fungicides the same mathematical reasoning as one uses for sanitation.

Nature seldom draws lines without smudging them. Biologists waste time seeking the perfect classification and clean lines that often, in fact, do not exist. They certainly do not exist in sanitation, and one might just as well accept this.

In this chapter we shall concentrate on sanitation as it affects blight of potatoes caused by *Phytophthora infestans*. The rest of this chapter is written with *P. infestans* in mind, even when this is not expressly stated. The chapter after this deals with wheat stem rust. The change from potato blight to stem rust brings with it such a great change of stress so that it is better to discuss the two diseases apart.

11.2. How the Infection Rate Affects the Benefit from Sanitation

Suppose there are two fields. In one of them the initial inoculum has been reduced by sanitation; in the other it has not. Apart from this, one supposes that the fields are identical in every respect: in time of planting, variety, fertilizer, cultural methods, aspect, and so on. Suppose further that the fields are large enough for us to ignore random variation.*

Suppose the proportion of tissue initially infected without sanitation to be x_0 and with sanitation, x_{0s}. We are simplifying the problem by letting the initial proportion of infected tissue represent the initial inoculum, which might, e.g., be sporangia from potato cull piles. But with potato blight there is good experimental support for this. In Section 7.10 it was shown from experimental data that the number of lesions is proportional to the number of spores that formed them. See also Section 13. 9.

Consider (to simplify the discussion temporarily) the logarithmic stage of an epidemic. The compound interest equation gives

$$x = x_0 e^{r_l t}$$

and

$$x_s = x_{0s} e^{r_l t}$$

in which x is the proportion of infected tissue after time t, r_l is the average infection rate during that time, and the subscript s refers to the field with sanitation. Division gives

$$\frac{x}{x_s} = \frac{x_0}{x_{0s}}$$

The effect of sanitation remains proportionately the same at all times in the logarithmic stage. If $x_0/x_{0s} = 5$, then $x/x_s = 5$ at all times.

In the previous paragraph we make use of the fact (see Section 7.2) that in the logarithmic stage r_l is independent of x. For example, other

* The supposition that there are two fields, one with and one without sanitation, is not very apt when the initial inoculum comes from large concentrated sources, e.g., when *Phytophthora infestans* comes from potato cull piles. Instead, suppose that there are two large populations of fields, all alike except that initial inoculum has been reduced by sanitation in one population but not in the other.

things being equal, it would take the same time for x to increase from 0.0001 to 0.001 as from 0.001 to 0.01. In the logarithmic stage lesions act independently of one another, so that the number of lesions is irrelevant.

One can estimate the benefit of sanitation as the delay in reaching any given level of disease. This delay is the time taken for the proportion

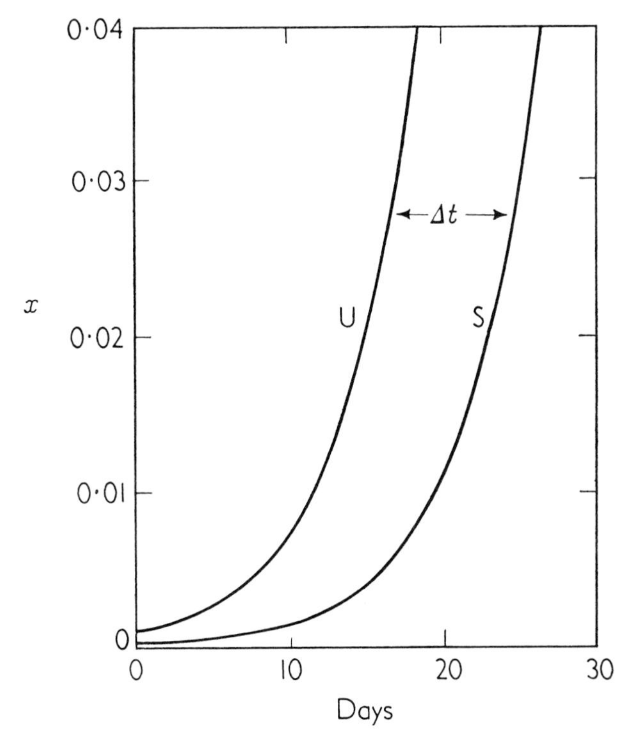

FIG. 11.1. The delay of disease brought about by sanitation. The sanitation ratio $x_0/x_{0s} = 5$; and $r_l = 0.2$ per unit per day. The delay Δt is 8 days at all levels of disease.

of diseased tissue to increase from x_s to x; and (because $x/x_s = x_0/x_{0s}$) this is the time taken for the proportion to increase from x_{0s} to x_0 at the same rate. Thus

$$x_0 = x_{0s}\, e^{r_l \Delta t}$$

where Δt is the delay. Hence

$$\Delta t = \frac{1}{r_l} \log_e \frac{x_0}{x_{0s}}$$

$$= \frac{2.3}{r_l} \log_{10} \frac{x_0}{x_{0s}} \tag{11.1}$$

For example, if the sanitation ratio $x_0/x_{0s} = 5$ and $r_l = 0.2$ per unit per day then

$$\Delta t = \frac{2.3}{0.2} \log_{10} 5$$
$$= 8 \text{ days}$$

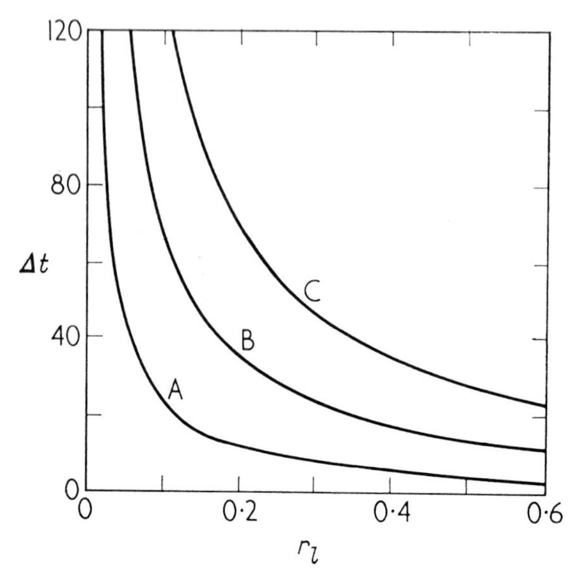

FIG. 11.2. The relation between the delay Δt, the logarithmic infection rate r_l, and the sanitation ratio x_0/x_{0s}. In A, B, and C the ratios are 10, 10^3, and 10^6, respectively.

Figure 11.1 illustrates this numerical example. The curve U is for the untreated field, and S for the field with sanitation. At all times curve U shows 5 times as much disease as curve S. At every level of disease curve S is $\Delta t = 8$ days behind curve U. An optical illusion may obscure this on the graph, but it can be verified by direct measurement.

Figure 11.2 shows how the delay Δt is inversely proportional to r_l. Three different curves A, B, and C show the effect for three different sanitation ratios, 10, 10^3, and 10^6, respectively. For all ratios, the delay is long when r_l is small, and short when r_l is large.

11.3. The Effect of the Sanitation Ratio

Equation (11.1) shows that the delay varies with the logarithm of the sanitation ratio. For every value of r_l, the values of Δt for curves A, B, and C in Fig. 11.2 are as 1: 3: 6.

A sanitation ratio x_0/x_{0s} of 10 means that 90% of the initial inoculum has been destroyed, $100/(100-90)=10$. A ratio of 10^3 means that 99.9% of the inoculum has been destroyed; and a ratio of 10^6 that 99.9999% has been destroyed. Destroying 99.9% of the inoculum is no easy task. Destroying 99.9999% is still more difficult by far. Yet the gain in sanitation by destroying 99.9999% is only double that of destroying 99.9%. The logarithmic basis of sanitation means that the returns diminish fast as sanitation is increased.

Short of complete eradication, there normally seems little to be gained by trying to destroy more than 99% of the inoculum. If at that stage control of blight is unsatisfactory, further effort is likely to be more profitable if it is directed at reducing r_l. In particular, one is likely to gain little by carrying sanitation far when the infection rate is fast and there are other sources of inoculum not being controlled by sanitation.

11.4. The Effect of Sanitation on Disease after the Logarithmic Phase

For simplicity we have been discussing the logarithmic stage of an epidemic as though a delay of the epidemic in that phase brought about an equal delay after that phase. In other words, it has been implicitly assumed that the value of r_l averaged over the period Δt at the end of the logarithmic phase is also the value of r after this phase. Most of the loss from blight (apart from tuber infection) comes about as disease increases beyond the logarithmic phase to the point where from 50 to 75% of the foliage is destroyed. What one really wants to know, therefore, is the delay Δt at this part of the epidemic.

If the five epidemics of *Phytophthora* blight illustrated in Fig. 4.1 are guides, r often stays roughly constant from the end of the logarithmic phase for the rest of the period of observation. Our implicit assumption is not unreasonable. When r stays constant in this way there is no difficulty of calculation. The delay Δt at the end of the logarithmic phase is the delay at all later levels of disease, and one can apply Eq. (11.1) simply by substituting r for r_l,

$$\Delta t = \frac{2.30}{r} \log_{10} \frac{x_0}{x_{0s}} \qquad (11.2)$$

Although Eq. (11.2) deals with events after the logarithmic stage of the epidemic, the correction factor $1-x$ does not appear unless x_0 and x_{0s} are themselves outside this stage.

Equation (11.2) is valid only if r is independent of x, as when r is constant. In Sections 6.10 and 8.8 it was seen that, as a good approximation, r is independent of x until $x = 0.35$ and, as a rough approximation, until $x = 0.5$. But if r varies after $x = 0.5$, we have no means yet of knowing what error this introduces into Eq. (11.2). We shall be dealing with problems in which r is considered to be constant, so no difficulty arises immediately. But the difficulty remains to be solved in future.

11.5. The Use of Eqs. (11.1) and (11.2) when Disease Is in Foci

Foci of *Phytophthora* blight start early in the season in potato fields near infected cull piles or around infected tubers planted as seed. They are local concentrations of disease in fields predominantly healthy. Early in the season the focus is conspicuously the unit of disease.

One may apply Eqs. (11.1) and (11.2) to epidemics that start from foci, provided that the epidemic area is large enough to contain hundreds of foci. One may apply them to potato blight in Aroostook County, Maine, but not to a single potato farm.

How to treat epidemics when disease is in foci was a topic of Chapter 7.

11.6. *Phytophthora infestans* from Potato Cull Piles

P. infestans overwinters in infected tubers. Some of these tubers are planted as seed the following spring. Occasionally, but not commonly, the fungus makes its way up the new shoots from the seed and releases sporangia above ground to infect other plants and initiate infection. Sometimes the fungus grows directly through the soil to infect leaves near it without invading stems from below ground (Hirst and Stedman, 1960b).

The number of seed pieces that initiate infection in either of these ways is small. In a careful study in the Netherlands, van der Zaag (1956) found that about one piece per square kilometer initiated infection in fields planted to the very susceptible variety, Duke of York. With less susceptible varieties the number was even less.

In North America more stress is put on another path. At the end of winter when potato stores are opened partially rotten tubers are culled and dumped on waste piles. Here the tubers form sprouts, often blighted. From these piles sporangia are released early and blown into fields of potatoes before the farmer starts to spray. Infected cull piles are dangerous compared with infected seed planted in the field, because they are not covered with soil and because the shoots grow in a dense tangled mass in which the ecoclimate is humid and favorable for spore production.

Special attention has been paid in Aroostook County, Maine, to the cull pile danger. Aroostook is the leading potato county of the United States and grows 130,000 acres of potatoes annually. The danger from cull piles was investigated there by Bonde and Schultz (1943, 1944), who published much useful information.

This is the history of an epidemic from one of these piles. In April and May 1938 about 25 barrels of cull potatoes were dumped on a pile about 100 ft. from a potato field. Blight was first seen on this pile on June 15, and by June 25 the plants on the pile were badly infected. Aroostook County then had a week of cloudy weather followed by light wind and rain from the southeast. On July 12 the field was examined and records taken of the prevalence of late blight at different distances from the infected cull pile.

Results are summarized in Table 11.1. The field was heavily infected near the cull pile, and hardly infected at all 600 ft. away. The evidence makes it perfectly clear that the cull pile was in fact the source of the fungus.

TABLE 11.1

INFECTION BY *P. infestans* OF A POTATO FIELD AT VARYING DISTANCES FROM AN INFECTED CULL PILE[a]

Distance from cull pile (ft.)	Plants infected (%)	Lesions per 100 plants
100	98	293
200	55	98
300	21	31
400	6	9
500	0	0
600	1	1

[a] Data from Bonde and Schultz (1943).

Observations of this sort were repeated. The pattern of spread of disease from infected pile to field was abundantly clear. Photographs show how the potato foliage was devastated in the fields near the piles in July, long before any general epidemic of blight occurred in Aroostook

County. In 5 years 417 cull piles were examined. Of these 232, or more than half, were infected; and from 52 of them, or one-eighth, infection had already established itself in neighboring fields before July 10. On most of the 6000 potato farms in Aroostook County as well as in the yards of many potato dealers there are one or more cull piles where potatoes are discarded each spring. The number of fields infected from cull piles before July 10 each year must thus be about 1000.

In 1943 the Farm Bureau and the Maine Agricultural Experimental Station started a campaign to destroy cull piles. The campaign continued for some years. It was highly successful in keeping blight out of fields until late in the season. It coped with the local epidemics which otherwise would have been seen spreading early in the season into fields from adjacent infected cull piles. But it did not stop a general epidemic from developing later. Indeed, as Bonde and Schultz (1944) noted, because blight seemed to be absent early in the season, many farmers were lulled into relaxing their spraying programs. Their farms were poorly sprayed. *Phytophthora infestans* spread quickly into them when the general epidemic occurred, and losses from blight were heavy.

The lesson was learnt, and potato farmers of Aroostook County, Maine, have the reputation for spraying their fields with fungicides more often and more diligently than anywhere else in the world.

11.7. Focal Outbreaks and General Epidemics of Potato Blight

What went wrong? Cull piles were the main source of initial inoculum. Infection spread from them to fields early in summer. Fields infected at this time suffered badly. Destroying cull piles saved fields near them from early infection. These were facts observed beyond dispute. But, on the evidence, destroying cull piles neither retarded subsequent general epidemics enough nor reduced their severity enough.

As Hirst and Stedman (1960b) stressed, potato blight in a season goes through two phases. The first phase is one of focal outbreaks around infected cull piles or (where Hirst and Stedman worked) infected shoots from infected seed pieces. The second phase is one of general epidemics. The change from the one phase to the other is often dramatic—the adjective is Hirst and Stedman's. After being confined to conspicuous foci, *P. infestans* starts fairly abruptly to spread over whole fields and from field to field over the whole countryside. This is the time recorded as the date on which an epidemic starts—the time blight forecasting systems are designed to determine. It is a time clearly defined in the minds of those studying potato blight epidemics.

Bonde and Schultz studied the spread of infection from potato cull piles in the phase of focal outbreaks. The amount of infection in this phase may be severe locally, but it is very small spread over the fields as a whole. *P. infestans* still has to multiply for a long time before it can be seen to cause a general epidemic. Foci are clear in the first half of July in Aroostook County. But, from the survey of Cox and Large (1960), only in 3 of the 13 years, 1943–1955, did general epidemics in Aroostook start in July. In 4 they started in August, and in 6 they started in September or no general epidemic started at all.

In the phase of focal outbreaks, the foci stand clear against a background of healthy fields. The clearness is itself a sign that general damage is small; and in Chapter 22 it will be seen that foci are most sharply defined when disease has multiplied little. The danger of general damage begins only after foci cease to be clearly discernible and start to merge into a background of general disease.

In this phase of focal outbreaks losses from blight are proportional to the number of foci. Two thousand foci separate from one another do twice as much damage as one thousand, random variation apart. So, too, loss from blight is reduced in direct proportion to the destruction of infected cull piles or other source of initial inoculum. If 99% of the infected cull piles or other sources are destroyed, loss from disease is reduced to 1% of what it otherwise would be.

But when the change occurs to a general epidemic, loss from blight stops being proportional to the number of infected cull piles or other sources of initial inoculum. Instead (to anticipate a later calculation) destroying 99% of the infected cull piles and other sources of inoculum saves only 60% of the loss of crop. The 40% left unsaved compares with the 1% left unsaved in the phase of focal outbreaks.

This difference, of 40% from 1%, explains what went wrong. It was hoped that benefits would be proportional to the cull pile destruction. They are not.

To the onlooker the difference may seem even greater than the figures, 40% and 1%, perhaps suggest. The 1% left unsaved is unsaved from very small losses. The general impression is of having blight well under control, and, as we have seen, farmers were lulled into a false sense of security and did not spray. The 40% left unsaved is unsaved from extensive losses. Despite this saving, blight is still seen to destroy the foliage completely if it is unsprayed. One gets the impression of general destruction, and is inclined to overlook the very real, but unseen, saving that has been effected.

11.8. The Change from Focal Outbreaks to General Epidemics

Nowadays, when one thinks of epidemics of blight in unsprayed potato fields, it is better to switch one's mind away from Aroostook County, Maine, where the unsprayed field is almost an anachronism. But it was not always so there. Nor is it at present so elsewhere; it seems clear from the survey of Cox and Large (1960) that the bulk of the world's potatoes are grown without fungicidal protection against blight.

Two changes turn focal outbreaks into general epidemics.

The first, and probably more important, is the increase of disease with time. Blight progress curves slope upward. As the amount of disease increases the boundaries of foci become vaguer. Finally the foci become blurred in the background of general disease. The process has been inadequately studied; a little will be said about it when we come to examine spread of disease.

The second contributes to the first. As the season advances there is normally an increase in the relative infection rate. Early in the season, before the potato foliage meets in the rows, *Phytophthora infestans* spreads slowly and multiplies slowly. Later it spreads swiftly and multiplies swiftly.

There are rival and probably complementary theories about this.

Grainger (1956, 1962) inquired why potato plants are less susceptible to blight in the middle stages of their growth than when they approach maturity. Chemical analyses showed that this was most clearly connected with the total carbohydrate in the plant. High carbohydrate content brings greater susceptibility. To obtain a suitable index, he used the ratio of total carbohydrate to dry matter other than carbohydrate. Under conditions in Scotland he found this ratio to be very low during June and July. It rose in August, and reached a maximum at the end of the growing season, in August or September according to the lateness of the variety. So too the susceptibility of the plants to blight was lowest in June and July, and rose to a maximum in August or September.

Hirst and Stedman (1960a) see the matter differently. As we have already noted, they saw that the activity of *P. infestans* during the growing season of potato crops is divided into two phases: an early phase of slow spread and focal development, and a later phase of a general epidemic. The date on which the epidemic became general varied from year to year, but did not normally range over much more than a month.

The date was not accompanied by lasting changes in climate thought sufficient to account for the dramatic change in the multiplication of the fungus. They turned their attention from the climate as measured in louvered screens above the fields to the ecoclimate measured by instruments within the crop itself. During the early phase when the plants are young and the soil exposed, ventilation is little impeded. The air at midday is warmer and drier within the crop than 4 ft. above ground, especially in dry sunny weather. But when the foliage grows and the canopy becomes dense, humid periods persist within the crop after rain long after the air has become dry above it. Climate now no longer dominates ecoclimate. Conditions within the canopy after rain allow abundant sporulation on foliage that stays wet long after the rain has stopped. The potato crops commit suicide by producing as they grow an ecoclimate that allows blight to destroy them.

It is during the second phase—the phase of high carbohydrate and humid ecoclimate—that the bulk of the potato crops are attacked. It is during this phase that we must calculate the effect of destroying infected cull piles.

11.9. The Delay of the General Epidemic as a Result of Destroying Cull Piles

As a figure for the infection rate during this phase we take $r = 0.46$ per unit per day. This is the value estimated in Exercise 8 at the end of Chapter 3. It agrees well with a value estimated for a number of crops of susceptible varieties during epidemics. Some of these values are reported in later chapters. With this value one estimates from Eq. (11.2) that destroying 90% of the infected cull piles delays a general epidemic 5 days. Because r often stays nearly constant throughout an epidemic (see Figs. 3.1 and 4.1), one can take the delay of 5 days to refer to all stages of the epidemic: to the epidemic when there is 0.1% blight, or 75% blight, or any other percentage one wishes to name.

This delay of 5 days is the most that one could get from thorough and conscientious destruction of all detected cull piles if 10% of the initial inoculum came from undestroyed sources. These might be infected seed tubers or undetected cull piles dumped near woods or in fields of cereals or clovers. Or the inoculum might be sporangia blown in from other areas.

If undestroyed sources were only 1% of the initial inoculum, the maximum delay (with $r = 0.46$ per unit per day) is 10 days.

General blight epidemics often start a month or more before potato fields are normally mature. A delay of 5 or 10 days is a real gain. But it does not stop the epidemic from developing until practically 100% of the foliage is destroyed. It is not seen to delay the epidemic. There is no standard by which a farmer can tell that he gained anything from cull pile destruction. If there is a gain he cannot weigh it, because there is no standard for comparison. If the epidemic is delayed, he cannot time it, because epidemics do not come on regular dates from year to year. The research worker is not much better placed to judge. The usual technique of field experimentation fails him, because blight epidemics range county-wide and delays from sanitation cannot be measured in small plots. In fact there is no way of determining the benefit of sanitation except by Eqs. (11.1) and (11.2).

It is no wonder that the farmers of Aroostook County, who take pride in keeping fields free from blight until the end of the season, relied on their spray machinery and showed little interest in sanitation.

Sanitation by destroying infected tubers or sterilizing them has been advocated off and on since the last century. It needs more quantitative investigation. But increasing labor costs and improving fungicides are moving against the method.

11.10. Cull Piles and Blight Forecasts

It can be argued that if sanitation delays an epidemic 5 or 10 days it will at least save the first weekly spray with fungicide. The argument is all very well provided that one can forecast when an epidemic will start.

Bonde *et al.* (1957) have devised a system of forecasting blight in Aroostook County. Blight is forecast at a time when there have been 10 consecutive days on which both temperature and rainfall were favorable to infection, and when the weather forecast at that time is for continued favorable conditions to come. (What constitutes favorable conditions does not concern us here.) A blight epidemic can be expected within 1 or 2 weeks after it is forecast.

The forecast allows a play of a week or two. To illustrate what this means, assume a play of 7 days. From Eq. (11.2) if, as before, $r = 0.46$ per unit per day, a 7-day delay of an epidemic can be expected if 96% of the cull piles or other sources of initial inoculum are destroyed. A play of 7 days in a forecast is equivalent to taking no notice of a 96% destruction of initial inoculum. Similarly, a play of 14 days in a forecast is equivalent to taking no notice of a 99.8% destruction of initial inoculum. A forecast with a large play has the virtue, if one may call it that, of being unaffected by variation, within large limits, in the amount of initial inoculum. That is, one can make forecasts on weather data alone, without bothering about the amount of initial inoculum. But by the same token the forecast is not precise enough to allow a fungicide spray program to be shortened safely by sanitation. Indeed it follows logically that spray programs can be safely cut as a result of sanitation only after blight forecasting systems have been evolved for which one needs to know the amount of initial inoculum as well as the weather. So far no systems of this sort have been devised in any country.

11.11. The Increase in Yield as a Result of Sanitation

Large (1945, 1952) and Cox and Large (1960) have explained how to estimate losses of yield from attacks of *Phytophthora infestans*. The curve for the progress of blight with time is read in conjunction with the curve for the progress of tuber development with time.

The top graph in Fig. 11.3 shows how tuber yield increases from the middle of August on. The data are those of Akeley *et al.* (1955) for crops kept free from blight in Maine. The graph uses an average of the results for all varieties (Chippewa, Green Mountain, Katahdin, Kennebec, Mohawk, Sebago, and Teton) other than the early variety, Irish Cobbler. Vines were killed on September 18 in preparation for harvesting. The yield on this date, an average of 597 bushels per acre, is taken as 100%. Yields on earlier dates are plotted as percentages of the final yield.

The lower graph in Fig. 11.3 shows the progress of two hypothetical blight epidemics. The earlier epidemic, shown by the curve on the left, is for conditions without sanitation. The curve on the right is for blight progress after 99% of the cull piles and other sources of initial inoculum have been destroyed. As calculated previously, for $r = 0.46$ per unit per day, the delay from 99% sanitation is 10 days. The two curves are 10 days apart on the graph.

Cox and Large estimate losses from blight on the basis that plants stop forming tubers when 75% of the foliage is destroyed by blight, i.e., when $x = 0.75$. Any slight increase of the tubers after this stage is offset by the slowing down of tuber growth before it. Figure 11.3 shows that without sanitation blight caused 75% defoliation on August 26; with sanitation, 10 days later on September 5. On August 26, 75% of the tuber yield had been developed; the loss through blight was 25% of 597 bushels, which is 149 bushels. On September 5, 90% of the yield had been developed; the loss was 10% of 597 bushels, which is 60 bushels. Of the 149 bushels, 60 bushels or 40% of the loss from blight was unsaved by sanitation. This is the value we used for illustration in Section 11.7.

It is possible that Cox and Large's criterion, that tubers stop developing when 75% of the foliage is destroyed, may have to be modified slightly. The results of Radley *et al.* (1961) for the varieties Majestic and Ulster Torch suggest that tubers stop developing rather abruptly when from 40 to 50% of the foliage is destroyed. Most probably, there is no fixed

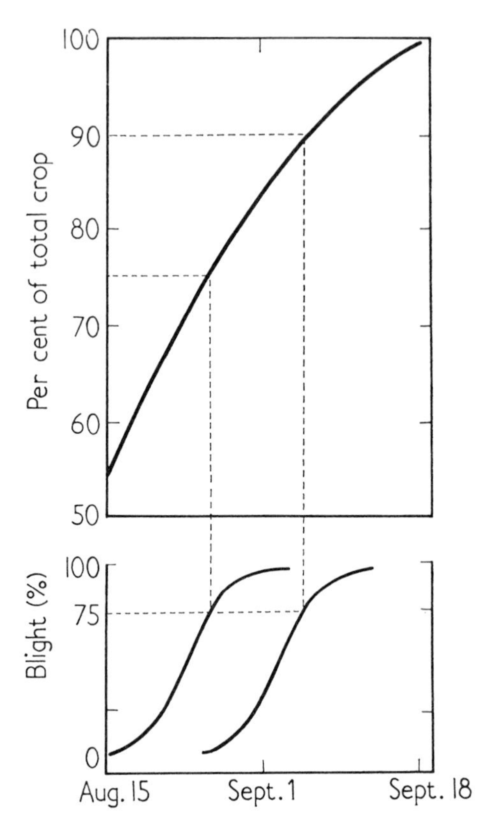

FIG. 11.3. Top graph—the development of tubers in potato crops in Maine. The values are percentages of the final crop in fields in which blight is fully controlled. Bottom graph—blight progress curves. The curve on the left is for conditions without sanitation; on the right, for conditions with sanitation. (For details, see text.)

criterion, and the percentage defoliation taken to mark the end of tuber bulking varies with the infection rate. The higher the value of r, the higher the percentage of defoliation one should take as the criterion. The reason for this will be evident in Sections 12.7 and 12.9 when infection rates and losses of yield from wheat stem rust are discussed.

CHAPTER 12

Sanitation with Special Reference to Wheat Stem Rust

SUMMARY

The chapter deals with barberry eradication, the effect of sanitation on the amount of stem rust, and the effect of the amount of stem rust on yield. Discussion is based on disease progress curves. Despite differences that are outwardly great, wheat stem rust and potato blight are treated essentially alike. The change from Chapter 11 to Chapter 12 is primarily in stress, because the epidemic process commonly goes further in unsprayed potato fields than in wheat fields.

12.1. Potato Blight and Wheat Stem Rust Contrasted and Compared

Potato blight caused by *Phytophthora infestans* and wheat stem rust caused by *Puccinia graminis tritici* are similar in many ways. Both fungi multiply fast if they are given a susceptible host and weather favorable to them. On data at present available one cannot decide which can multiply faster. But the results of Rowell (1957) and Asai (1960) show that what we have taken as a fast rate for potato blight ($r = 0.46$ per unit per day) is a fast rate for wheat stem rust as well. This does not exclude the possibility of faster rates with either disease.

As epidemics mount, the curve for the progress of disease with time is shaped as a rough S. For potato blight this is shown in Figs. 4.1 and 7.7; and S-shaped curves are familiar to all who have read Cox and Large's (1960) work on blight throughout the world. With wheat stem rust, the results of Rowell (1957), Underwood *et al.* (1959), and Asai

(1960) suggest an *S*-shaped curve. The *S* may not be perfect, but it is not a bad *S* either.

In relation to sanitation, the great difference between the diseases is that with potato blight the *S* is commonly complete, whereas wheat stem rust commonly writes only a fragment of the *S*. Typically, *Puccinia graminis tritici* reaches wheat fields too late to do more than start the *S*. The difference between the diseases has its roots as deep in economics as in host–pathogen relations.

Blight is common in areas of high yields of potatoes. The Netherlands have the highest average yields of potatoes per acre of any country. Blight epidemics writing the full *S* can be expected in most seasons. Ten years ago there were more unsprayed than sprayed fields. Nowadays the use of fungicides is becoming the rule. But even so epidemics mostly complete the *S* or would do so if the farmer did not intervene to destroy the foliage to prevent further tuber infection. The Netherlands farmer uses fungicide to delay the blight epidemic rather than to prevent it. In this he differs from the farmer in Aroostook County, Maine.

It has been shown by Cox and Large (1960) that in England and Wales potato yields are higher in years of severe blight than in years of little blight. In "blight years" the average yield is about 8.0 tons per acre; in other years about 7.1 tons. Heat and drought are worse enemies of the potato than blight, and weather that favors the potato also favors blight. (We except here countries where potatoes are grown under irrigation.) In England, as in the Netherlands, fungicides are used primarily to postpone epidemics, not to control them absolutely; and unless the farmer destroys the vines to prevent tuber infection, the epidemic completes the *S* in most years.

In Aroostook County, Maine, a climate good for potatoes is also a climate good for *Phytophthora infestans*. But fungicides now obscure the relation. In 1954, a year of very severe blight in unsprayed plots, blight was kept completely under control by fungicides in 40% of the fields, and only in 14% of the fields did the percentage of defoliation exceed 75% (Cox and Large, 1960).

In most wheat producing areas there is no comparable tendency for a full *S* with stem rust. In the spring wheat fields of the United States, there were years with a fairly full *S*: 1904, 1916, 1935, 1953, and 1954. But these years were the exception rather than the rule.

Lambert (1929) gives information about "the average year" in the western Mississippi Valley. In southern and northern Texas wheat ripened with 10% stem rust (i.e., $x = 0.1$); in Oklahoma, Kansas, and Nebraska, with 5%; and in South Dakota, North Dakota, and Minnesota, with 60%. These figures are for the middle 1920's. They show that the

S was usually incomplete, even while many barberry bushes remained undestroyed and before the common wheat varieties were specially bred for resistance against stem rust.

There are at least two reasons for this contrast between potato blight and wheat stem rust.

Phytophthora infestans clings to the potato. Where blight occurs on potatoes in summer, *P. infestans* survives on potatoes in winter. Any large tract of potatoes supplies its own inoculum in spring, and when one considers epidemics in, say, Aroostook County one need not look elsewhere for a source of inoculum. This is one reason why blight forecasters take little notice of initial inoculum and confine attention almost entirely to summer weather.

With wheat stem rust it is otherwise. *Puccinia graminis* does not overwinter on wheat north of Texas, except very occasionally. The spring wheat in the Upper Mississippi Valley must get inoculum either from barberries or as spores blown up from the south. Both sources are likely to be the limiting factor in epidemics; the inoculum is likely to be too little or arrive too late to start a general epidemic.

The other reason is economic. The potato plant sets tubers early. Even if blight destroys the foliage before it is ripe, the chances are good that there will be a crop beneath adequate in quantity and quality. The wheat plant fills out the grain late. If it is killed before it is ripe, the chances are that the crop will be light and the grain shriveled and unfit for milling and baking. A farmer cannot survive if his wheat is attacked regularly by wheat stem rust. Areas where stem rust would complete the epidemic S in most years doubtlessly exist. They are to be found among the blank parts of the world wheat map. Economics confines wheat to where calamitous epidemics of stem rust are rare.

12.2. Stem Rust and Barberry Eradication

Against *Puccinia graminis* sanitation has been practiced by destroying the fungus's alternate host, the barberry. Successful control of stem rust was achieved in north-western Europe by barberry eradication. Massachusetts, Connecticut, and Rhode Island passed barberry eradication laws before the Revolutionary War, but they were not applied thoroughly and fell into disuse.

Barberry eradication started in earnest in the United States as a result of devastation by the great stem rust epidemic of 1916. State and Federal legislation against the barberry began in 1917. In 1918, Congress passed a Federal appropriation to support the destruction of barberries, and the support has continued since. A barberry eradication area was established by agreement between the Bureau of Plant Industry of the United States Department of Agriculture and the enforcement

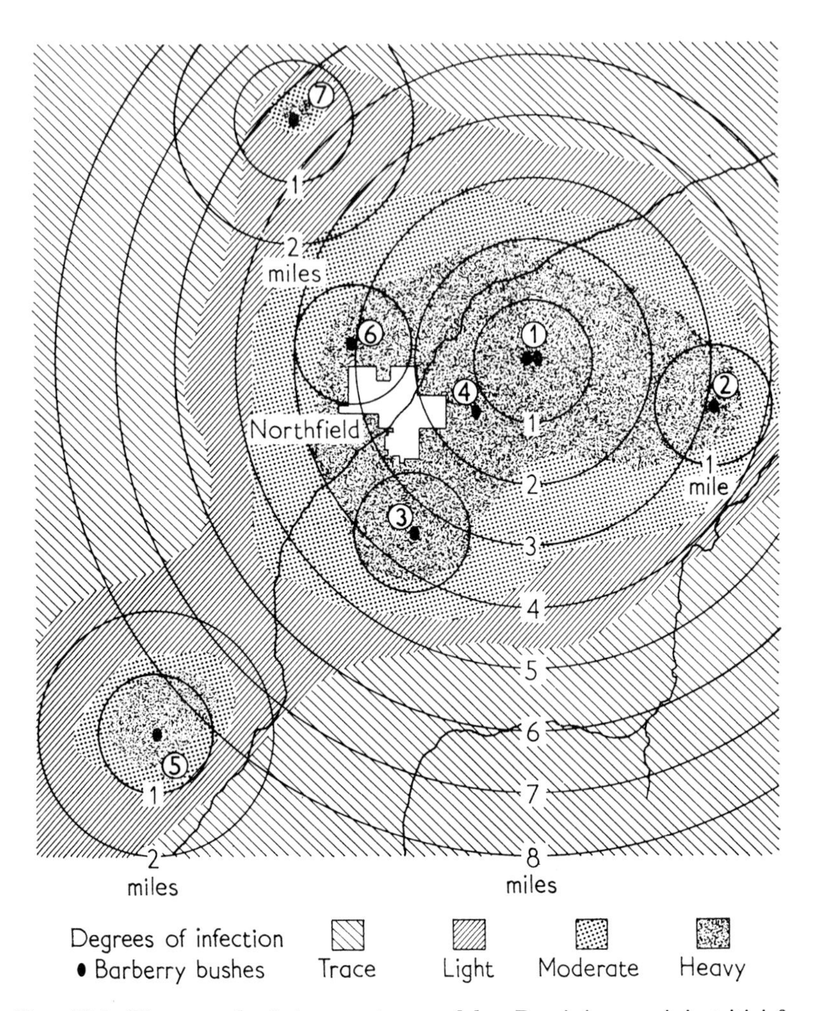

FIG. 12.1. The spread of stem rust caused by *Puccinia graminis tritici* from 175 large barberry bushes in seven plantings near Northfield, Minnesota, in 1922. The numbers from 1 to 7 indicate the location of the plantings. After Stakman and Fletcher (1930).

agencies in Colorado, Illinois, Indiana, Iowa, Michigan, Minnesota, Montana, Nebraska, North Dakota, Ohio, South Dakota, Wisconsin, and Wyoming. It was attempted to destroy every bush of the common, susceptible barberry and, through reinspection, to destroy new shoots and seedlings that grew later.

The literature of the early years of the campaign against barberries showed complete faith in the benefits to be gained. It is essential now to accept the facts as they were seen then, and not allow ideas to be colored by later disillusionment. The direct visual evidence was clear and unambiguous; and writer after writer noticed how stem rust outbreaks occurred near barberry bushes and ceased when the bushes were killed. The early *Supplements of the Plant Disease Reporter* make useful reading. But we shall confine attention to the evidence presented by Stakman and Fletcher (1930).

"Near Northfield, Minnesota [they wrote], in the summer of 1922 there were 175 large heavily rusted bushes in one planting. By May 26 the rust had spread 100 ft. from the bushes; by June 6 more than 1½ miles; by June 12, it could be traced 2¼ miles; by June 17, 4 miles, and eventually it spread at least 10 miles from the bushes." (See Fig. 12.1.)

"Not only did barberries cause local epidemics but they often caused destructive regional epidemics. For instance, in North Dakota in 1924 rust was discovered over a large area. Survey showed the area to be fan shaped. The narrow but heavily rusted handle of the fan was near Jamestown, and the rust epidemic became lighter and lighter as the fan spread out to the north, northwest, and northeast for a distance of about 100 miles. It was suggested that rusted bushes might be found near Jamestown. The search was made and 86 rusted bushes were found. No one suspected that they were there. The effect produced by the bushes was found first and the bushes were found later by searching near the most heavily infected grain. The heavy rust infection near the bushes and the gradually diminishing infection as the distance from these bushes increased clearly indicated that these bushes were the cause of this widespread infection, which eventually extended northward into Canada."

Again, observations in Wisconsin during 1924 on wheat and oats showed a very direct regional correlation between the infected barberry areas and regions of severely rusted grain. "The spread of rust from the Trempealeau County wild barberry area extended northward in a fan-shaped area over the most productive oat area in Wisconsin for a distance of about 100 miles. The infections were found within a few miles of the barberry early in the spring and spread gradually northward from these early infections until at the end of August practically all the fields in this entire area were infected. Fields within 25 or 30 miles of the barberry area were very severely damaged by rust, because of the early appearance of the disease. On the other hand, the region north of Dodgeville, extending to La Crosse, where the barberry had been removed, was relatively free from stem rust on oats. It was possible to trace most of the stem rust infection in the State directly to regions where the barberry was rusted during the early spring months."

Spore movements from the south, from Mexico and Texas, had been mapped and analyzed since 1917 and especially 1923 (Stakman and Harrar, 1957). Yet the way Stakman and Fletcher (1930) sum up all the known evidence shows that they were convinced that most infection in the eradication area in the past had come from local barberry bushes and that barberry eradication had markedly reduced stem rust despite the uredospores coming in from the south. In 1925, for example, the mapped data from spore traps showed that during the first few days of June stem rust spores disseminated over an area of about a quarter of a million square miles, over South and North Dakota, practically up to the Canadian border. The year 1925 has been cited in the literature as a classic example of spread of inoculum from the south. But 1925 was a year of light stem rust infection in the eradication area of the United States, in South and North Dakota as in the other States. Loss from stem rust in the eradication area was estimated to be 12 million bushels of wheat against 184 million in 1916. Stakman and Fletcher (1930) thought that the weather in 1925 had been just as favorable to stem rust as in 1916, and attributed the marked reduction in damage to all the barberry bushes that had been destroyed between these years. Although the rust spores from the south arrived early in 1925, they were, it seems, too few to offset the good work of barberry eradication.

Stakman and Fletcher also analyzed data to bring out the correlation between the decline of stem rust losses and the progress of the eradication campaign. In the 5 years, 1915 through 1919, when eradication had hardly begun, the average annual loss of wheat was 50 million bushels in the eradication area. From 1920, by which time 4 million barberry bushes had been destroyed, to 1924, when 11 million had been destroyed, the average annual loss of wheat was 26 million bushels. From 1925, when 12 million bushes had been destroyed, to 1929, when 18 million had been destroyed, the average annual loss of wheat was 11 million bushels.

Chester (1950, p. 285) disputed the correlation, and attributed falling losses to the introduction of resistant varieties. History is not on Chester's side. Ceres, the first resistant variety to be widely grown, was released only in 1926, and became available in adequate bulk too late greatly to affect Stakman and Fletcher's analysis. In 1929 susceptible Marquis was still far the most popular variety in the barberry eradication area.

Commenting on this correlation, Stakman and Fletcher warned that there were still many barberry bushes and seedlings undestroyed in the eradication area: enough to cause severe epidemics in conditions favorable for stem rust. Still better results, one infers, could be expected from better eradication.

But despite all this evidence, enthusiasm for barberry eradication waned when the great epidemics of 1935, 1953, and 1954 showed that all the cumulative killing of barberry bushes failed to stop the devastation. Indeed it was authoritatively written in 1960 that the eradication campaign was being continued "on a new and somewhat reduced basis —primarily to limit the number of new genetic races evolving out of sexual reproduction on barberry."

This summing-up is too harsh. Barberry eradication did everything it was seen to do. When outbreaks start at barberry bushes—and there were very many observations to show that outbreaks did, in fact, start there—destroying the bushes destroys the outbreaks. For many years barberry eradication substantially reduced stem rust infection. One need not question for a moment the validity of the correlation brought out by Stakman and Fletcher's figures: that with the start of eradication the average annual losses of wheat from stem rust fell from 50 million bushels to 26 million and then to 11 million in successive 5-year periods.

Nevertheless it is evident that the hopes of 30 or 40 years ago are no longer held so high. The reason is now clear. The first years of the barberry eradication campaign were essentially years of focal outbreaks. But in 1935, and again in 1953 and 1954, inoculum swept in from the south and undid much of what barberry eradication had done.

12.3. Focal Outbreaks and General Epidemics of Wheat Stem Rust

In Chapter 11 it was seen how *Phytophthora infestans* multiplies in foci near initial sources of inoculum, and then spreads widely to start a general epidemic. Lambert (1929) has described the same process for *Puccinia graminis* in wheat fields. For four "generations" the fungus tends to remain in a distinct focus, typically with a heavily rusted area 5 to 10 ft. in diameter. With further "generations" the rate of spread becomes faster, and the fungus disperses practically evenly over the field and moves from field to field. Finally the fungus spreads far and fast.

Chapter 22 deals with the general epidemic process, common to potato blight, wheat stem rust, and other diseases. As disease multiplies, disease gradients flatten. As the epidemic *S* develops, increase of disease locally is accompanied by a more than proportional spread of disease away from its initial source.

The benefits of barberry eradication are greatest when the epidemic process does not go far either in the eradication area itself or south of it. A smaller epidemic process to the south means less spores from the south to interfere. A smaller process in the area itself means less overshadowing of the good work of eradication. (See Section 13.6.)

The scenes that Stakman and Fletcher (1930) describe are typical of years with a relatively small epidemic process. It is not just coincidence that their most vivid examples of the benefit of barberry eradication are drawn from 1922 and 1924 when stem rust did little damage.

The contrast came in 1935. In May rainfall on wheat in Texas was almost treble the normal. Wheat ripened late and became heavily rusted. Spores moved north to Kansas and Nebraska. Here, too, wheat was late and became heavily rusted. Spores continued to move and reached the barberry eradication area early and in great masses. Conditions in the area itself again favored rust. Wheat was late, July wet, and the stem rust epidemic explosive and general. From south to north the epidemic process went far, and in the north, despite barberry eradication, stem rust devastated the wheat fields. The melancholy moral is that all the accumulated effort of barberry eradication helps least when it is needed most. Far from smoothing losses from stem rust, barberry eradication has probably sharpened the contrast between "rust years" and others.

One must extend the moral to barberry bushes as a source of new races of *Puccinia graminis*. Periods of weather that make for a full, or nearly full, epidemic S favor the spread of uncommon races as much as that of common races. Races uncommon a few years earlier are abundant when conditions favor a general epidemic. Ceres was the popular variety of spring wheat in the early 1930's. It was somewhat resistant to the races then common, but not to race 56. This race was first found in 1928 near barberry bushes. In 1935 it was there to take advantage of the wet seasons and late crops in Texas, Kansas, Nebraska, and the spring wheat area. It destroyed Ceres. Somewhat similarly, race 15B, first found on barberry in 1939, was uncommon before 1950. It was there in the spring wheat area to destroy the bread wheats and durums in 1953.

12.4. Barberry Eradication and Wheat Stem Rust in Northwestern Europe

Centuries back in Europe, long before the reason was known, the association of stem rust and barberries was observed. Most barberry bushes have since been destroyed, some under compulsion of laws passed relatively recently, and stem rust is now not a major problem in northwestern Europe. It is pertinent to ask why eradication was so successful here.

Uredospores come up from the south in Europe as well as in North America. But they come in smaller numbers, which enhances the benefit of sanitation in the north.

Temperature is probably another reason. In the Upper Mississippi Valley temperatures are about optimal for stem rust epidemics, or slightly below the optimum because Stakman and Lambert (1928) and Lambert (1929) found from a study of the records that epidemics tended to be somewhat commoner in years hotter than average. Summer temperatures in northwestern Europe are from 4 to 6°C. below those in the Upper Mississippi Valley, and considerably below the optimum for stem rust epidemics. One reaches the same conclusion in another way. The epidemic rust disease of wheat fields of northwestern Europe is stripe rust caused by *Puccinia striiformis*. Stripe rust is an epidemic disease of wheat near the cool limit of wheat production. For example in Kenya, on the equator, wheat is grown at altitudes from 6000 to 9000 ft. above sea level. At 6000 ft. stem rust is epidemic, at 9000 ft., stripe rust. The rule holds generally elsewhere: where stem rust becomes destructively epidemic, stripe rust is relatively unimportant; where stripe rust is epidemic, stem rust is relatively important.

Probably because of suboptimal temperatures for stem rust the change from focal outbreaks to general epidemics goes less far in northwestern Europe than in the Upper Mississippi Valley. Less of the epidemic S is written except locally; geographical spread of *Puccinia graminis* within a season is more limited; and it was possible for Denmark to have an eradication scheme on her own and for England successfully to control stem rust by barberry eradication while there were still many barberry bushes in Wales.

12.5. The Relation between Sanitation and the Percentage of Stem Rust in Ripe Fields

Figure 12.2 shows the effect of sanitation on the progress of stem rust up to the time when fields are ripe. The percentage of disease is plotted against the number of days still needed for the fields to ripen. The sanitation ratio is $x_0/x_{0s} = 10$, i.e., 90% of the initial inoculum is destroyed, e.g., by destroying barberry bushes. With time scale A, $r = 0.46$ per unit per day. Successive curves show the effect of sanitation. The top curve ends with 99% of rust at ripeness; the next with 91%. This means that sanitation, with a sanitation ratio of 10, would reduce rust from 99 to 91% at ripeness. Similarly, it would reduce it from 91 to 50%, or from 50 to 9%, or from 9 to 1%.

The curves are drawn on the assumptions that the infection rate is constant (i.e., $r = 0.46$ per unit per day for time scale A) and that disease

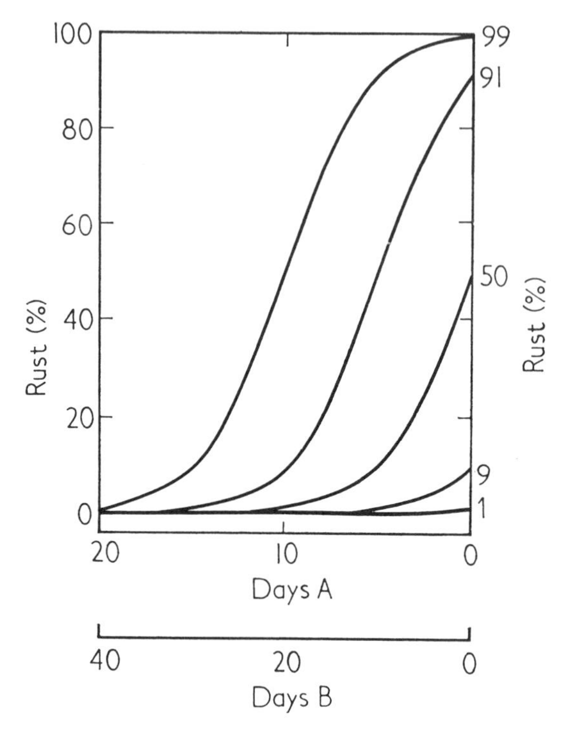

FIG. 12.2. The relation between sanitation and stem rust in wheat fields. The sanitation ratio is 10. Time is measured as days from ripeness. Time scale A is for $r = 0.46$ per unit per day; time scale B is for half this rate. (See text for other details.)

starts from inoculum arriving all at once at a given date. The first assumption is unlikely, and the second erroneous. (Apart from all else, inoculum would have to arrive from day to day in the proportions needed to keep r constant. See Chapter 6.) But neither affects the issue, and we can leave them as they are.

To avoid the empiricism implicit in curves for high percentages of disease, we shall look mainly at percentages of 50 or less.

If one substitutes time scale B for time scale A, r then becomes 0.23 per unit per day. But the rust progress curves shown in Fig. 12.2 remain unchanged. This illustrates a fundamental point. Sanitation, at sanitation ratio of 10, reduces disease from 50 to 9% or from 9 to 1%, irrespective of the infection rate at which these figures were reached. The ratio of percentage disease without sanitation to the percentage disease with sanitation is determined by the sanitation ratio and not the infection rate. (It is, of course, assumed that the infection rate, whatever it may be, is the same with as without sanitation.)

Sanitation reduces $\log [x/(1-x)]$ by $\log [x_0/x_{0s}]$, i.e., by log (sanitation ratio). One may use logarithms either to the base 10 or to the base e. Suppose that without sanitation the amount of disease at ripeness is 50%. That is, $x=0.5$, $\log [x/(1-x)]=0$. With a sanitation ratio of 10, $\log_{10} 10=1$. With this degree of sanitation $\log_{10} [x/(1-x)]=0-1=-1$ at ripeness. Hence $x=0.09$. The amount of disease with sanitation is 9%. With a sanitation ratio of 100 (i.e., with 99% of the initial inoculum destroyed) disease at ripeness would fall from 50 to 1%, corresponding with a change in $\log_{10} [x/(1-x)]$ from 0 to -2. With the same ratio of 100, it would fall from 31 (for which $\log_{10} [x/(1-x)]=\bar{1}.65$) to 0.45% (for which $\log_{10} [x/(1-x)]=\bar{3}.65$).

12.6. The Reduction by Sanitation of Loss in Yield

The figures in the previous section show an impressive response from sanitation if the final amount of disease is not high. But what does a change in the percentage of stem rust mean in terms of bushels of wheat per acre?

TABLE 12.1

KIRBY AND ARCHER'S (1927) TABLE FOR COMPUTING THE PERCENTAGE LOSS OF YIELD OF WHEAT BECAUSE OF STEM RUST

State of development of the crop:						Loss from stem rust (%)
Boot[a]	Flower[a]	Milk[a]	Soft dough[a]	Hard dough[a]	Mature[a]	
—	—	—	—	Tr.	5	0.0
—	—	—	Tr.	5	10	0.5
—	—	Tr.	5	10	25	5
—	Tr.	5	10	25	40	15
Tr.	5	10	25	40	65	50
5	10	25	40	65	100	75
10	25	40	65	100	100	100

[a] Average percentage of stem rust in the field; Tr. = trace.

To assist observers in the United States plant disease survey, the Office of Cereal Crops and Diseases of the Federal Department of Agriculture worked out a table to estimate yield loss when stem rust attacked wheat at different stages of development. It was published by Kirby and Archer (1927), and is reproduced here as Table 12.1. Presumably it distilled the Office's experience over the years.

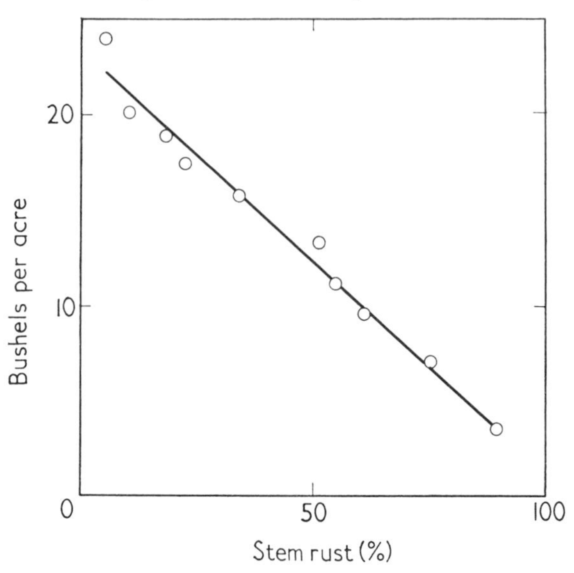

FIG. 12.3. The relation between yield of wheat, in bushels per acre, and the percentage of stem rust at maturity. The data are from the 1937 experiment of Greaney *et al.* (1941).

In Germany, Gassner and Straib (1936) studied the effect of rust on yield of wheat. They worked mainly with stripe rust caused by *Puccinia striiformis*, but also with stem rust. They introduced an injury coefficient, which was the percentage loss of yield brought about by each week's duration of rust attack. Thus a moderately severe attack of stem rust 2 weeks before the crop ripened reduced the yield by 24% in one variety and 17.6% in another. The injury coefficient for moderately severe stem rust was therefore 12 for the one variety and 8.8 for the other. They did not lead evidence to show that there is a linear relation between loss of yield and duration of attack, which their coefficient implies.

In Canada, Goulden and Greaney (1930) and Greaney *et al.* (1941) made use of sulfur to estimate loss of yield from stem rust. They found that sulfuring did not affect the yield of plots from which rust was absent. By altering the number of sulfur applications they altered the amount of stem rust and were able to estimate how rust affects yield.

Figure 12.3 analyzes data from their experiment with Marquis wheat in 1937 when stem rust was abundant (the plots were inoculated with stem rust in June) and leaf rust scarce. A linear regression line for yield on percentage rust fits the points reasonably well. There is a loss of 2.2 bushels per acre, or 9.4% of the crop, for each increment of 10% of stem rust. This loss is paralleled by a reduction of weight per 1000 grains. With loss of yield goes a concomitant loss of quality.

A linear relation between yield and the percentage of stem rust was found in all their experiments. If one were to apply this relation to the curves in Fig. 12.2, one would have to conclude that stem rust does its damage immediately before the field ripens, i.e., the stem rust present when the crop is, say, milk ripe does no harm. This is against all experience. One concludes that these results from sulfuring experiments are applicable to gains in yield in sulfuring, but not to gains from sanitation. The difference is this. Figure 12.2 was constructed for a constant infection rate: $r = 0.46$ per unit per day for time scale A or $r = 0.23$ for time scale B. Disease with, and disease without, sanitation increases at the same infection rate, whatever this may be. But sulfuring reduces the infection rate, and final percentages are approached at very different rates. Sanitation reduces the initial inoculum, not the infection rate; sulfur reduces the infection rate, not the initial inoculum (added artificially in these experiments).

Kingsolver *et al.* (1959) in Maryland inoculated winter wheat when it was jointing on April 20 and 22. Different levels of inoculum were used in different plots, and disease assessments were made in order to follow the progress of infection. These assessments are shown in Fig. 12.4. During the course of the experiment rust spread from plot to plot, and even the uninoculated plots were later swamped by infection from the rest of the experiment. (The vast spread of inoculum and interference of plots with one another made the terminal percentages of disease differ by less than the amounts calculated on sanitation theory.) To assess yield loss, Kingsolver *et al.* had to make use of a comparable field 400 m. upwind from the experimental area. This field had only a trace of stem rust when it ripened. It yielded 43.1 bushels per acre, whereas the uninoculated plots in the experimental area yielded only 11.6 bushels per acre because of the disease that spread into them. Other yields were 8.6 bushels per acre on plots inoculated with 0.1 gm. of a uredospore–talc mixture per acre; 7.5 bushels per acre on plots inoculated with 1 gm.; 4.1 bushels per acre on plots inoculated with 10 gm.; and 0.9 bushels per acre on plots inoculated with 100 gm.

The most heavily inoculated plots had 56% stem rust at flowering and 87% at the milk stage. Their yield was only 2% of that of the control field. This suggests that grain development stops when stem rust infection reaches a level somewhere between 56 and 87%. In later discussion we shall take a provisional figure of 75%. The heavy infection at flowering severely interfered with reproduction, and an analysis of the data shows that far fewer grains were set in the heavily inoculated

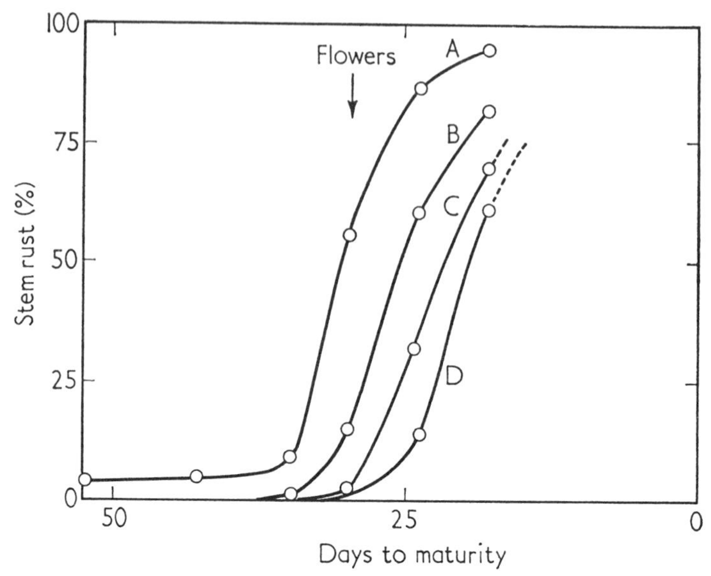

FIG. 12.4. The development of stem rust in wheat after artificial inoculation on April 20 to April 22. Time is measured in days before the crop ripened on July 2. Curve A is for plots receiving 100 gm. of inoculum (a uredospore–talc mixture) per acre; curve B, for plots receiving 10 gm.; curve C, for plots receiving 1 gm.; curve D, for plots receiving 0.1 gm. and plots not artificially inoculated, there being no difference between the two. Data of Kingsolver *et al.* (1959).

plots. At lower levels of inoculation, stem rust also interfered with reproduction but to a smaller extent.

Kingsolver *et al.* related yield loss to stem rust in two different ways. They related loss to the host stage when 1% stem rust was recorded; and they related loss to the logarithm of the percentage of rust at the dough stage.

12.7. The Relation between Loss of Yield and the Area under Stem Rust Progress Curve

To try to solve the relation between stem rust and yield loss, we suggest the two simplest possible hypotheses.

First, injury is proportional to the amount of disease. Two pustules do twice as much damage as one pustule, other things being equal. If, as the evidence in the previous section suggests, grain development stops when about 75% stem rust is present, this hypothesis is necessarily only approximately true. Alternatively, it should be applied only when the final amount of rust in the ripe crop is less than 75%.

Second, injury is proportional to the duration of the disease. A pustule developed 2 weeks before the crop ripens does twice as much damage as a pustule developed 1 week before, other things being equal. This is also the idea behind Gassner and Straib's injury coefficient.

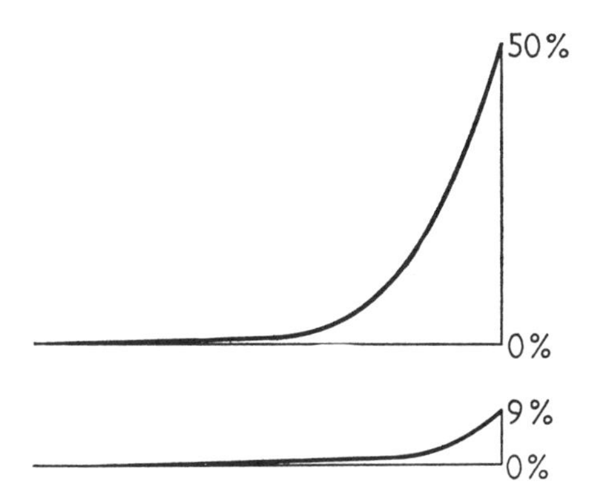

FIG. 12.5. The areas under disease progress curves. The upper area is for disease without sanitation, ending with 50% stem rust when the fields are ripe. The lower area is for disease reduced by sanitation and ending with 9%. The progress curves were taken from Fig. 12.2; and details for Fig. 12.2 apply to Fig. 12.5.

On these hypotheses, injury is proportional to the area under the disease progress curve. Figure 12.5 illustrates this for the reduction by sanitation of disease from 50% when the crop is ripe to 9%. The two areas shown in the figure are taken from Fig. 12.2; they are in the ratio of approximately 15:2. Sanitation, on these hypotheses, cut losses to 2/15 of what they otherwise would have been.

Inevitably, factors (such as soil fertility or wheat variety) other than area are assumed equal.

For the suggestion to hold it is necessary that the wheat plant should continue to increase the dry weight of the grain until the plant is ripe. The results of Percival (1921) suggest that it does. He measured the change in weight of grain as the plant ripened. His results are given in Fig. 12.6. The dry weight as a percentage of the final dry weight is plotted against the number of days before the crop ripened on August 4.

Unfortunately Percival did not say when the plants flowered, so the time of zero weight is unknown. The rate of ripening is for summer conditions in England, and relatively slow. There is no suggestion of a falling-off of the rate toward maturity.

To test the area hypothesis, Kirby and Archer's table (Table 12.1) was analyzed. Results are given in Fig. 12.7. The loss of yield recorded by them is plotted against the area under the appropriate disease progress curve. This was drawn on the assumption, obviously intended by Kirby and Archer, that the intervals between the stages (boot, flower, milk, soft dough, hard dough, mature) are equal. The area is expressed in

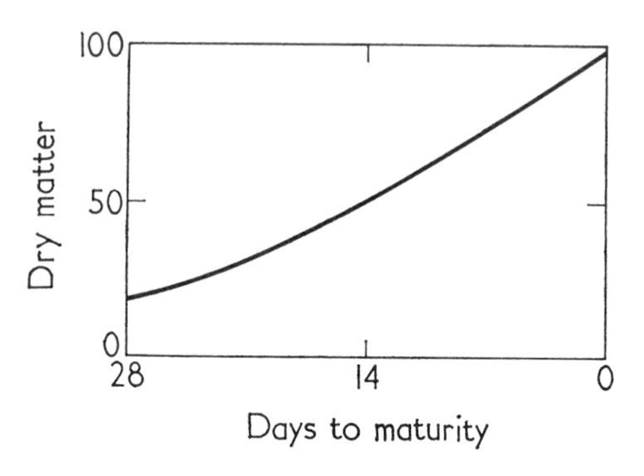

FIG. 12.6. The rate of development of wheat grain in England. The dry weight as a percentage of the final dry weight is plotted against time before the plants ripened. Data of Percival (1921).

arbitrary units. The last entry in their table, for 100% loss from disease, has been excluded because it is irrelevant. (There is nothing in their table to preclude the possibility of 100% loss from less disease than they enter.) A straight line fits the points reasonably well, as it should on the area hypothesis. Only one point is much out of line (the point for 50% loss of crop). Since Kirby and Archer gave no details about how they constructed the table, one cannot now judge whether the fault is in the table or the area theory.

Figure 12.8 uses the data of Kingsolver *et al.* (1959) presented in Fig. 12.4. Just as Fig. 12.7, Fig. 12.8 also plots the recorded loss from stem rust against the area under the disease progress curve, but with the restriction that the area is estimated only below the line for 75% disease. That is, Fig. 12.8 adopts the suggestion made in the previous section that grain development stops when stem rust is 75%. (This restriction is unnecessary in the analysis of Kirby and Archer's table, which is concerned with lower percentages of stem rust. Also, the restriction is not of very great importance to the analysis of the results in Fig. 12.8; it is included just as an improvement.) A regression line is fitted to the points except to the one for the highest amount of disease and abnormal reproduction. The line is not very informative one way or the other. If it gives no strong support for the area hypothesis, it also gives no evidence against it.

At one stage at least, the area hypothesis can be seen to be correct: the logarithmic stage. If the crop ripens with disease still at this stage, reducing the initial inoculum to 1/10 reduces the area under the disease

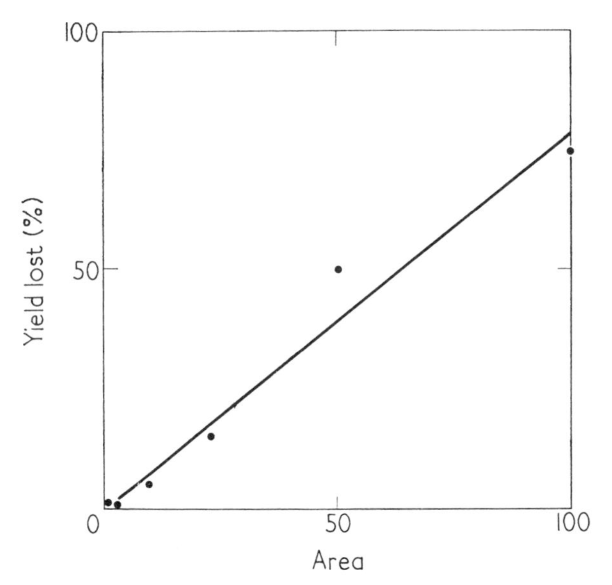

FIG. 12.7. An analysis of Kirby and Archer's table (Table 12.1) to relate loss of yield of wheat to the area under the stem rust progress curve. Area is shown relatively, in arbitrary units.

progress curve to 1/10, which on the area hypothesis reduces yield loss to 1/10. At the logarithmic stage the initial foci (e.g., near barberry bushes) do not interfere with one another. (See Chapter 7.) Sanitation that reduces the number of foci to 1/10 would, other things being equal, reduce yield loss to 1/10, as the area hypothesis predicts. The hypothesis has a sound base to stand upon.

We can now return to the two areas in Fig. 12.5. The ratio 15 : 2 of these two areas is independent of the time scale. The curves on which they are based are taken over from Fig. 12.2, and can be used with either time scale, A or B, in that figure. That is, the ratio is independent of the infection rate. It is determined by the sanitation ratio (10 in this example), which in turn determines the final percentages of stem rust (50 and 9% in this example).

But, although the ratio does not vary with the infection rate, the actual areas do. Changing from time scale A to time scale B doubles the area under each curve. The area (measured in any absolute units one cares to choose) is inversely proportional to r.

One must reconsider Kirby and Archer's table in this light. The table is for a slow infection rate, slower than that in the experiments of Rowell

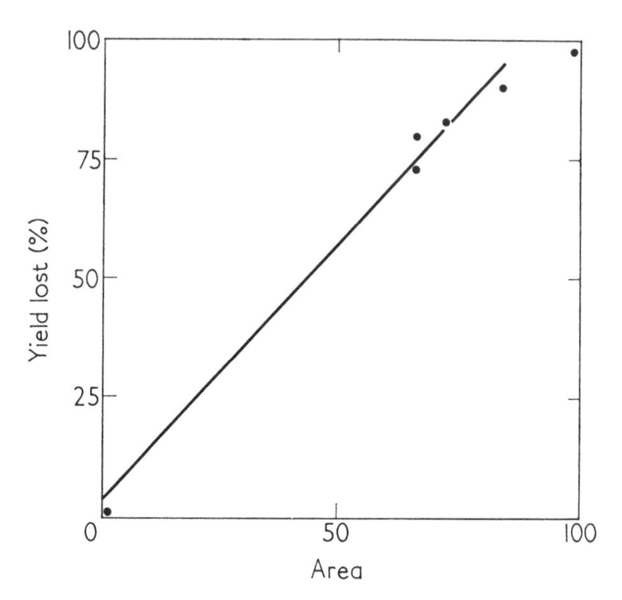

FIG. 12.8. An analysis of the data of Kingsolver *et al.* (1959) to relate loss of yield of wheat to the area under the stem rust progress curve. Area is shown relatively, in arbitrary units.

(1957), Kingsolver *et al.* (1959), or Asai (1960). In the table, 40% of stem rust in the ripe crop means a 15% loss of yield. But, if the infection rate had been doubled, 40% of stem rust in the ripe crop would mean only a 7.5% loss of yield if the area hypothesis is correct.

Any table that relates loss of yield to terminal stem rust infection holds only for some particular infection rate. That much is clear, even if one does not wish to accept a simple area relation without further evidence. Naumov (quoted by Chester, 1950) reports that Kirby and Archer's table does not apply to losses from wheat stem rust in Russia. One does not expect it to do so, unless infection rates there are precisely those Kirby and Archer had in mind. (Other variables, such as variety or temperature, may also affect differences between results in different countries, but they do not concern us here.)

12.8. The Infection Rate in Relation to Benefits Gained from Sanitation

In Section 12.6 infection rates were discussed in a way that is true only to a limited extent. To simplify the argument attention was directed at the percentage of rust when the crop was ripe: e.g., at 50% of rust without sanitation or at the corresponding 9% with sanitation at a sanitation ratio of 10. With time scale A in Fig. 12.2, the 50% and the 9% levels were reached as a result of a relatively fast infection rate following the relatively late arrival of the initial inoculum. With time scale B, they were reached with a lower infection rate and a correspondingly earlier arrival of the initial inoculum.

All this is artificial. There is no compensating correlation between infection rates and times of arrival of the initial inoculum. So now let us free the argument from this correlation. Only one conclusion needs changing. The rest remain.

One needs to change the conclusion that the infection rate does not affect the benefit to be gained from sanitation.

For a given time of arrival of inoculum, the higher the infection rate, the higher the percentage of stem rust will be when the field is ripe. For the purpose of calculation assume throughout that the sanitation ratio $x_0/x_{0s} = 10$. If the proportion of rust never increases beyond the logarithmic phase of an epidemic (i.e., if inoculum arrives very late or in very small amounts or r_l is very small), sanitation in this ratio will reduce the area under the disease progress curve to 1/10. On the hypothesis that loss of yield is proportional to this area, sanitation in this ratio will cut yield loss to 1/10 of what it would have been without sanitation. If (because of a higher value of r or more abundant or earlier inoculum) there was 50% of rust in the ripe fields without sanitation, there would be 9% with sanitation, and sanitation would cut the yield loss to 2/15.

If without sanitation there was 91% of rust and with sanitation 50%, sanitation would cut yield loss to 3/10. If the amounts were 99% and 91% of rust, respectively, sanitation would cut yield loss to 1/2. (One can calculate these results from the areas under the disease progress curves in Fig. 12.2, by making use of the fact that the ratio between any two of these areas is independent of r.)

For a given time of arrival of inoculum, the greater value of r, the higher the amount of disease in the ripe crop and the less, proportionately, sanitation cuts loss in yield.

How sanitation cuts the absolute loss of yield (measured in, say, bushels per acre) depends on soil fertility and other factors that affect yield.

12.9. Potato Blight and Wheat Stem Rust Compared Again

The quantitative treatment of the effect of sanitation in Chapters 11 and 12, on potato blight and wheat stem rust respectively, seems very different. Actually, the epidemic process is similar, and the treatment seems different only because the chapters concentrate on the opposite ends of the process.

Loss of yield of wheat because of stem rust is assumed to be proportional to the area under the rust progress curve. Loss of yield of potatoes because of blight is estimated by Large's method, which assumes tuber bulking to stop when 75% of the foliage is destroyed. The methods look dissimilar but are in reality similar. If one used for potato blight a figure of 50% defoliation instead of 75% (a minor change and, as we saw in Section 11.11, possibly a justified change) in a sigmoid blight progress curve and if the rate of tuber bulking were constant, the methods would be the same. In other words, Large's method in effect equates yield losses with the area under the blight progress curve, but with a neat correction for change in rate of tuber bulking with time.

Our treatment of potato blight and wheat stem rust is consistent. What varies is the stress; and that varies largely because the two diseases tend to write different amounts of the epidemic curve S.

CHAPTER 13

Sanitation and Two Systemic Diseases. Sanitation when Other Things Are Not Equal

SUMMARY

This is the last of three consecutive chapters in which the relation between disease and initial inoculum is probed.

The chapter is in two parts.

The first deals with the effect of inoculum on bunt of wheat and fusarium wilt of tomatoes. Both are "simple interest" diseases—the pathogens do not multiply in the sense of the word given in Chapter 4—and sanitation in one form or another (including the use of vertically resistant varieties) is normally effective.

The second part deals with the way in which the good effects of sanitation may be neutralized by an increase in the infection rate in seasons unusually favorable to disease. They are least likely to be neutralized if rt is usually small, or, what comes to the same thing, if the disease usually starts from relatively high amounts of initial inoculum. Control by sanitation is safer with "simple interest" diseases than with "compound interest" diseases, especially if the infection rate of the compound interest disease is high and the epidemic lasts long. On the other hand, simple interest diseases are less efficiently controlled by reducing the infection rate.

13.1. Common Bunt of Wheat and *Fusarium* Wilt of Tomatoes

So far sanitation has been discussed mainly around *Phytophthora* blight of potatoes and stem rust of wheat. Epidemics start from small amounts of inoculum and the pathogens multiply fast in a form similar to compound interest.

To strike a balance we now discuss two systemic diseases: common bunt (stinking smut) of wheat caused by *Tilletia caries* and *Tilletia foetida*, and fusarium wilt of tomato caused by *Fusarium oxysporum* f. *lycopersici*. Bunt and, for most practical purposes, fusarium wilt of tomato do not spread from plant to plant during the course of a single season. Increase is in a form similar to simple interest. The relation between initial inoculum and disease in the crop is therefore close and direct.

Both bunt and fusarium wilt are largely controlled by sanitation if under sanitation one includes vertical resistance. Common bunt is controlled by crop rotation, by using clean seed or treated seed, and by planting vertically resistant varieties. Fusarium wilt is controlled by using vertically resistant tomato varieties (see Chapter 17).

We shall be mainly concerned with the relation between initial inoculum and infection of the crop.

13.2. The Relation between the Number of Spores and Infection of Wheat by Bunt

Heald (1921) investigated the relation between the load of spores of *Tilletia* spp. on the grains planted as seed and the percentage of bunt in the crop that grew from them. The results presented here are from experiments in which he artificially infected the susceptible variety Jenkins Club. In his experiments farm infected seed produced very little bunt in the crop, too little for analysis. Heald explains this as the effect of the lessened viability of spores that have passed the winter on the surface of the seed.

Figure 13.1 shows the relation between the number of spores per grain and the proportion x of smutted plants. To allow for multiple infections $\log_e [1/(1-x)]$ is plotted against the spore load. Figure 13.2 is similar, except that the results are taken from Heald's data for the percentage of smutted heads instead of smutted plants. Linear regression lines are fitted; they pass practically through the origin, and there is nothing in the data to demonstrate a departure from a linear relation.

It is unfortunate that Heald's publication seems to contain a misprint. With the lowest spore load (104 spores per grain) the percentage of smutted plants is recorded as 0.49 and of smutted heads as 1.80. This rules out a detailed analysis of points near the origin.

The highest spore load recorded in Figs. 13.1 and 13.2 is 37,000 spores per grain. Heald used up to 164,000 spores per grain in his experiments. His results show very little increase of infection with this extra load. The points for higher loads are excluded from Figs. 13.1 and 13.2 because they are irrelevant to our discussion. In a survey of naturally infected grain Heald found loads in excess of 10,000 spores per grain only in 6 out of 84 samples. The top spore load of 37,000 spores per grain shown in Figs. 13.1 and 13.2 is already very high. Higher loads may interest the physiologist, but it is doubtful whether they bear much on epidemiology. Section 7.10 deals with the caution needed about high loads.

Figures 13.1 and 13.2 show no evidence for a departure from a simple proportional relation between the number of spores and the number of infections. Heald interpreted his results differently, and others have followed him in looking for complications.

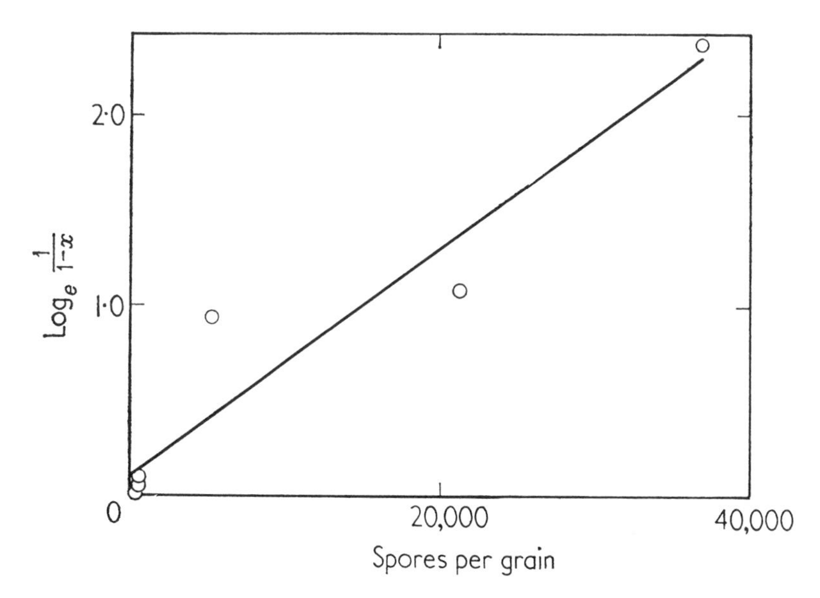

Fig. 13.1. The relation between the proportion x of smutted plants and the number of spores of *Tilletia* spp. per grain. In order to correct for multiple infections, $\log_e[1/(1-x)]$ is plotted instead of x. Data of Heald (1921) from artificial inoculation of the wheat variety Jenkins Club.

"It seems to have been [Heald wrote] the general opinion of plant pathologists that infection of a wheat plant with bunt might be accomplished by a single spore, but the results seem opposed to such an idea. Food for thought should be found in the fact that a considerable number of spores per grain may not be sufficient to cause any infection. Two possible explanations may be suggested. Either an infection occurs in which a number of spores participate; or there is a chemical mass effect due to numbers of spores, and infection may then be from a single infection thread."

Future experiments may justify Heald's speculations; his own do not. There is nothing to suggest interaction between spores at the loads recorded in Figs. 13.1 and 13.2. All that is evident is that the proportion of spores causing infection is small. Whereas (in approximate numbers) 1 uredospore in 100 leads to the development of a stem rust pustule on wheat under suitable conditions, only about 1 bunt spore in 10,000 led to infection of the very susceptible variety of wheat, Jenkins Club. There is a marked quantitative difference, but no evidence for a qualitative difference.

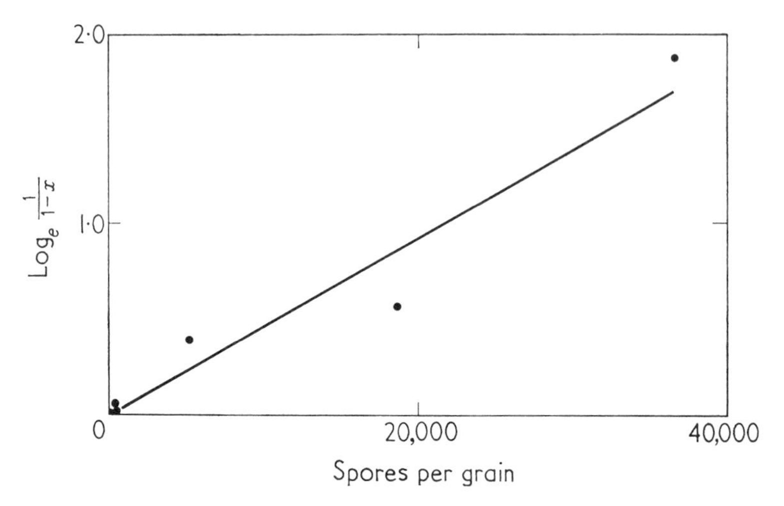

FIG. 13.2. The relation between the proportion x of smutted heads and the number of spores of *Tilletia* spp. per grain. Other details are the same as those for Fig. 13.1.

13.3. The Relation between the Number of Spores and Infection of Tomatoes with *Fusarium* Wilt

Haymaker (1928) grew tomato plants in sterilized soil. When they were in the two-leaf stage he inoculated them with spores of *Fusarium oxysporum* f. *lycopersici*. Thirty days later he counted infected plants.

Figure 13.3 reproduces his results graphically, with $\log_e [1/(1-x)]$ plotted against the amount of inoculum. A linear regression line fitted to the points goes practically through the origin. The evidence suggests that infections are proportional to the number of spores.

Fusarium oxysporum f. *lycopersici* is like *Tilletia* spp., and *Puccinia graminis* and *Phytophthora infestans* discussed in Chapter 7. There is no evidence of interaction between spores, provided that the concentration of spores does not exceed the limits likely to be found in epidemics. (Log $[1/(1-x)]$ corrects for a plant being infected more than once. The use of log $[1/(1-x)]$ has no known bearing on whether spores interact or not.)

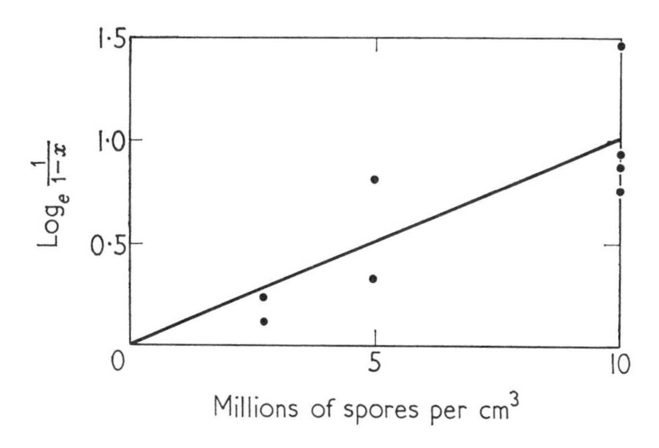

FIG. 13.3. The relation between the proportion x of tomato plants that become infected with fusarium wilt and the number of spores of *Fusarium oxysporum* f. *lycopersici* per $cm.^3$ of inoculum. In order to correct for multiple infections, $\log_e [1/(1-x)]$ is plotted instead of x. Data of Haymaker (1928).

13.4. The Effect of Sanitation on Disease of the "Simple Interest" Type

When one deals with the simple interest type of disease, one plots $\log_e [1/(1-x)]$ against the amount of initial inoculum. (See the remarks on transformations in the Appendix.) This was done in Figs. 13.1, 13.2, and 13.3. It shows how to estimate the effect of sanitation on the amount of simple interest disease.

Suppose that without sanitation there is 62% disease. If nine-tenths of the initial inoculum is destroyed by some method of sanitation, how much disease will there be? From Table 3 in the Appendix, we read that when $x = 0.62$, $\log_e [1/(1-x)] = 0.968$. When $\log_e [1/(1-x)] = 0.968/10 = 0.0968$, $x = 0.092$. Destroying nine-tenths of the inoculum would reduce disease from 62% to an estimated 9.2%.

This method for the simple interest type of disease differs from the method for the "compound interest" type described in Section 12.5. A given degree of sanitation affects the simple interest disease more than the compound interest type. But a comparison of the two types of disease in relation to sanitation is conveniently left until later in this chapter.

13.5. The Benefit of Reducing Systemic Disease by Sanitation

We discuss now systemic disease, not simple interest disease.

In the preceding section we estimated by how much sanitation reduced disease. But by how much does sanitation reduce loss of yield from disease? What is the relation between the amount of disease and the loss of yield?

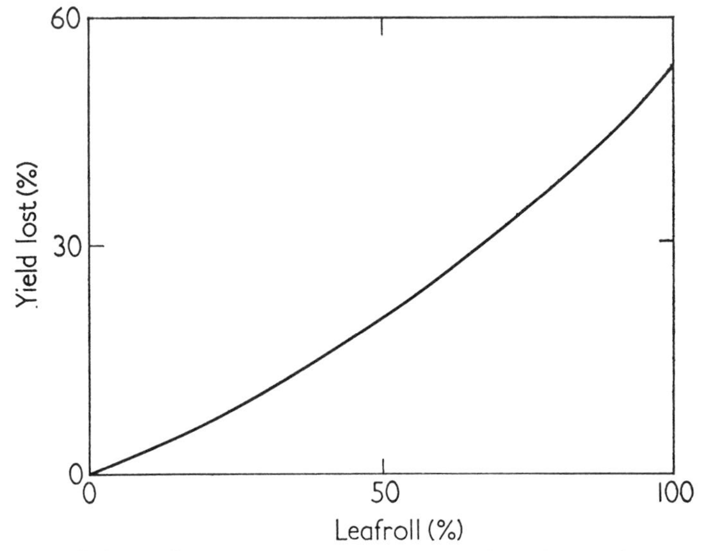

Fig. 13.4. Relation between the percentage of leaf roll disease in potato fields and the percentage loss of crop. Data of Tuthill and Decker (1941) for the variety Irish Cobbler.

With bunt, Heald and Gaines (1930), Flor *et al.* (1932), Leukel (1937), and Slinkard and Elliot (1954) obtained results which suggest that loss of yield is proportional to the percentage of disease. The percentage loss of yield is only slightly less than the percentage of disease.

When therefore sanitation reduces the amount of bunt it reduces almost proportionately the loss of yield from bunt.

Our treatment, in this and preceding chapters, of loss of yield from potato blight, wheat stem rust, and bunt has been consistent. (It is easily seen, from the experimental results given two paragraphs back, that loss of yield from bunt is nearly proportional to the area under the bunt progress curve, which is horizontal after the seedling stage.) We can treat the three diseases consistently, because with none of them is there much compensation. New healthy growth does not compensate for diseased tissue. But with many systemic diseases it does. A healthy plant between two diseased plants grows better and yields better than a healthy plant competing against two healthy neighbors. A healthy plant compensates partially for disease in its neighbors. When this happens, sanitation reduces loss of yield proportionately more than it reduces disease. There is an extra gain from sanitation.

A healthy potato plant yields better when it is next to a plant with leaf roll disease, and compensates somewhat for loss from leaf roll. Blodgett (1941), Tuthill and Decker (1941), and Kirkpatrick and Blodgett (1943) neatly investigated this in potato fields in which plants with leaf roll disease are randomly scattered. Figure 13.4, drawn from the data of Tuthill and Decker, shows how compensation makes the curve for the relation between yield loss and percentage disease become steeper as disease increases. The curvature depends much on the variety, the number of plants per acre, the fertility of the soil, the nature of the disease, and other factors.

13.6. Dependability of Sanitation as a Method of Disease Control in Relation to *rt*

We now come to the second part of the chapter.

Any reduction in the amount of initial inoculum inevitably reduces the final amount of disease, other things being equal. But other things are not always equal; and the effect of changes in the amount of initial inoculum may be small compared with the effect of other changes.

For example, by referring to data in Chapter 11, one can see that even a careful campaign against potato cull piles and other sources of *Phytophthora infestans* is unlikely to postpone a blight epidemic by more than a week or two. In Maine and England, however, a blight epidemic can be expected at any time from July to September. If the weather is very favorable for blight, an epidemic will start in July in unsprayed fields. The effect of weather on blight is much more striking than that of sanitation. In other words, with potato blight the average value of r can change so much from season to season that the change overshadows any change that sanitation is likely to effect in the amount of initial inoculum. In a season with a high average value of r, sanitation will be inadequate to control blight. In a season with a low average value of r, sanitation will be unnecessary to control blight.

The tendency for other things not to be equal inevitably influences the choice of sanitation as a method of control. And the tendency is largely determined by the value of rt. The higher the value, the greater is the tendency for other things to obscure the effect of sanitation. A fast infection rate and a long period over which infection can mount are the primary factors that make sanitation inadequate in some seasons and perhaps unnecessary in others.

Consider this example. Potato plants stop developing tubers when 75% of their foliage is destroyed by blight. (See Chapter 11.) In the Netherlands van der Zaag (1956) investigated the danger from blighted tubers planted as seed. In the very susceptible variety, Duke of York, he found an average of one blighted shoot, growing from an infected tuber, per square kilometer of potatoes. This was enough to cause an epidemic that destroyed the foliage. There are about 4 million plants per km.^{2} of potato fields, and about 750 lesions per plant, or 3 billion lesions per km.^{2}, when 75% of the foliage is destroyed, i.e., when $x = 0.75$. To keep the figures simple we assume that a blighted shoot—the initial inoculum—is equivalent to 3 lesions. This makes the increase one billion-fold in a season. For a very simple model assume that this increase occurs in 88 days, from June 1 to August 28. Assume too that the increase is according to Eq. (3.6). In this equation we put $t_2 - t_1 = 88$ days; $x_2/x_1 = 10^9$; $1 - x_1 = 1$; $1 - x_2 = 0.25$. From these figures $r = 0.251$ per unit per day, averaged over the 88 days.*

From 1 diseased shoot per km.^{2}, blight increased from June 1 to destroy 75% of the foliage on August 28 with $r = 0.251$ per unit per day. But suppose a season arrives with weather more favorable to blight, and r averages 0.30 per unit per day. If blight starts spreading on the same date, June 1, then, if it is not to reach the 75% level before August 28, the initial inoculum must be reduced to 1/74, or 1.3%, of what it was before. That is, it must be reduced to 1 diseased shoot in 74 km.^{2}.

Consider now a hypothetical disease. It has an infection rate only one-tenth that of potato blight. At an average rate of $r = 0.0251$ per unit per day the disease, as before, increases to 75% defoliation after

* One can change the model without substantially changing the result. Our model does not apply to single fields. *P. infestans* moves from field to field. Blight is severe in the first field (the field with the initial inoculum) before it is severe in the next field. If one thinks of blight in individual fields, r for each individual field would average more than 0.251 per unit per day. Alternatively, with blight scattered throughout the country, one can picture blight increasing according to Eq. (3.6) throughout the country as a whole. (See Sections 7.8 and 22.3 for examples.) As before, the average for the country as a whole, $r = 0.251$ per unit per day, is lower than in the individual fields.

88 days. If a season arrives to raise the average rate to $r = 0.030$ per unit per day, the initial inoculum would have to be reduced by one-third to keep defoliation down to 75% after 88 days.

With $t_2 - t_1 = 88$ days, a change from $r = 0.251$ to $r = 0.30$ per unit per day would neutralize sanitation that reduced the initial inoculum by 73/74. But a change from $r = 0.0251$ to $r = 0.030$ per unit per day would neutralize only a very mild degree of sanitation that reduced initial inoculum by about a third.

One can make another calculation, to put the matter differently. With $t_2 - t_1 = 88$ days, sanitation that reduces initial inoculum to 1/74 would be offset by an increase of r from 0.251 to 0.30 per unit per day (an increase by one-fifth in the infection rate) or from 0.0251 to 0.074 per unit per day (a threefold increase in the infection rate). The effect of a given degree of sanitation is more easily nullified by a given relative increase in the infection rate, if the disease normally has a fast infection rate.

What holds for a normally fast infection rate r holds also for a normally long time of multiplication t. Doubling t (i.e., $t_2 - t_1$) has the same effect as doubling r. When there is a long interval of time, sanitation can be trusted to give adequate control only if the infection rate is very small.

13.7. Sanitation in Relation to the Absolute Amount of Initial Inoculum

Sanitation is a reduction of initial inoculum. One may estimate the effect of reducing the initial inoculum by, say, 99%. It can be asked, 99% of how much? Does the absolute amount of inoculum not bear on the sanitation problem?

Consider the extremes. An epidemic may start from very small initial amounts of inoculum. For example, relatively few pustules of stem rust surviving the winter on wheat in Texas in 1935 started an epidemic that destroyed millions of acres of wheat in North Dakota, Manitoba, and elsewhere in the spring wheat area. Such an epidemic can be destructive only if disease multiplies fast or for a long time. At the other extreme, disease may multiply slowly or for only a short time. If it does, an epidemic can be destructive only if it starts from a high initial amount of inoculum.

We can now rewrite what we deduced in the previous section, namely, that fast and prolonged multiplication makes it hazardous to rely on sanitation to control disease. We now write: Sanitation is a safe method of control of epidemics that normally start from a relatively large amount of initial inoculum.

The rewritten form sometimes is the more useful, because more is often known about initial inoculum than about infection rates.

From evidence already discussed, a potato blight epidemic can develop destructively when the initial inoculum is not more than 1 infected shoot per km.^{2} of potato fields. Root knot nematodes (*Meloidogyne* spp.) may occur abundantly in each cubic centimeter of infected soil. With no more evidence than this, one may reasonably surmise that a given degree of sanitation against nematodes (e.g., by destroying 99% of the inoculum by soil fumigation) is less likely to be neutralized by seasonal weather changes than an equal degree of sanitation against potato blight (e.g., by destroying 99% of the diseased seed or diseased cull piles).

The more abundant the inoculum, the more safely the disease is controlled by sanitation. This statement does not imply that it is easier to destroy, say, 99% of the inoculum, if the inoculum is abundant. How to destroy the inoculum by biological, physical, or chemical methods and the ease with which the inoculum is destroyed do not enter into the statement.

13.8. A Comparison between "Simple Interest" and "Compound Interest" Disease in Relation to Sanitation

Our discussion has not been apt for diseases such as bunt of wheat or fusarium wilt of tomatoes because r is not an apt measure of infection rate when the disease is of the simple interest sort. With this type of disease, there is a simpler relation between sanitation and factors such as seasonal changes of climate.

Equation (4.1) applies to the simple interest type of disease. The equation shows that to maintain a constant absolute infection rate dx/dt, QR must stay constant, where Q is, for example, the number of bunt spores per grain of wheat, and R is the relative infection rate. If R is increased by 1/5, this increase can be balanced by sanitation that reduces Q by 1/6. For example, reducing the number of bunt spores per grain by 1/6 would balance and neutralize the harm done by an increase of 1/5 in the relative infection rate of bunt.

This relation, that Q must decrease by 1/6 to balance an increase of R by 1/5, holds for all values of R. That is, the efficiency of sanitation against the simple interest type of disease does not depend on the infection rate being small. In this, the simple interest type of disease contrasts sharply with the compound interest type. With the compound interest type, as we have seen, the benefits of sanitation are much more easily neutralized by a given relative increase of r when r is great.

The same relation, that Q must decrease by 1/6 to balance an increase of R by 1/5, holds also for all values of t. In this, too, the simple interest type of disease contrasts sharply with the compound interest type.

An increase of infection rate can balance and neutralize the benefit of reducing initial inoculum by sanitation. How do simple interest and compound interest diseases compare? To answer this quantitatively one must return to the mathematics of Chapter 6. We shall just give a general answer. If an increase in R just neutralizes the benefit of a given amount of sanitation during the first generation of the pathogen, it more than neutralizes this benefit during the second generation. The harmful effect of increased R—the overneutralizing of sanitation—increases with each succeeding generation. (In the language of Chapter 6, the harmful effect increases as t/p increases.) The first generation of the pathogen is the simple interest phase of the disease. Subsequent generations make the disease a compound interest disease. In other words, a given increase of infection rate (as a result, e.g., of weather becoming very favorable to disease) is more likely to neutralize the benefit of sanitation if the disease is of the compound interest type, other things being equal. The greater the infection rate and the longer the duration of the epidemic, the greater becomes the difference between compound interest and simple interest diseases.

This is one side of the picture: A simple interest form of increase of disease puts a premium on sanitation as a method of control. The other side is that, by the same argument, a simple interest disease is more difficult to control by reducing the infection rate. This is discussed further, in relation to horizontal resistance, in Section 20.7. Again, with a compound interest disease, if the epidemic lasts long and the infection rate is high, there is the hazard that the benefits from sanitation will easily be neutralized by a change in the rate if the season is favorable to disease. This is one side of the picture. The other side is that, by the same argument, a compound interest disease, if the epidemic lasts long and the infection rate is high, is especially amenable to control by reducing the infection rate, even relatively slightly. This, for example, is the theme in Chapter 16 for defense against wheat stem rust.

There is an element of mutual exclusiveness about desirable control measures. The characteristics that make a disease safe to control by sanitation necessarily make the disease less easily controlled by methods of reducing the infection rate. The characteristics that make a disease easy to control by reducing the infection rate necessarily make the disease less safely controlled by sanitation.

Because of the element of mutual exclusiveness one may have to stress one form of control rather than another. If the disease is of the simple interest type, stress sanitation; if it is of the compound interest type, stress reducing the infection rate if the rate is fast and the epidemic lasts long. The faster the rate and the longer the epidemic, the more the stress should shift from sanitation to reducing the infection rate.

In terms of controlling disease with resistant varieties of the host plant, for sanitation use vertical resistance; and for reducing the infection rate use horizontal resistance.

These rules about stress say nothing about the biological, chemical, and physical difficulties of implementing any particular method of control. The rules may suggest that sanitation is a desirable method of control; they do not say how one should set about achieving sanitation or whether sanitation can be easily achieved.

The wheat variety Elmar was released in the Pacific Northwest of the United States in 1949. (See Section 17.5.) It had vertical resistance to the races of the bunt fungus common there. Sanitation, by means of this vertical resistance, was effective. Then the races changed, and after some years Elmar was badly smutted. This does not contradict what we have been saying. During a single season bunt is a simple interest disease; but if increase is considered from year to year (i.e., over several seasons) bunt is a compound interest disease. When Elmar was smutted in later years, the fault lay not in sanitation, but in lack of it because Elmar was not vertically resistant to the races that later became abundant. If a variety is vertically resistant to the bunt races present at the beginning of a season, the benefit of this resistance is unlikely to be neutralized during the season by weather or other conditions favorable to infection. This is the gist of this chapter. But what will happen some years later, if the variety continues to be grown, is not relevant to simple interest disease.

13.9. Two Kinds of Initial Inoculum

In the last three chapters we have been discussing initial inoculum without closely defining it. To conclude these chapters, we now discuss the two sorts of inoculum, for which the symbols x_0 and Q have been used.

The initial inoculum may be ordinary lesions, or infected plants if the disease is systemic. For example, in Section 5.4, we considered the initial inoculum to be the stem rust pustules developed directly by artificial inoculation. These lesions or systemically infected plants beget others of the same kind; the infection chain is homogeneous. The proportion x_0 of initial infection is then just x at whatever time we choose to be zero time.

The other sort of initial inoculum is not in a homogeneous infection chain. For example, in Section 4.4, we considered the initial inoculum to be *Fusarium oxysporum* f. *vasinfectum*, present in the soil or added to the soil in a bran culture. We used the symbol Q for it.

The distinction between the two sorts of initial inoculum is sometimes important. But in calculations we can often conveniently ignore it because we often deal with ratios. For example, in the next chapter we apply Eq. (11.2) to data on potato blight without distinguishing between x_0 and Q because, for reasons given in Chapter 7, $x_0/x_{0s} = Q/Q_s$ during the logarithmic phase of an epidemic.

EXERCISE

This exercise considers details from Section 13.6.

Blight increased 1 billion times in 88 days to the stage $x = 0.75$. From Eq. (3.6),

$$r = \frac{2.30}{88} \log_{10} 4 \times 10^9$$

where

$$4 = \frac{1}{1 - x_2}$$

Hence $r = 0.251$ per unit per day.

If r is increased to 0.30 per unit per day,

$$0.30 = \frac{2.30}{88} \log_{10} \frac{4x_2}{x_1}$$

and

$$\log_{10} \frac{4x_2}{x_1} = \frac{0.30}{0.251} \log_{10} 4 \times 10^9$$

$$= \frac{0.30 \times 9.60}{0.251}$$

$$= 11.47$$

In the foregoing calculation x_2 is constant ($x_2 = 0.75$). Hence

$$\log_{10} \frac{x_1 \text{ when } r = 0.251 \text{ per unit per day}}{x_1 \text{ when } r = 0.30 \text{ per unit per day}} = 11.47 - 9.60$$

$$= 1.87$$

Hence

$$\frac{x_1 \text{ when } r = 0.251 \text{ per unit per day}}{x_1 \text{ when } r = 0.30 \text{ per unit per day}} = 74$$

For the hypothetical disease, when $r = 0.0251$ per unit per day,

$$0.0251 = \frac{2.30}{88} \log_{10} \frac{x_2}{x_1(1-x_2)}$$

and

$$\log_{10} \frac{x_2}{x_1(1-x_2)} = \frac{0.0251 \times 9.60}{0.251}$$

$$= 0.960$$

When $r = 0.030$,

$$\log_{10} \frac{x_2}{x_1(1-x_2)} = \frac{0.030 \times 11.47}{0.30}$$

$$= 1.147$$

In these calculations x_2 is constant. Therefore

$$\log_{10} \frac{x_1 \text{ when } r = 0.0251 \text{ per unit per day}}{x_1 \text{ when } r = 0.030 \text{ per unit per day}} = 1.147 - 0.960$$

$$= 0.187$$

$$\frac{x_1 \text{ when } r = 0.0251 \text{ per unit per day}}{x_1 \text{ when } r = 0.03 \text{ per unit per day}} = 1.54$$

$$= \tfrac{3}{2} \text{ approximately}$$

Much of this calculation is superfluous, because if the first ratio is the antilogarithm of 1.87, the second ratio is the antilogarithm of 0.187. If the infection rate of the hypothetical disease was one-fifth of that of blight, the ratio would be the antilogarithm of 1.87/5.

CHAPTER 14

Vertical and Horizontal Resistance against Potato Blight

SUMMARY

Vertical resistance in a host is directed against some races of a pathogen but not others. In potato foliage vertical resistance is conferred by R genes. It amounts almost to immunity from some races of *Phytophthora infestans*. Vertical resistance reduces the initial inoculum x_0.

Horizontal resistance operates against all races. It reduces the infection rate r. The rate is reduced because the host resists the establishment of lesions by spores, and because the fungus in lesions produces fewer sporangia and takes longer to produce them.

Vertical resistance exists only when there is a mixture of genotypes of the host. Vertical resistance helps to defeat itself. *P. infestans* is plastic, and forms new races easily. If vertical resistance makes a genotype commercially popular, the races able to attack the genotype become more abundant, and much of the advantage of vertical resistance wears off. Vertical resistance is at its best when the popular varieties belong to several different R-types.

A variety benefits from vertical resistance directly in proportion to the amount of horizontal resistance it has. By shielding varieties from full selection pressure, vertical resistance has significantly reduced the horizontal resistance of late maturing varieties of continental Europe. Methods for assessing horizontal resistance in the presence of vertical resistance are available. Unless they are used, it would be better if breeders left vertical resistance alone.

Most commercial varieties have only a fraction of the horizontal resistance potentially available in *Solanum tuberosum*. Incorporating more of this resistance into new varieties should be the first aim of potato breeders. Apart from the merits it has on its own, horizontal resistance enhances the benefit from vertical resistance, general sanitation, and fungicides.

14.1. The Relation between Races of *Phytophthora infestans* and Resistance Genes in the Potato

Table 14.1 shows how races of *P. infestans* and "major" genes in the potato are named. The system was proposed by Black *et al.* (1953) and

TABLE 14.1

INTERNATIONAL SYSTEM OF DESIGNATING INTERRELATIONSHIPS OF GENES AND RACES OF *Phytophthora infestans*[a]

Genotype	(0)	(1)	(2)	(3)	(4)	(1,2)	(1,3)	(1,4)	(2,3)	(2,4)	(3,4)	(1,2,3)	(1,2,4)	(1,3,4)	(2,3,4)	(1,2,3,4)
r	S	S	S	S	S	S	S	S	S	S	S	S	S	S	S	S
R_1	—	S	—	—	—	S	S	S	—	—	—	S	S	S	—	S
R_2	—	—	S	—	—	S	—	—	S	S	—	S	S	—	S	S
R_3	—	—	—	S	—	—	S	—	S	—	S	S	—	S	S	S
R_4	—	—	—	—	S	—	—	S	—	S	S	—	S	S	S	S
R_1R_2	—	—	—	—	—	S	—	—	—	—	—	S	S	—	—	S
R_1R_3	—	—	—	—	—	—	S	—	—	—	—	S	—	S	—	S
R_1R_4	—	—	—	—	—	—	—	S	—	—	—	—	S	S	—	S
R_2R_3	—	—	—	—	—	—	—	—	S	—	—	S	—	—	S	S
R_2R_4	—	—	—	—	—	—	—	—	—	S	—	—	S	—	S	S
R_3R_4	—	—	—	—	—	—	—	—	—	—	S	—	—	S	S	S
$R_1R_2R_3$	—	—	—	—	—	—	—	—	—	—	—	S	—	—	—	S
$R_1R_2R_4$	—	—	—	—	—	—	—	—	—	—	—	—	S	—	—	S
$R_1R_3R_4$	—	—	—	—	—	—	—	—	—	—	—	—	—	S	—	S
$R_2R_3R_4$	—	—	—	—	—	—	—	—	—	—	—	—	—	—	S	S
$R_1R_2R_3R_4$	—	—	—	—	—	—	—	—	—	—	—	—	—	—	—	S

[a] Notation: — = resistant; S = susceptible.

is now internationally accepted. To give examples, the gene R_1 makes a potato variety resistant to all races of *P. infestans* that do not have the number 1 in them: races (0), (2), (3), (4), (2,3), (2,4), (3,4), and (2,3,4). It leaves the variety susceptible to races with the number 1 in them: (1), (1,2), (1,3), (1,4), (1,2,3), (1,2,4), (1,3,4), and (1,2,3,4). Similarly a potato variety with the three genes R_1, R_3, and R_4 is susceptible to races with the three numbers 1, 3, and 4 in them: races (1,3,4) and (1,2,3,4), but resistant to all races without these three numbers.

This system was proposed at a time when only the four genes R_1, R_2, R_3, and R_4 had been much studied. At least two more genes, R_5 and R_6, are now known. The system is readily extended to include them, e.g., $R_1R_2R_3R_4R_5R_6$-types are susceptible only to race (1,2,3,4,5,6). The system refers only to resistance in the normal foliage. (Senescent leaves lose a little of their resistance.) Tuber resistance does not follow the pattern in the foliage exactly, and tubers may be susceptible to races that cannot attack the foliage. We shall be concerned mainly with the foliage.

All R genes introduced into commercial potato varieties probably come from *Solanum demissum*, a hexaploid species of Central America. This is certainly true where the breeding material was of known origin. And there is little doubt that the W-race of potatoes, widely used in Germany and the United States, was also derived from *S. demissum* (Müller, 1951).

14.2. The Mutability of Races of *Phytophthora infestans*

When genes for resistance from *S. demissum* were first incorporated in the common potato, they conferred resistance amounting almost to immunity from *P. infestans* for a while. In the breeders' nurseries the new varieties with *S. demissum* genes behaved excellently. But as soon as the varieties had been increased to a few acres, blight began to appear conspicuously. From the blight lesions, Schick (1932) and O'Connor (1933) isolated a new race of *P. infestans*. Since then it has been realized that, in conditions that favor blight, new derivatives from *S. demissum* invariably succumb before long to appropriate races of the blight fungus.

Gallegly and Galindo (1958) showed that the blight population of Mexico comprised two intercompatible groups of races that intercrossed freely in nature. Here there is sexual recombination to produce new races. Elsewhere in the world only one of the compatibility groups has been found (Smoot *et al.*, 1958), and there is no evidence for sexual recombination. But asexual mutability in *P. infestans* is quite enough to account for the origin of new races.

From race (0) Mills and Peterson (1952) developed races (1), (2), (4), (1,4), and (2,4) simply by serial passage through senescent leaves of the appropriate genotypes. Black (1960) gives examples of changes from simple to more complex races, particularly those involving the gene R_4 (see accompanying tabulation).

Race (0) to race (4)
Race (1) to race (1,4)
Race (2) to race (2,4)
Race (3) to race (3,4)
Race (4) to race (1,4)
Race (1,2) to race (1,2,4)

Eide *et al.* (1959) report that races (1,2,3,4,5), (1,2,3,4,6), and (1,2,3,4,5,6) were obtained by inoculating senescent or juvenile leaves with race (1,2,3,4). The same authors found that races (2), (3), and (4) appear frequently in single-spore isolates of race (0). Graham *et al.* (1961) record several examples of change: e.g., from race (1,2,4) to (1,2,3,4) and from race (1,2,3) to (1,2,3,5).

Black (1960) also cites examples of the reverse process in which some complexity is lost. This happens in the field or when tubers lacking R genes are used as media for keeping cultures. (See accompanying tabulation.)

Race (1,4) reverted to race (4)
Race (1,3,4) reverted to race (3,4)
Race (1,2,3,4) reverted to race (1,3,4)
Race (1,2,4,6) reverted to race (2,4,6)

P. infestans is an adaptable organism. In the presence of R-genotypes it extends its host range; in their absence it tends to revert to a simpler host range.

14.3. Vertical and Horizontal Resistance

When a variety is resistant to some races of a pathogen we shall call the resistance vertical or perpendicular. When the resistance is evenly spread against all races of the pathogen we shall call it horizontal or lateral.

Figure 14.1 shows the behavior of two varieties, Kennebec and Maritta, both with the gene R_1. This gene confers vertical resistance to races (0), (2), (3), (4), (2,3), (2,4), (3,4), and (2,3,4). This is shown in Fig. 14.1, by making the level of resistance to these races complete, irrespective of the variety.

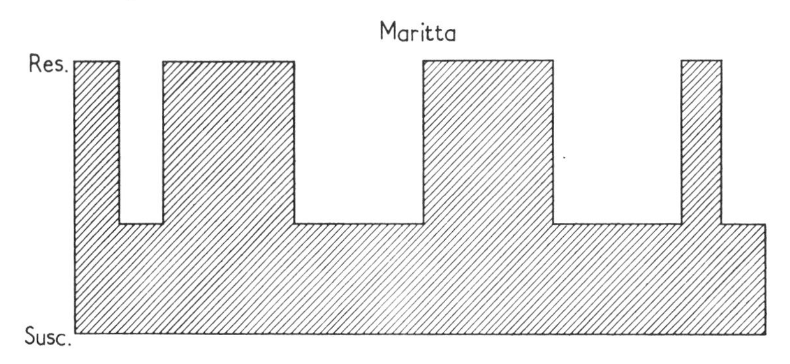

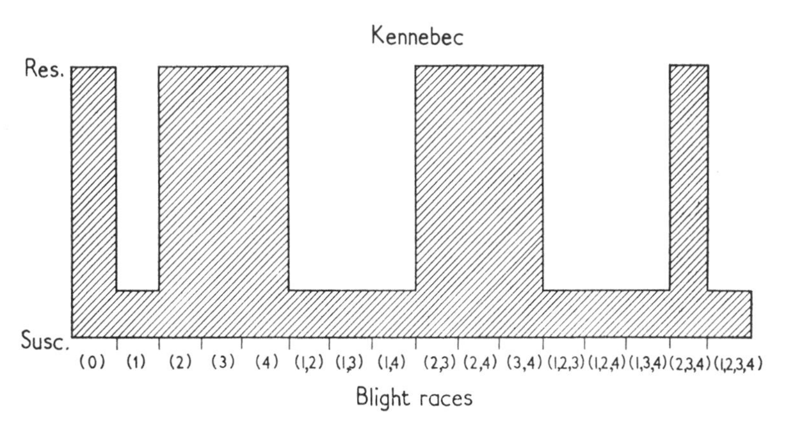

FIG. 14.1. Diagram of the resistance to blight of the foliage of two R_1 varieties: Kennebec and Maritta. The resistance is shown shaded to 16 races of blight. To races (0), (2), (3), (4), (2,3), (2,4), (3,4), and (2,3,4), the resistance of both varieties is vertical and complete. To races (1), (1,2), (1,3), (1,4), (1,2,3), (1,2,4), (1,3,4), and (1,2,3,4), resistance is horizontal, small in Kennebec and moderate in Maritta. Res. = resistant; susc. = susceptible.

Against the other races, to which the gene R_1 does not confer resistance, Kennebec and Maritta behave differently. Kennebec succumbs faster to blight than Maritta; the infection rate is faster in Kennebec. Grown side by side, Kennebec is blighted brown whereas Maritta is still mainly green. Maritta has more "field resistance," to use the current term; Maritta has more horizontal resistance, to use our term. This is shown diagrammatically in Fig. 14.1. Where there is no vertical resistance against a race, the horizontal resistance is shown higher for Maritta than for Kennebec.

Figure 14.2 shows how two varieties without R genes, Katahdin and Capella, behave. No races are known that can attack other varieties but not these. They are, therefore, shown in the diagram to lack vertical resistance. (This point is taken up in more detail in Section 14.4.)

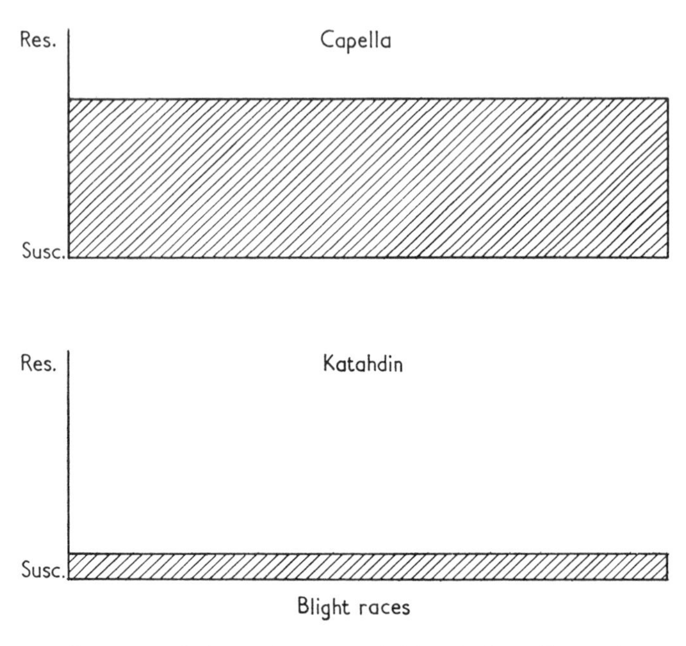

FIG. 14.2. Diagram of the resistance to blight of the foliage of two varieties without R genes: Katahdin and Capella. The resistance, to undefined races, is shown shaded. It is horizontal resistance, small in Katahdin and considerable in Capella. Res. = resistant; susc. = susceptible.

Katahdin is very susceptible to blight in the field. In weather favorable to blight the foliage is easily attacked, and (if unprotected by fungicides) succumbs fast. The infection rate in favorable weather is high. Katahdin is therefore shown in Fig. 14.2 to have little horizontal resistance. However, there are varieties in which blight increases even faster, so Katahdin is shown not altogether devoid of resistance.

The other variety Capella has possibly more horizontal resistance than any other in common cultivation. A blight attack increases slowly in the foliage; and fungus sporulates sparsely in the lesions; and leaves and stems succumb tardily. The infection rate is slow, even in weather favorable to blight and even if the foliage is unprotected by fungicides. It is, therefore, shown in Fig. 14.2 to have much more horizontal resistance than Katahdin.

Lines are drawn horizontally in Fig. 14.2. This indicates that there is no vertical resistance, and anticipates the conclusion come to in the next section.

One can compare Capella not only with Katahdin but also with Maritta. As we shall see (in Table 14.2) Capella has more horizontal resistance than Maritta. This is shown in Figs. 14.1 and 14.2 by making the level of resistance of Capella greater than that of Maritta to those races [race (1), etc.] not controlled by Maritta's gene R_1.

To sum up, against race (0)—or other race controlled by the gene R_1—Kennebec and Maritta are immune, Capella considerably resistant, and Katahdin very susceptible. Against race (1)—or other race not controlled by the gene R_1—Capella is considerably resistant, Maritta moderately susceptible, and Kennebec and Katahdin very susceptible.

All these remarks, it should be remembered, refer to the behavior of the foliage, not of the tubers.

14.4. The Unimportance of Vertical Resistance in Varieties without R Genes

By definition, horizontal resistance is spread evenly against all races. Does such resistance against *Phytophthora infestans* occur in a pure state, unmixed with vertical resistance? In varieties with R genes, vertical resistance is usually accompanied by at least some horizontal resistance, as in Maritta. In varieties without R genes, is horizontal resistance accompanied by vertical resistance? Because one cannot always distinguish changed resistance of the host from changed aggressiveness of the pathogen, the question is of little importance at present except in this form: Can races of *P. infestans* arise that are better adapted to some potato varieties without R genes than to others?

It should be interpolated here that isolates vary considerably in aggressiveness on varieties without R genes. Fresh isolates from the field seem sometimes to differ from one another; cultures that have been kept on artificial media or on tubers lose some of their ability to infect foliage; and many specialized races seem less able than races (0) and (4) to attack varieties without R genes (see Section 14.15). We can call these different isolates strains, races, or what we will. But their existence has nothing to do with our present problem unless there is a differential response to them among varieties without R genes. This section is concerned purely with differential response.

With this said, let us come back to the question: Can races arise that are better adapted to some varieties without R genes than to others?

Toxopeus (1956) believes they can. He cites two examples.

The potato variety Champion was introduced in Ireland in 1876. It withstood blight epidemics so well that from 1882 to 1894 it accounted for 80% of the potato acreage of Ireland. Later, its popularity declined. It lost its resistance to blight, and Toxopeus quotes Pethybridge as saying in 1921 that "nowadays, probably, it would be difficult to find a variety which is more susceptible to blight." It had also become severely infected with potato virus *A*. When in the early 1920's a search was made in Ireland for healthy Champions to start new virus-free stocks, only two healthy tubers could be found (Davidson, 1926). From these tubers (and others from Scotland), a healthy stock was built up. Davidson (1928) found this stock to have all the old vigor and blight resistance of Champion when it was introduced.

Toxopeus interprets Champion's history thus: When Champion was first introduced, it met races of *P. infestans* adapted to the older varieties then popular. These races were not specially adapted to Champion. Because of this, Champion was resistant. As Champion became popular and dominated the potato fields of Ireland, *P. infestans* survived on Champion and became specially adapted to it. Because of this, Champion became susceptible. Later, as Champion lost popularity, *P. infestans* survived on the varieties that replaced it, and lost its special adaptation to Champion. So when Davidson introduced his virus-free stock in the 1920's, it met ill-adapted races of *P. infestans* and was resistant. The virus part of the history Toxopeus regards as irrelevant. Indeed, he quotes with approval the finding of Müller and Munro (1951) in the laboratory that virus infection increases resistance to blight.

The difficulty is to reconcile the theory of Toxopeus with what Pethybridge saw in 1921. According to Toxopeus' theory, the virus-infected Champions which Pethybridge saw had two features that should have made them resistant to blight. They were virus infected. The variety had long ago lost its popularity and been replaced by other varieties. But the variety was in fact very susceptible.

The other example Toxopeus cites is Voran. In 1940 the planting of wart-susceptible potatoes was prohibited in the reclaimed peat soils of the northern provinces of the Netherlands. Voran soon became the popular wart-immune variety. Within a few years it covered nearly 80% of the 40,000 hectares in the region. In the beginning [Toxopeus reports] Voran was highly resistant to *P. infestans*, and for several years blight was never seen in the tubers. This original resistance gradually declined, especially in the tubers, and in 1954 and 1955 the tubers were severely attacked.

Toxopeus explains the history of Voran in the same way as he explains the history of Champion. When Voran was first introduced it met races of *P. infestans* adapted to other varieties and not to it. Voran was resistant. Later, after Voran had become very popular, *P. infestans* survived on, and became adapted to this variety. Voran consequently lost some of its original resistance.

The first difficulty with Toxopeus' theory is the conflict of evidence. The Netherlands has a State Commission whose duty it is to prepare every year an official list of varieties of all agricultural crops. For each potato variety the list records behavior toward blight, on a scale from 3 = very susceptible to 10 = immune. The 1946 list can be taken to describe Voran in its early prime, when the acreage under this variety was still expanding fast. In 1961 Voran was still by far the most popular variety of the region. The 1961 list can, therefore, be taken to show what popularity had done to Voran. In 1946 Voran was given a rating of 7 for tuber resistance (i.e., it was regarded as somewhat susceptible). In 1961 it got a rating of 6.5, a small change. One can assess what the change amounts to in this way. Of the medium-late or late maturing varieties in the 1946 list, seven, including Voran, still survived in the 1961 list. In 1946, four of these (Alpha, Bevelander, Libertas, and Noordeling) were rated as having tubers more resistant than those of Voran, and two (Furore and Industrie) as having tubers more susceptible. The identical order of rating reappeared in 1961.

The second difficulty is that even if for argument's sake one assumes a small change in the behavior of Voran toward blight, there is still no evidence that the change was in *P. infestans* itself and not in other variables. Since the early 1940's there have been great changes in general agricultural practice. Did they not affect Voran's behavior? And Voran itself may have changed. Mutation in potato clones, i.e., asexual variation, has long been recognized. So important do the Netherlands authorities consider clonal variation to be that the production of certified seed potatoes is founded on clonal selection. Every year new clones must be selected. Starting from one plant, clones are multiplied and selected before being handed over for mass production. After (normally) 9 years from start to finish the clone is discarded and replaced. If one grants that the Netherlands authorities have reason to believe in the importance of clonal selection, one must also grant that clonal selection pressure over the years may have changed Voran. Why does Toxopeus assume that selection pressure, presumably in the direction of higher yield and earliness, has no awkward side-effects in the form of slightly changed horizontal blight resistance? The onus is on him to exclude this possibility and every other possibility of change except in *P. infestans.* He has not done so.*

To leave Champion and Voran out of further discussion, the general evidence that vertical resistance does not exist in varieties without R genes is that these varieties have a record of stability in their behavior toward blight. Suppose that there are blight races specially adapted to some potato varieties without R genes but not to others. A variety grown near a source of *P. infestans* adapted to it would be rated as susceptible. The same variety grown near a source of *P. infestans* not adapted to it would be rated as resistant. Because of this, if blight races are specially adapted to some varieties and not to others, one would expect much chopping and changing of the order in which varieties are ranked for resistance. In point of fact, observers agree generally in the way they rank varieties without R genes. There seems to be no evidence of a considerable change of rank from place to place or from observer to observer; and the writer knows of no current dispute about rankings of varieties without R genes.

One cannot on available evidence ignore the possibility that blight races arise that are more adapted to some varieties without R genes than to others. But the evidence at least shows that the ability of *P. infestans* to adapt itself in this way is small.

14.5. The Distribution of Races

Surveys in many countries have shown that races (0) and (4) are the common races on varieties without R genes. Of these, race (4) tends to be the more common. Records of the surveys have been summarized by Black (1957).

Races (1), (2), (3), and higher races are common only on the appropriate genotypes or on other varieties growing near them.

14.6. The Effect of Popularity on the Behavior of R-Types toward Blight

Consider the events of some 30 years ago. Potato breeders were just beginning to bring out R_1 varieties. (The gene R_1 was the first to be

* He must show that genetic changes, if there were any, were in the pathogen and not in the host, although the host as well as the pathogen is genetically unstable. One can better understand the relevance of this if one remembers that a potato variety is a clone. Introducing a new clone of an old variety is, in principle, the same as introducing a new variety.

used of the R genes.) There was much excitement, because these varieties seemed immune from blight. Later, as plots of the new varieties in the breeders' nurseries became bigger, a few blight lesions were noticed toward the end of the growing season. Races able to attack the new varieties had appeared. There were comforting theories about them. One theory had it that the new races would overwinter less easily than the old. Another had it that the new races were primarily adapted to old foliage.

But events soon showed that the new races were quite normal in their behavior and arrived with distressing regularity whenever varieties with R genes began to make their mark commercially.

The history of Kennebec is typical. Kennebec has the gene R_1. It was the first, and so far is the only, variety with an R gene to be included in the top six United States varieties. Kennebec was introduced in 1948. At first it was much discussed as a resistant variety. Production increased fast. By 1954 Kennebec and other R-types accounted for 6.3% of the certified seed grown in Maine. (Kennebec, 1 million bushels; Cherokee, $\frac{1}{4}$ million; and small amounts of Canso and Pungo. These compare with 20 million bushels of varieties without R genes, Katahdin contributing 16 million. All these figures are for 1954.) In 1954 blight was severe. At the experiment station on the Aroostook Farm, Presque Isle, Maine, blight caused over 90% defoliation of Kennebec before September 4 (Stevenson *et al.*, 1955). This was in unsprayed plots. (Katahdin suffered even more severely. But no comparison can be made. The plots were artificially inoculated in mid-July with race (0), which immediately biased the experiment against Katahdin and in favor of Kennebec.) Elsewhere in Maine the behavior of Kennebec varied. Some fields were severely damaged (Webb and Bonde, 1956), but on the whole Kennebec withstood blight better than Katahdin.

Ominously, in 1954 slight to moderate infection by blight was found in widely separated fields of Kennebec as early as the beginning of July (Webb and Bonde, 1956). This suggested that Kennebec was already grown widely enough for appropriate inoculum to overwinter in substantial amounts. The suggestion was confirmed by Webb and Bonde (1956). They tested the races of *P. infestans* from 15 cull piles in 1955. From 11 of them, races were found able to attack Kennebec. Most of the piles had two or more races. Of the 56 isolates of *P. infestans* from these piles, 30 were of races (1), (1,2), (1,3), (1,4), and (1,2,4) that can attack Kennebec. Unfortunately it was not reported to what varieties the blighted potato plants in the cull piles belonged, so that finer analysis is not possible.

Since 1954 the position has changed little in Maine. Kennebec and other R_1 types now account for about 10% of the certified seed in Maine.

Blight is feared, and Kennebec receives a series of protective fungicidal sprays similar to those given to varieties without R genes (Cox and Large, 1960).

Let us return to the general problem. Consider the introduction of R_1 varieties into a region where they, or other varieties with R genes, never existed before. At first, the races of blight to which they are exposed are those found in areas with no R gene varieties, i.e., almost entirely races

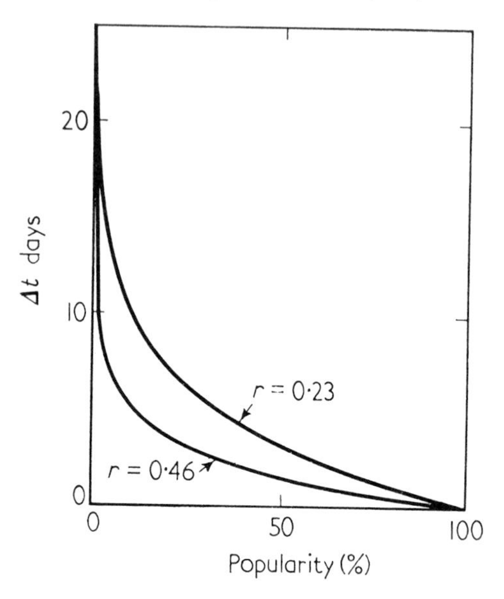

FIG. 14.3. The relation between the benefit from the R_1 gene and the popularity of varieties with this gene. The benefit is measured as the number of days Δt by which an epidemic is delayed. Popularity per cent is the percentage of the potato acreage in a region planted to R_1 varieties. The fast infection rate, $r = 0.46$ per unit per day, can be taken to represent a "blight year"; and the slow rate, a year relatively unfavorable to blight.

(0) and (4). Races able to attack an R_1 variety are scarce, and one expects no more than light infection that becomes noticed only at the end of the season. But as the R_1 varieties become more popular, they provide their own inoculum, from their own cull piles or from their own infected tubers planted as seed. Eventually in a hypothetical country in which R_1 varieties have replaced all others and cover 100% of the land under potatoes, the inoculum that overwinters would all be of races able to attack R_1 varieties. The farmer would then be just as badly off as if nobody had ever grown an R_1 variety at all.

The change is illustrated loosely in Fig. 14.3. Abundance of inoculum is assumed to be proportional to the abundance of R_1 varieties. On this assumption one can use Eq. (11.2) to calculate how much an epidemic of blight is delayed by the gene R_1.

If only about 1 or 2% of the potato acreage is down to R_1 varieties, epidemics in them will occur from 1 to 3 weeks later than in varieties without R genes. This accords with experience: all experience showed that at first blight tended to occur toward the end of the season in R_1 varieties. Also Fig. 14.3 shows that the delay is greatest when r is small. That, too, accords with experience. It was commonly noticed that the delay was great in seasons with little blight in ordinary varieties, but that resistance tended to "break down" in "blight years."

As R_1 varieties become increasingly popular, the delay in the epidemic is reduced. Finally, when farmers plant nothing but R_1 varieties, there is no advantage in the gene R_1 at all.

The harm popularity does is sharper than Fig. 14.3 suggests. The assumption, that the amount of inoculum is proportional to the amount of R_1 varieties planted, implies that inoculum is evenly distributed. It is not. Thus, cull piles next to Kennebec fields are more likely to be of Kennebec and to produce inoculum that can attack Kennebec. So, too, infected Kennebec tubers planted as seed bring into the field a potential source of inoculum of races that can attack Kennebec. The inoculum to which an R_1 variety is exposed is not a random sample from the country as a whole, but a biased sample likely to contain more of the appropriate aggressive races.

Because initial inoculum is not evenly distributed, there is not an even epidemic throughout the fields. Instead, some fields with R_1 varieties are attacked earlier than Fig. 14.3 suggests; others escape until later. This was the situation in Maine in 1954, for example.

Nevertheless, although Fig. 14.3 understates, possibly grossly, the harm popularity does, it does at least illustrate the inevitable frustration from relying on the gene R_1 to control blight in the absence of adequate horizontal resistance. A breeder inevitably hopes that his variety will maintain a reputation for blight resistance. It will do this only if R_1 varieties fail to become popular among the farmers. He is in the ludicrous position that his variety, as a blight resister, will be defeated by its own popularity. Fortunately for breeders' reputations, R_1 varieties have made relatively little headway in most countries; and it is still possible to describe R_1 varieties as "resistant to the common races of blight" without being altogether misleading. Only in Germany have R_1 varieties topped the popularity list. By 1958 Maritta was the most popular variety in Western Germany. It lost its original reputation for blight resistance (Cox and Large, 1960).

What holds for varieties with the gene R_1 holds for varieties with any other gene. So far, no other R gene has been incorporated in a variety of much commercial importance. But there is no reason to believe that if all farmers in a country switched over entirely to $R_1R_2R_3R_4$ varieties, they would be any better off than if they had kept to varieties without R genes. All inoculum would then be of races able to attack $R_1R_2R_3R_4$ varieties.

What is likely to happen if farmers switched to a diversity of genotypes is discussed in Sections 14.14, 14.15, and 14.19.

14.7. The Place of Vertical and Horizontal Resistance in Epidemiology

One chooses definitions because they are apt for some special purpose. For example, a geneticist may think in terms of monogenic or polygenic resistance, and a physiologist in terms of resistance through hypersensitivity. We define resistance as vertical or horizontal because it suits our particular epidemiological approach to the problem of controlling disease. Definitions interlock. Vertical resistance to potato blight is often monogenic and expressed as hypersensitivity. Horizontal resistance is polygenic and expressed in the ways discussed in Section 14.16. This book stresses epidemiology, and uses definitions that make discussions on epidemiology quantitative.

As was pointed out in Chapter 10, vertical resistance reduces the initial inoculum. It reduces x_0.

Horizontal resistance reduces the infection rate r. Consider the three varieties, Bintje, Eigenheimer, and Voran. On the official Netherlands scale, with 3 = very susceptible and 10 = immune, foliage resistance against blight is rated as 3 for Bintje, 5 for Eigenheimer, and 7 for Voran. Bintje is one of the most susceptible varieties; Voran is mildly resistant; and Eigenheimer lies between. None of these three varieties has an R gene. What resistance there is, is horizontal. Figure 14.4 reanalyzes the data (Anonymous, 1954) given in Fig. 7.7 for 117 unsprayed fields in the sand area of the Netherlands in 1953. It will be remembered from Exercises 8, 9, and 10 at the end of Chapter 3 that the regression coefficient of $\log_e [x/(1-x)]$ on time measures r. From Fig. 14.4 we estimate r to be 0.42, 0.21, and 0.16 per unit per day for Bintje, Eigenheimer, and Voran, respectively. These are estimates averaged for the whole period of the graph. More accurately, one should compare varieties at the same time, when they share the same weather. To confine the comparison to July, r is estimated as 0.42 and 0.11 per unit per day for Bintje and Voran, respectively.

The horizontal resistance is clearly reflected by r.

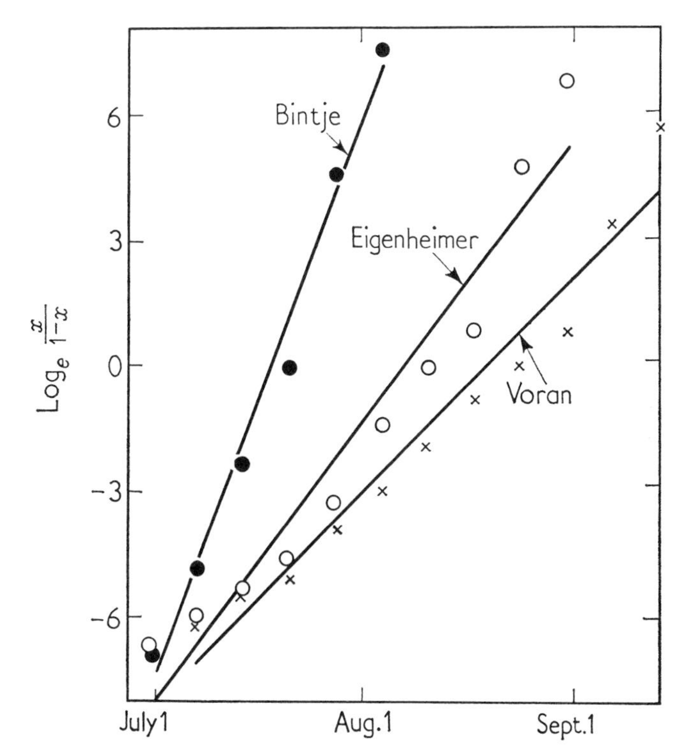

FIG. 14.4. The progress of blight in 117 fields of potatoes. The data are for the sand area of the Netherlands (Anonymous, 1954).

14.8. Analysis of Kirste's Findings: the Effect of Vertical Resistance

Table 14.2 reproduces some data that Kirste (1958) obtained at Celle, Western Germany, in 1955 when blight was severe. Kirste rated the blight attack in five severity groups, from 0 = no blight to 5 = completely blighted.

For each variety he recorded the severity at successive examinations. Table 14.2 shows when that severity was first reached. Thus, blight on Augusta was first rated as severity 1 on July 30, as severity 2 on August 7,

TABLE 14.2

PROGRESS OF AN EPIDEMIC OF *Phytophthora infestans* ON VARIOUS POTATO VARIETIES[a]

Variety	Maturity[b]	Major genes[c]	Date of reaching blight rating[d]			
			1	2	3	4
Bona	ME	0	July 25	—	July 30	Aug. 16
Concordia	ME	0	July 25	July 30	Aug. 13	Aug. 25
Heideniere	ME	0	July 25	—	July 30	Aug. 13
Forelle	ME	R_1	July 30	Aug. 7	Aug. 13	Aug. 25[e]
Lori	ME	R_1	July 30	Aug. 7	Aug. 10	Aug. 19
Augusta	ME	R_1	July 30	Aug. 7	Aug. 13	Aug. 19
Cornelia	ME	R_1	July 30	Aug. 7	Aug. 10	Aug. 22
Luna	ME	R_1	July 30	Aug. 7	Aug. 13	Aug. 19
Suevia	ME	R_1	July 25	July 30	Aug. 13	Aug. 19
Heida	ML	0	July 30	Aug. 7	Aug. 13	Aug. 28
Voran	ML	0	Aug. 7	Aug. 13	Aug. 28	—
Maritta	ML	R_1	Aug. 7	Aug. 16	Aug. 22	Aug. 28
Margot	ML	R_1	Aug. 7	Aug. 13	Aug. 16	Aug. 22
Benedikta	ML	R_1	Aug. 7	Aug. 13	Aug. 22	—
Oda	ML	R_1	Aug. 3	Aug. 13	Aug. 19	—
Apta	ML	R_1	Aug. 3	Aug. 13	Aug. 22	Aug. 25
Nova	ML	R_1	Aug. 10	Aug. 16	Aug. 25	Aug. 28
Urtica	ML	R_1	Aug. 7	Aug. 13	Aug. 25	—
Virginia	ML	R_1R_4	Aug. 7	Aug. 10	Aug. 19	Sept. 8
Lerche	L	0	Aug. 3	Aug. 7	Aug. 25	—
Carmen	L	0	Aug. 7	Aug. 25	Sept. 4	—
Heimkehr	L	0	Aug. 3	Aug. 16	Aug. 28	—
Ackersegen	L	0	Aug. 3	Aug. 13	Aug. 28	—
Capella	L	0	Aug. 13	Aug. 16	Sept. 20[f]	—
Ronda	L	0	Aug. 10	Aug. 16	Sept. 1	Sept. 4
Ancilla	L	R_1	Aug. 13	Aug. 16	Sept. 1	—
Monika	L	R_1	Aug. 13	Aug. 19	Sept. 1	—
Adelheit	L	R_1	Aug. 13	Aug. 22	Aug. 28	—
Herkula	L	R_1	Aug. 13	Aug. 19	Sept. 4	—
Panther	L	R_1	Aug. 13	Aug. 19	Aug. 25	—
Vertifolia	L	R_3R_4	Aug. 22	—	Aug. 28	Sept. 1

[a] Adapted from a table of Kirste (1958).

[b] ME = medium early, ML = medium late, L = late in maturity. Classification of Schick *et al.* (1958a).

[c] No major gene present = 0.

[d] Ratings: 1 = very mild attack, only occasional lesions found; 2 = mild attack, lesions on about one leaf per plant; 3 = medium infection, several leaves per plant attacked; 4 = all leaves attacked, but the plants still generally green.

[e] Forelle was still at stage 3 when last examined on August 25.

[f] Capella was still at stage 2 when last examined on September 20.

as severity 3 on August 13, and as severity 4 on August 19. From his descriptions, given at the foot of Table 14.2, one cannot estimate the percentage of disease. One cannot therefore estimate r. But this does not prevent a detailed analysis from being made.

As background to the analysis, Table 14.3 records observations of Schick *et al.* (1958b) on the prevalence of races of *P. infestans* in Germany

TABLE 14.3

OCCURRENCE OF VARIOUS RACES OF *Phytophthora infestans* IN 1956[a]

	Frequency (%)		
Race	Aug. 3	Aug. 29	Oct. 5
(4)	85.4	64.6	58.5
(0)	—	5.4	3.2
(1)	10.4	21.2	27.4
(1,4)	4.2	7.6	9.7
(1,2)	—	1.2	—
(1,3,4)	—	—	1.1

[a] Data of Schick *et al.* (1958b) for 209 isolates of *P. infestans* from potato varieties without R genes.

in 1956. (Unfortunately, appropriate data for 1955 are not available.) Schick *et al.* identified 209 isolates of *P. infestans* taken from plants of varieties without R genes. This is important, because varieties with R genes necessarily limit the number of races that can be found on them. The main races were (4), (1), and (1,4). The proportion of race (4) dropped as the season advanced. No reason was given. But the change is perhaps related to the fact that the most popular R_1 varieties are, like Maritta, moderately late in maturing. This must influence the race population, even that reaching fields of varieties without R genes.

We can use Table 14.3 to estimate how vertical resistance conferred by the gene R_1 delays the start of a blight epidemic. Consider the figures for August 3. Races (1) and (1,4) were, together, 14.6% of the isolates. These races can attack R_1 varieties as well as varieties without R genes. The remainder 85.4% was race (4) which cannot attack R_1 varieties. We, therefore, write $x_0/x_{0s} = 100/14.6$. With this ratio, Eq. (11.2) estimates $\Delta t = 4$ days if $r = 0.46$ per unit per day, which is a fast infection rate. With a fairly slow rate, $r = 0.23$ per unit per day, $\Delta t = 8$ days. We do not know what r was, but it is likely to have been within the stated limits.* On this assumption, we estimate that varieties with the gene R_1 would have been 4 to 8 days behind varieties without R genes in blight development.

If we use the figures in Table 14.3 for August 29 instead, $x_0/x_{0s} = 100/30$; and the corresponding delays would have been 3 and 5 days, respectively.

From Table 14.2 one can determine what delay was actually observed.

Consider first the nine medium-early varieties. The three (Bona, Concordia, and Heideniere) without R genes all reached blight severity 1 on July 25. Five (Forelle, Lori, Augusta, Cornelia, and Luna) of the six with the gene R_1 reached severity 1 on July 30. The sixth (Suevia) reached severity 1 on July 25. This gives an average for the six of July 29. On an average, the vertical resistance conferred by the gene R_1 delayed blight from reaching severity 1 by 4 days, from July 25 to July 29.

If one considers all the varieties without R genes and all with the gene R_1 in Table 14.2, the average delay is 5 days. That is, from Eq. (11.2) one expects a delay of from 3 to 8 days; and the observed delay is 5 days.

Other comparisons can be made.

Vertifolia, a late variety, has the genes R_3 and R_4. Table 14.3 shows that Schick *et al.* (1958b) found 1 isolate out of 209 of a race [race (1,3,4)] able to attack it. With $x_0/x_{0s} = 209$ in Eq. (11.2) one estimates $\Delta t = 12$ days if $r = 0.46$, and $\Delta t = 23$ days if $r = 0.23$ per unit per day. In another larger survey in 1954 and 1955 (of varieties of unspecified genotypes) Schick *et al.* obtained three isolates of race (3,4) and seven of race (1,3,4) out of a total of 979 isolates. With $x_0/x_{0s} = 97.9$, Δt is 10 days and 20 days, respectively, for $r = 0.46$ and $r = 0.23$ per unit per day. The observed delay, found by comparing the date (August 22) in Table 14.2 when Vertifolia reached blight severity 1 with the average date (August 7) when the corresponding late varieties (Ackersegen, etc.) without R genes reached severity 1, was 15 days.

So too one can compare Virginia with genes R_1 and R_4 with other medium-late varieties.

These comparisons show no noteworthy discrepancy between theory and observation.

* These are estimated likely limits for r in a severe blight year, which 1955 was. Higher values than 0.46 per unit per day have been recorded. (An example is given in Section 21.3.) Lower values than 0.23 per unit per day (such as the estimated value of Voran in the previous section) are common, but not relevant to a blight year. Stretching the limit in either direction would not alter the substance of our analysis.

14.9. Analysis of Kirste's Findings: the Effect of Horizontal Resistance

From Kirste's data reproduced in Table 14.2 one can estimate relative values of r. Equations (3.5) and (3.6) show that r is inversely proportional to the time taken for disease to increase from any given proportion x_1 to any other given proportion x_2. Blight on Bona and Heideniere took 5 days, from July 25 to July 30, to increase from severity 1 to severity 3. On the third medium-early variety without an R gene, Concordia, the time was 19 days. This gives an average for the three varieties of 9.7 days. For the six medium-early varieties with the gene R_1, the average time was 13.8 days. Therefore r for the varieties without the gene R_1 was 13.8/9.7 times as great as r for the varieties with the gene R_1.

Apart from being statistically insignificant, this difference between varieties is inappropriate for a comparison of horizontal resistance. In the varieties without the gene R_1 most of the increase of blight from severity 1 to severity 3 took place at the end of July; in the varieties with the gene R_1 most took place in August. One cannot, therefore, necessarily ascribe a difference in r to a difference in resistance; it might have been caused by a change in the weather.

But when one examines the twelve late varieties a better comparison is possible. On an average it took 15.5 days for blight to increase from severity 1 to severity 3 in the six varieties with R genes. But in the six varieties without R genes it took as much as 26.7 days.* That is, r in the varieties with R genes was 26.7/15.5 times as great as r in the varieties without R genes. This difference is highly significant statistically ($P < 0.01$).

The difference is interpreted to mean that the varieties without R genes (i.e., without vertical resistance) had considerably more horizontal resistance. The same weather affected both groups of varieties. On an average blight in the six varieties without R genes reached intensity 1 between August 6 and 7 and intensity 3 on September 2. On an average blight in the six (including Vertifolia) with R genes reached intensity 1, 8 days later and intensity 3, 3 days earlier than in the six without R genes. The period for varieties without R genes straddles the period for varieties with R genes. The varieties without R genes were exposed to all the blight weather to which varieties with R genes were exposed. One can as an illustration consider the extremes. In Capella, without an R gene, blight reached intensity 1 on August 13 and was still at

* In Capella, severity 3 was never reached at all, so the interval was taken from severity 1 to the date of the final examination.

intensity 2 when examinations were discontinued on September 20. In Vertifolia with genes R_3 and R_4 blight reached intensity 1 on August 22, intensity 3 on August 28, and intensity 4 on September 1. Weather that allowed Vertifolia to become quickly blighted in the last third of August failed to bring about comparable blighting of Capella. The difference between the varieties was not in the weather, but in their horizontal resistance.

14.10. Other Evidence for a Difference in Horizontal Resistance

The evidence from Kirste's data is that potato varieties with vertica resistance have less horizontal resistance. Evidence from another source confirms this.

Rudorf and Schaper (1951) investigated the so-called "incubation resistance" in potato varieties. In some varieties lesions form sporangia more abundantly and more quickly than in others. Because of more abundant sporangia, blight multiplies faster in these varieties. That is, they have less horizontal resistance. (See Section 14.16.) Rudorf and Schaper inoculated nine varieties with race (1) on August 16, and a few days later measured the abundance of sporangia. Eight of the varieties had the gene R_1; the ninth was Ackersegen, with no R gene. Their results are given in Table 14.4.

TABLE 14.4

INTENSITY OF SPORANGIAL DEVELOPMENT OF RACE (1) OF *Phytophthora infestans* ON VARIOUS POTATO VARIETIES[a]

Variety	Maturity[b]	Major genes[c]	Intensity[d]	
			Aug. 19	Aug. 20
Erica	L	R_1	1	3
Falke	L	R_1	1	3
Aquila	ML	R_1	2	4
Maritta	ML	R_1	2	4
Roswitha	L	R_1	3	4
Pommernbote	ML	R_1	3	5
Panther	L	R_1	5	5
Robusta	L	R_1	5	5
Ackersegen	L	0	1	2

[a] Data of Rudorf and Schaper (1951).
[b] ML = medium late, L = late in maturity. Classification of Schick *et al.* (1958).
[c] No major gene present = 0.
[d] Few sporangia = 1, abundant sporangia = 5.

Ackersegen had significantly less sporangia than the varieties with the gene R_1. Yet within its own class, i.e., of medium-late and late varieties without R genes, Ackersegen is no better than a good average, if one can judge by the production of sporangia in another experiment of Rudorf and Schaper that is not reported here. One also assesses Ackersegen as a good average within its own class from Kirste's data in Table 14.2.

The results of Kirste and of Rudorf and Schaper seem to be the only evidence available. For late varieties the evidence is consistent. With vertical resistance there is on an average less horizontal resistance. For early maturing varieties there is no significant evidence at all, either in one direction or the other.

14.11. Some Published Assessments of Blight Resistance

In northwestern Europe early maturing varieties often escape serious blight injury even though they are very susceptible. Indeed history has shown that great susceptibility is no barrier to the success of an early variety that adequately meets other requirements. But late maturing varieties must stand the full blast of blight epidemics. Failure to stand up reasonably well to blight would bar them from being commercially accepted. Late varieties have therefore been under considerable selection pressure for resistance to blight.

With this as background, we can inquire how the resistance of late maturing varieties has been assessed by competent observers.

Schick *et al.* (1958a) kept more than 370 varieties under observation in the field, most of them for 8 years, from 1950 to 1957. The purpose was to compare the field resistance of varieties that have R genes with the resistance of those that have none. Field resistance is accepted by workers dealing with potato blight to mean what we call horizontal resistance. Schick *et al.* leave one in no doubt that this is how they interpret field resistance. It is characterized, they say, by smaller susceptibility to infection, by slower growth of mycelium through the plant, by delayed and reduced sporing. These are all characteristics of horizontal resistance, to be discussed in Section 14.16.

They concluded that varieties with R genes were on the whole more field resistant.

How does this conclusion stand up to analysis? Table 14.5 shows how Schick *et al.* (1958a) assessed the late maturing varieties studied by Kirste and included in Table 14.2. In the varieties without R genes that Schick *et al.* classed as very resistant, blight took on an average 32 days to increase from severity 1 to severity 3 in Kirste's experiment. The corresponding figure for varieties with R genes was 16 days.

Their classification was far too lenient for varieties with R genes. They classed these varieties as highly resistant even though *r* for them was (in Kirste's experiment) twice as great as *r* for the corresponding varieties without R genes.

One comes to the same conclusion about the varieties they classed as moderately resistant. Here, too, *r* was twice as great in varieties with R genes as in varieties without these genes.

TABLE 14.5

THE HORIZONTAL RESISTANCE TO BLIGHT OF LATE MATURING POTATO VARIETIES. A COMPARISON BETWEEN THE CLASSIFICATION OF SCHICK *et al.* (1958a) AND THE FINDINGS OF KIRSTE (1958)

Resistance class of Schick *et al.*[a]	Days from blight severity 1 to severity 3[b]	
	Varieties with R genes	Varieties without R genes
1	16	32
2	12	25
3	—	22

[a] Very resistant = 1, very susceptible = 5.
[b] Averages from Kirste's data in Table 14.2.

Schick *et al.* (1958a) did not allow for differences in the amount of initial inoculum. They made the common mistake of assuming that as long as some inoculum is present it does not greatly matter how much is present. For example, they rated both Capella and Vertifolia as very resistant. Capella is attacked by all races of *Phytophthora infestans*. Vertifolia is attacked only by races which all surveys have shown to be rare at present and which can be expected to remain rare while Vertifolia and other varieties with the gene R_3 remain relatively unimportant commercially.

Only on one condition can one compare varieties for horizontal resistance in the field just by watching how severely blight attacks them. The condition is that the only races present should be those that can attack every single variety present. (See Section 14.17.)

Before leaving the assessments of Schick *et al.* (1958a), it is worth interpolating that within each of the two groups of varieties, i.e., within the group of varieties with R genes and within the group without R genes, their assessments agree with the results of Kirste. The lower they assessed the resistance of varieties, the shorter was the time taken from blight severity 1 to severity 3 in Kirste's experiments. Thus, in varieties without R genes resistance classes 1, 2, and 3 corresponded with intervals of 32, 25, and 22 days, respectively.

TABLE 14.6

THE RESISTANCE TO BLIGHT OF LATE MATURING POTATO VARIETIES. A COMPARISON BETWEEN THE CLASSIFICATION OF HOGEN ESCH AND ZINGSTRA (1957) AND THE FINDINGS OF KIRSTE (1958)

Resistance class of Hogen Esch and Zingstra[a]	Days from blight severity 1 to severity 3[b]	
	Varieties with R genes	Varieties without R genes
9	16	—
8	6	30
7	—	—
6	—	24

[a] Immune = 10, very susceptible = 3.
[b] Averages from Kirste's data in Table 14.2.

Hogen Esch and Zingstra (1957), in the Netherlands, also examined a large collection of potato varieties and classed them for resistance to blight. In Table 14.6 their assessments are also compared with the findings of Kirste. Only three (Monika, Panther, Vertifolia) of the six late maturing varieties with R genes in Kirste's experiments appear in Hogen Esch and Zingstra's list. The table is therefore somewhat incomplete. But even with this incompleteness the difference between the varieties with R genes and those without them is statistically significant. Table 14.6 shows the same features as Table 14.5. Just as Schick *et al.* (1958a), so also Hogen Esch and Zingstra greatly overrated the resistance of varieties with R genes. They too assessed Capella and Vertifolia as being equally resistant.

Again we may interpolate that Hogen Esch and Zingstra, like Schick *et al.*, made assessments within each of the two groups of varieties, those with and those without R genes, that tally with the results of Kirste.

To return to the problem of comparing varieties that have R genes with those that have none, there is more realism about the assessment of blight resistance in the recent annual lists of varieties grown commercially in the Netherlands. (These lists were discussed in Section 14.4.) The same classes are used as Hogen Esch and Zingstra used: 10 = immune and 3 = very susceptible. Varieties without an R gene are classed with a single figure, e.g., Ackersegen is classed as 8, i.e., as having considerable resistance. But varieties with R genes have a second number in parentheses. This number shows the susceptibility when the variety is infected at a late stage. Thus, Maritta's resistance is shown as 9(7). That is, Maritta is classed as very resistant early in the season, but only moderately resistant later on. This is a good empirical description of Maritta's behavior recorded by Kirste and reproduced in Table 14.2. According to our explanation, Maritta has vertical resistance to the commonest races—races (0) and (4)—that delays a blight attack. But it has only moderate horizontal resistance to slow down the attack when once it has started. Vertifolia is not a commercial variety in the Netherlands and is not on the list. But Ambassadeur also has the gene R_3; its resistance is classed as 9(5). That is, Ambassadeur is very resistant early but markedly susceptible late in the season, if it is attacked. This would be a good empirical description of Vertifolia's behavior recorded by Kirste. Ambassadeur and Vertifolia have vertical resistance to all but races rare at present, which greatly delays a blight attack. But when the attack comes they have little horizontal resistance to slow it down.

14.12. The Vertifolia Effect

When blight first struck the potato fields of Europe in the 1840's, *P. infestans* met a host that had existed for centuries in a blight-free world. The potato varieties were very susceptible, and great havoc was wrought. Since then conscious and unconscious selection has raised the level of resistance considerably. Modern late maturing European varieties are resistant enough to yield fairly well in blight epidemics even when the fields are unprotected by fungicides.

The change that has come about since the 1840's can be seen by comparing modern varieties with the old varieties that have persisted in the mountains of Basutoland (van der Plank, 1960). These old varieties were introduced in 1833. They are probably of European origin; i.e., they are probably relics from the pre-blight era of Europe. There is little blight in the Basutoland mountains, so the varieties have been able to continue without severe infection. Most of these old varieties mature late. Nevertheless they are, as a group, intensely susceptible to blight. They show, it seems, what the European varieties were like before they were selected in an environment of blight, i.e., before there had been selection in favor of blight resistance.

The horizontal resistance of modern European late varieties was accumulated by selection. Horizontal resistance is polygenic. Many genes are involved, none of them important enough for it to have been identified. In crossing varieties the genes are dispersed and resistance easily lost. This is specially true when a resistant is crossed with a susceptible variety. Many modern late varieties have susceptible ancestors, e.g., Ackersegen was bred from Allerfrüheste Gelbe, a very susceptible variety. These late varieties are resistant because they were exposed to blight in the breeders' nurseries and in commercial fields. They were selected, whereas very susceptible varieties fell by the wayside.

Varieties with vertical resistance were protected from the full pressure of selection. All the R_1 varieties listed in Tables 14.2 and 14.4 were bred when even races (1) and (1,4) were rather uncommon. The races to which Vertifolia and Ambassadeur are susceptible are even today still uncommon by comparison with races (0), (4), (1), and (1,4). The varieties were selected under conditions of blight attacks delayed because initial inoculum was reduced. Selection was therefore less severe. The inevitable result of less severe selection is seen in Tables 14.2 and 14.4. Varieties very susceptible horizontally were accepted into commerce just because they were resistant vertically.

Vertical resistance reduces selection pressure. Reduced selection pressure reduces the average horizontal resistance of varieties that survive the selection process. Vertical resistance reduces the average horizontal resistance of varieties that survive the selection process. This sequence we shall call "the Vertifolia effect", naming it for a conspicuous example.

What Kirste found for late varieties was not fortuitous, but the result of a Vertifolia effect. But one cannot assume a Vertifolia effect for the early varieties he studied, because one cannot assume that resistance to blight plays a decisive part in the selection of these varieties.

Great selection pressure and great vertical resistance (i.e., resistance to the great bulk of the prevalent races of the pathogen) are needed for a great Vertifolia effect.

14.13. Blight on *Solanum demissum*

Solanum demissum has been widely used by potato breeders as a source of R genes for commercial varieties. *S. demissum* is an uncultivated species prevalent as a weed at high altitudes in Central Mexico.

Many clones (varieties) exist. They differ considerably in resistance to *P. infestans*. The natural clone S-434 has no R gene at all (Graham *et al.*, 1959), and is moderately susceptible to blight. Other clones have various R genes. But none has been found that is immune from all races of *P. infestans* (Niederhauser *et al.*, 1954). *P. infestans* has been able to match all genes in *S. demissum*.

Nevertheless, even though foliage lesions are produced on them, some clones of *S. demissum* show great resistance to blight. The lesions are slow spreading, often turn necrotic, and sporulate sparsely (Niederhauser *et al.*, 1954). Petiole and stem lesions tend to be superficial and do not kill the distal portions of the affected leaf or stem. Relatively few lesions develop on a plant, and these are mostly on the older leaves.

These clones have great horizontal resistance; the infection rate is low.

Horizontal resistance enhances the effect of vertical resistance. Calculations from Eq. (11.2) have been made in previous chapters. We make another here for convenience. Suppose that a variety has vertical resistance to 80% of the initial inoculum, i.e., suppose that its R genes confer immunity from 80% of the sporangia of *P. infestans* that reach it to start the epidemic ($x_0/x_{0s} = 5$). If $r = 0.4$ per unit per day the epidemic will start 4 days later in this variety than in a variety with the same horizontal resistance but with no vertical resistance. But if $r = 0.04$ per unit per day, a figure that can be used to indicate great horizontal resistance, vertical resistance will delay the epidemic by 40 days.*

One cannot judge the value of vertical resistance without reference to the horizontal resistance. The vertical resistance of many *S. demissum* clones gives great protection against blight because it is coupled with great horizontal resistance. But breeders have taken R genes from these resistant clones and incorporated them into commercial potato varieties without also incorporating the horizontal resistance. They have taken the cart without the horse.

14.14. Blight on Mixtures of Clones

Solanum demissum exists in many different clones. If *P. infestans* occurred all as one single superrace, say, race (1,2,3,4,5,6), that could attack all clones, the diversity of clones would not protect against blight.

* The rate, $r = 0.4$ per unit per day, is reasonable for commercial varieties. But the rate, $r = 0.04$ per unit per day, is a shot in the dark. No data seem to have been published from which an infection rate for resistant clones of *S. demissum* can be calculated. It may well be that the rate, $r = 0.04$ per unit per day, is too fast, and the estimate, 40 days, too short.

All clones would be susceptible, and inoculum would be common to all of them. But *P. infestans* does not occur solely as a superrace. It tends to occur most abundantly as the simplest races that can attack the host. Thus, on *S. demissum* clone S-434 (without an R gene), Graham *et al.* (1959) collected only races (0) and (4); on clone S-406 only races (1) and (1,2). In line with this, of the 22 isolates of *P. infestans* which they collected from Mexican "Criolla" varieties of *Solanum tuberosum* susceptible to all races, 11 were of race (0), 1 of race (1), and 3 of race (4). Complex races attack complex genotypes, but are less fit to reproduce on simple genotypes.

Because complex races tend to be less common than simple races on simple genotypes, a diversity of genotypes within the same environment can create its own vertical resistance. This vertical resistance, like any other vertical resistance, derives its power from the horizontal resistance that backs it. A diversity of genotypes without much horizontal resistance would not thwart the pathogen for long. Diversity in *S. demissum* must be judged in relation to the very great horizontal resistance some clones of this species have.

14.15. What Vertical Resistance Implies

Horizontal resistance is an inherent quality of the host. It will always operate in a blight attack and influence the infection rate. It exists in any genotype on its own.

Vertical resistance is not an inherent quality of the host. The same genotype may have great vertical resistance, or it may have none at all. Its vertical resistance depends on what races of the pathogen reach it. In turn, the races of the pathogen depend on the genotypes of the host plants from which they come.

The existence of vertical resistance carries two clear implications. First, vertical resistance is a quality not of a single genotype of host plant, but of a mixture of genotypes. Second, vertical resistance implies that there is a stabilizing selection within races of *P. infestans* when directional selection stops. (These implications are discussed further in Chapter 17.)

Consider the first implication, that at least two genotypes of the host must be present before vertical resistance can exist. If genes make a variety vertically resistant to a race, i.e., if that race cannot reproduce on the variety, then it is certain that sporangia of that race must come from a blighted variety with different genes or with no genes for resistance at all. A variety with the gene R_1 is vertically resistant to race (0) or race (4) coming from a field of a variety without R genes, or to race (2) coming from a field of variety with the gene R_2, or to race (3) or (3,4) coming from a field of variety with the gene R_3. But a variety with the gene R_1 will not be resistant to any race coming from another field of a variety with the same gene R_1. What holds for the gene R_1 holds for any other gene or combination of genes.

Consider the second implication, that there must be stabilizing selection to make simple races of blight the most abundant on simple genotypes of potato. Suppose one introduces a potato variety that is vertically resistant by virtue of an R gene (except the gene R_4). All experience shows that, if weather favors blight, sooner or later appropriate races will attack the variety and become abundant on it. There has been natural selection of races fit to survive on the R variety. The R gene has directed selection. But these same races are not the fittest to survive on a variety without an R gene. All surveys have shown that, from varieties without R genes that are well isolated from varieties with R genes, one isolates mainly the simple races—race (0) or (4). These simple races are the most stable on simple varieties without R genes, and stabilizing selection makes them the most abundant.

Similarly, surveys of blight races on varieties with R genes show that one tends to find the relatively simple races that can attack these varieties. For example, races (1) and (1,4) are usually the commonest in commercial fields of varieties with the gene R_1.

Vertical resistance exists because of all this. If the stable fittest race on varieties without R genes had been race (1,2,3,4,5,6), the genes R_1, R_2, R_3, R_4, R_5, and R_6 would probably not have been discovered, and the vertical resistance conferred by them would certainly not have been used. It would not have existed.

14.16. The Manifestations of Horizontal Resistance

The leaflets, petioles, and stems of resistant varieties are more difficult to infect. The lesions grow more slowly. In lesions, *P. infestans* takes longer to start sporulating, and then sporulates less freely. These are ways in which horizontal resistance manifests itself and reduces the infection rate.

Müller (1953) sprayed a dilute suspension of zoospores on the upper surface of discs cut from leaves of different varieties without R genes. The proportion of discs that became infected was less when the discs came from varieties regarded as field resistant. Hodgson (1961) used the same technique with the same results. Van der Zaag (1959) inoculated leaflets by dipping them in a dilute zoospore suspension. Fewer lesions were formed in varieties regarded as field resistant. Lapwood (1961b) found differences between varieties in the resistance of the leaf lamina to infection, but thought other differences more important. Bigger differences were found when he inoculated varieties in the leaf axil and counted the number of axillary buds infected and destroyed.

Many workers have found that *P. infestans* advances more slowly through the lamina of resistant varieties. Lapwood (1961b) found important differences in the rate and extent of the advance through petioles and stems. In the resistant variety Arran Viking, the lesions usually remained small. In susceptible Up-to-Date and King Edward, the lesions often girdled the leaf petiole, the leaf collapsed, and after a while the stem was also girdled.

Vowinckel (1926), Kammermann (1950), Schaper (1951), Rudorf and Schaper (1951), and Lapwood (1961b) have studied the time after inoculation that *P. infestans* takes to start forming sporangia. Under favorable conditions sporulation is abundant after 4 days in susceptible varieties, but 5 or 6 days are needed in resistant varieties.

Sporulation is more abundant on susceptible varieties, and this is related to less necrosis in the lesions of these varieties (Kammermann, 1950; Lapwood, 1961b). Lapwood made a detailed study of this. The sporing zone surrounds a necrotic area. As the sporing zone advances, the necrotic area follows up behind. In susceptible varieties under conditions favorable to the fungus, the sporing zone of lesions on the lamina is about 1 day's growth wide. It is less in resistant varieties and in conditions not favorable to the fungus. If conditions change from adverse to favorable, sporulating activity recovers; but it recovers less fast in resistant varieties.

Lapwood (1961c) found one variety, Pimpernel, significantly better than his standard variety Majestic in all features he assessed: The proportion of spores that caused lesions to develop in lamina, petiole, and stem was less. Mycelium grew more slowly through the lamina. The width of the sporing zone was less, and so was the intensity of sporulation within the zone.

14.17. The Quantitative Determination of Horizontal Resistance when There Is Also Vertical Resistance

The features just listed influence the infection rate.

When one deals with varieties lacking vertical resistance it is not difficult to compare their horizontal resistance. The varieties that are blighted last are the most resistant.

Vertical resistance complicates the comparison, unless all the varieties being compared have the same vertical resistance. The amount of disease at any time depends on the initial inoculum x_0 and the average infection rate r. Vertical resistance determines x_0; horizontal resistance r. The problem is to assess r alone.

There are two ways of doing this. One can leave x_0 variable and assess r in spite of this. Or one can make x_0 the same for all varieties. Either way, one assesses r comparatively because weather, too, affects r.

The analysis of Kirste's data in Table 14.2 showed how r could be assessed even when x_0 varied from one genotype to another. The six late varieties without vertical resistance from R genes were compared with the six with vertical resistance. Blight in the varieties without vertical resistance took longer to increase from intensity 1 to intensity 3. Horizontal resistance in them was greater; and r less. The varieties were compared during the same time. Blight intensity 1 was reached on an average a little earlier and intensity 3 a little later in the varieties without vertical resistance. So weather was no excuse for the poorer performance of the varieties with vertical resistance.

In this experiment blight in the late varieties increased to the same extent (from intensity 1 to intensity 3) at dates that made an assessment possible. But it is not really necessary that blight should increase to the same extent in all varieties. One could fix the dates and use Eqs. (3.5) or (3.6) to determine r. That is, one could fix t_1 and t_2, and estimate x_1 and x_2 to find r. Provided that, for reasons given in Section 6.9, $x_2 < 0.35$, the different r-values for different varieties would indicate their comparative horizontal resistance. It would be convenient to include standard varieties in all tests.

This method of assessing horizontal resistance assumes that x is measured accurately. If a key is used to assess blight, it must be accurate for all values of x (up to 0.35).

The other way of assessing horizontal resistance is to ensure that x_0 is the same for all varieties. At the Alma Experiment Station, New Brunswick, on the Bay of Fundy, varieties have been compared by inoculating the experimental field with race (1,2,3,4) before any natural inoculum arrived. This allows all varieties, without R genes or with any of the genes R_1, R_2, R_3, and R_4, to be assessed for horizontal resistance by direct comparison in the field.

The essential point of this method is that all blight races must attack all the varieties present. One could, e.g., compare varieties without an R gene with varieties with the gene R_1 in the presence of race (1) or (1,4) or (1,2,4) or any other race with the number 1 in it. But for an exact comparison all other races, including race (0), would have to be absent.

The Rockefeller Foundation in Mexico has screened many thousands of clones since 1953 at Toluca. Research workers from many parts of the world have sent varieties there for testing. At Toluca the sexual process occurs. Blight is present abundantly; all races are there; and all varieties of *Solanum tuberosum* and *Solanum demissum* are attacked to some extent or other. Because of the abundance of inoculum and diversity of races, and because of weather favorable to blight, varieties with little horizontal resistance are screened out quickly. The tests at Toluca have been of the greatest value in eliminating varieties which in other countries might have been saved by their vertical resistance for years before their poor horizontal resistance became apparent.

Nevertheless, the tests are not perfect. The diversity of races allows horizontally susceptible varieties to be screened out quickly. But at the same time it makes comparisons of varieties inexact, especially if horizontal resistance is high. The varieties that so far have seemed most promising as blight resisters at Toluca have genes R_3 and R_4. In a survey made by Graham *et al.* (1959) of blight on varieties of *S. tuberosum* in the valley of Toluca only 8 isolates out of 69 were of races able to attack R_3R_4-types. There were 3 isolates of race (1,3,4) and 5 of (1,2,3,4) ($x_0/x_{0s} = 69/8$). On these results, if $r = 0.1$ per unit per day, which represents high horizontal resistance, a blight attack would be 22 days later on an R_3R_4 variety than on a variety with the same horizontal resistance but with no R genes.* The higher the horizontal resistance, the greater is the handicap on varieties with no R genes. A test that has proved to be of great value with very susceptible varieties must be interpreted with caution when it is applied to varieties with great horizontal resistance.

14.18. The Deficiency of Horizontal Resistance in Commercial Varieties

One may inquire, how far does the amount of horizontal resistance in commercial varieties fall short of what it might potentially be? How much of this resistance can be incorporated into a new variety?

The late maturing varieties of continental Europe without R genes (Ackersegen, Capella, etc.) have been much exposed to blight attacks.

* This infection rate, just as that in Section 14.13, is a shot in the dark. If inoculum is waiting for the plants almost from the time they emerge from the ground, as it often is in the valley of Toluca, the infection rate must be low if the plants are to give a satisfactory yield. The rate, $r = 0.1$ per unit per day, was chosen low in order to recognize this. Its exact value, provided that it is low, does not matter much.

Their level of resistance may yet be bettered; there are sources of resistance not yet broached. But let us take them to have a high resistance, and use them as a standard.

The late maturing varieties of continental Europe with R genes have been protected from full selection pressure. Varieties with the gene R_1 have been protected least, because varieties with this gene were the first to be released, and races that can attack them are now not uncommon. Nevertheless, in the experiment analyzed in Table 14.2, r was 26.7/17.4 times as great for varieties with the gene R_1 as for the corresponding varieties without R genes. Varieties with the gene R_3 have been better protected, and on the evidence of Table 14.2, r was more than 4 times as great for Vertifolia with the genes R_3 and R_4 as for the corresponding varieties without R genes. One should perhaps not attach too much meaning to this single result. But it is backed by observations in the Netherlands on the variety Ambassadeur, an R_3-type discussed in Section 14.11.

Early maturing varieties in northwestern Europe have been protected from full selection pressure because they are usually well on the way to maturity when blight comes. In their escape from full selection pressure there is little difference in principle between early varieties which escape the worst effects of blight and late varieties with enough vertical resistance to postpone blight epidemics for a few weeks. One suspects that the average horizontal resistance of early varieties is well below the best potential level of resistance for the group. In Kirste's experiment, r was 5 times as great in two medium early varieties (Bona and Heideniere) as in the late varieties without R genes. Some of this difference may have been caused by weather, because the readings were at different dates. Some may be owing to inherently smaller resistance in early varieties. But one suspects that much of the susceptibility can be ascribed to lack of selection for resistance. In confirmation of this suspicion one notes that in the third variety, Concordia, that Kirste studied in this group, r was approximately only one-quarter as great as in Bona and Heideniere. It is unlikely that this is wholly just a chance result because Hogen Esch and Zingstra (1957) assess Concordia as more resistant than average for its maturity class.

Other varieties have been protected from full selection pressure by fungicides. Katahdin, grown in Maine and protected by an efficient fungicide program, is an example.

Hogen Esch and Zingstra (1957) list forty varieties without R genes in the same maturity class as Katahdin's. [We use the assessments of Hogen Esch and Zingstra for illustration. The assessments of Schick *et al.* (1958a) would do just as well.] On the scale, 3 = very susceptible and 10 = immune, Katahdin's resistance in the foliage is assessed as 4. The average for all forty varieties in the class is 5.4. But two of the varieties were assessed over 7. This brings them above the average for the late maturing varieties without R genes that were tested in Kirste's experiment summarized in Table 14.2. One of them, Noordeling, is a well-known commercial variety. It has been studied by van der Zaag (1956, 1959). He found it much more resistant than Eigenheimer. This in turn is more resistant (in the foliage) than Katahdin, on the assessment of Hogen Esch and Zingstra. Van der Zaag dipped leaflets in a very dilute zoospore suspension. Twice as many lesions formed on Eigenheimer as on Noordeling. The fungus sporulates sparsely in lesions on Noordeling; he found from 5 to 10 times as many sporangia in lesions on Eigenheimer as in those on Noordeling. All this does not mean that r is from 10 to 20 times as great for Eigenheimer as for Noordeling. (Susceptibility to infection and abundance of spores affect the basic infection rate, discussed in Chapters 5 to 8, and through it the apparent infection rate r. This will be considered further in Chapter 20.) But it does mean that r is much lower for Noordeling than for Eigenheimer when they are grown under the same conditions.

Another variety with high resistance is Olympia. It matures somewhat earlier than Katahdin, but compares well with Noordeling in resistance. The lists of Hogen Esch and Zingstra and of Schick *et al.* (1958a) bring this out.

The evidence available is somewhat scrappy. But it is entirely consistent. Pieced together it leaves no doubt that varieties differ greatly in resistance, even in the same maturity group, and that the horizontal resistance of most varieties is much less than it needs to be.

14.19. Breeding Potato Varieties for Resistance to Blight

Blight may be controlled by: (*a*) horizontal resistance; (*b*) vertical resistance, and other forms of sanitation; and (*c*) fungicides.

Horizontal resistance on its own, i.e., straight field resistance, has long been used. It protects millions of acres a year, especially in continental Europe.

Horizontal resistance activates vertical resistance and other forms of sanitation, such as the destruction of infected cull piles.

Horizontal resistance determines how effective fungicides will be. This is discussed in Chapter 21. Replacing Katahdin by a variety with the full potential horizontal resistance of its maturity class would change the fungicide program in Maine as much as replacing copper by dithiocarbamate fungicides did.

The common ingredient of all these methods is horizontal resistance.

One could organize the use of horizontal resistance in two ways, one direct and uncomplicated, the other requiring restrictions on the planting of varieties.

The *uncomplicated* way is to carry on as potato breeders did for many years. Between the 1840's and the 1930's, when many breeders changed over to using vertical resistance, much horizontal resistance was added to new varieties. Now, with breeders disillusioned about vertical resistance, there is every reason to believe that they will again start adding horizontal resistance as steadily as ever before.

The *complicated* way, requiring restrictions, takes into account that a blight epidemic is not just in a single field, but ranges throughout the whole countryside. Van der Zaag (1956, 1959) believes that in the Netherlands epidemics start from inoculum in the early varieties and progress through to the late varieties. Table 14.2 illustrates the sequence. To consider only varieties without R genes, blight was recorded in the three medium early varieties on July 25, in the two medium late varieties on July 30 and August 7, respectively, and on the six late varieties from August 3 to August 13. On other evidence from the Netherlands (given later in Fig. 22.1), even in the same variety, blight swiftly spread from field to field.

The whole epidemic process would be slowed down most if it was slowed down in every field. This is true even of fields that themselves escape heavy infection. For example, in northwestern Europe, blight is normally not a serious economic problem in early maturing varieties. These varieties are harvested before damage is severe. Nevertheless, they contribute to the blight problem, because their great susceptibility speeds the first part of the epidemic process that eventually engulfs the more vulnerable, but less susceptible, later varieties.

With horizontal resistance, to keep blight to a minimum in any particular field not only should this field itself be of the most resistant variety available in its class but all other fields from which inoculum could come should also be of the most resistant varieties available in their classes. (By class we mean, for example, maturity class.) One would have to restrict the planting of varieties to the most resistant in their classes. These are precedents for restrictions. For example, some countries restrict the planting of potatoes to varieties immune from wart disease.

Restrictions would have to be applied with care. In many countries a policy of getting rid of very susceptible varieties would not be practical; some very susceptible varieties are old favorites with farmers and consumers.

Moreover, to restrict varieties to the most resistant is to restrict potato breeders in their primary aims of producing varieties with the greatest yields and the best quality. The restriction would amount to reducing the effective size of the breeding program (see Section 19.1) and, in most countries, might do more harm than good.

Nevertheless, there are countries, mostly at low latitudes, in which blight cripples potato production. In them, it may be just as important to weed out the very susceptible varieties as to search for more resistant new varieties.

Vertical resistance, like horizontal resistance, could be used in two ways: one direct and uncomplicated, the other requiring restrictions on the planting of varieties.

The popularity of vertical resistance has ebbed, probably because vertical resistance has allowed a loss of horizontal resistance to occur in those classes of varieties ordinarily subject to high selection pressure from blight. But, on condition that a Vertifolia effect can be avoided, vertical resistance should be encouraged.

In the uncomplicated way, breeders could use R genes as they like. They would automatically seek to use the less popular R genes because it is with the less popular genotypes that they would get the most vertical resistance (for reasons given in Section 14.6). In this, breeders would automatically create the diversity of genotypes which vertical resistance needs to manifest itself.

The complicated way of using R genes would be to restrict their use to late maturing varieties. If epidemics begin from inoculum in early maturing varieties, as van der Zaag (1956, 1959) believes, and if these early varieties were kept free from R genes, one could hope that stabilizing selection would keep complex races scarce and thus protect the late varieties. This suggestion is analyzed in more detail in Section 17.8.

CHAPTER 15

A Note on the History of Stem Rust Epidemics in Spring Wheat in North America

SUMMARY

Vertical resistance delays epidemics. Varieties with vertical resistance to stem rust were introduced into the hard-red spring wheat area and became popular some 30 years ago. They delayed epidemics enough to make stem rust almost harmless in most years, and in this way have markedly reduced losses from stem rust. But vertical resistance has been unable greatly to mitigate the devastation by the occasional first-class epidemic (to use Waldron's expression). First-class epidemics have been no rarer since vertically resistant varieties were introduced into the spring wheat area than they were before.

15.1. The Problem in the Spring Wheat Area of North America

Modern varieties in the spring wheat area of the United States and Canada have vertical resistance to stem rust.

In an analysis made some years after resistant varieties became available in Canada, Craigie (1944) computed that the resistant varieties had increased the wheat crop of Manitoba by 14 million bushels a year and that of eastern Saskatchewan by 28 million bushels. Primarily this increase was the result of protection against stem rust.

The protection that vertical resistance gives has repeatedly been noticed by other observers. It becomes apparent, e.g., by comparing the amount of stem rust in modern varieties with the amount in susceptible Marquis. There can be no doubt at all about the great benefit that vertical resistance has conferred on the spring wheat industry.

But all has not been well. A change in the relative abundance of races of *Puccinia graminis tritici* coincided with the great epidemics of 1935, 1953, and 1954. Race 56 was abundant in 1935; race 15B in 1953 and 1954. The spring wheat varieties widely grown at that time were not vertically resistant to these races, and were devastated by them.

What the evidence suggests is this: Vertical resistance is well able to control focal outbreaks and mild epidemics of stem rust. It controls them year after year to the satisfaction of all. But vertical resistance on its own has been inadequate against great epidemics. It puts out little fires, but lets great conflagrations roar.

This is in line with history which shows that great epidemics of stem rust have been as frequent in the United States spring wheat area since vertically resistant varieties were introduced as they were before. Modern literature sometimes gives the impression of a rust ravaged wheat industry waiting to be saved by resistant varieties. There were indeed frequent epidemics in the United States spring wheat area in the old days. But they were not great ravaging epidemics that could stop the industry from putting plump grain on world markets at slashing prices. Such epidemics were as infrequent then as they are now.

The purpose of this chapter is to look back a little on the history of stem rust in the barberry eradication area of the United States, to see how the coming of vertical resistance changed the frequency of great epidemics.

The history is of special importance, because of the extent of the wheat-growing area, the scale on which the work was done, the great number of pathologists and breeders who took part in it, and the duration of the work continuously for over half a century. The work has had the flattery of imitation throughout the world. It has set the pattern for breeding not only against stem rust, but also against other diseases of cereals and diseases of other crops. It has dominated thought on disease resistance to the extent that it is now most commonly assumed in the literature that resistance is vertical resistance. Within the limits of the strategy adopted, it has been among the greatest, the most consistent, the most sustained efforts in agricultural research.

15.2. Some Early History

One must go back to the 1870's, to the collapse of 1873–1874. A railroad panic, a drop in the price of manufactured goods, and a fall in wages drove thousands from the towns to farm the virgin soils of the West.

Wheat was the pioneers' crop. So successfully was it grown that America was soon pouring wheat on the British market in such quantity and of such quality that prices on the market slumped. In 1894–1895 wheat fell to the lowest price in England for 150 years. The wheat acreage in England dropped from 4 million acres to $1\frac{3}{4}$ million in the last half of the century, largely because of American competition. All too evidently it was not competition from an ailing industry.

Little was known about how severe rust was in the early years. To remedy this the Division of Vegetable Physiology and Pathology of the United States Department of Agriculture decided in 1892 to investigate what was happening. It sent a circular to crop reporters in the principal wheat-growing States and asked for information about the distribution and abundance of the different cereal rusts.

In 1894 the Division appointed Carleton to take over the work. Carleton was later to become the authority on small grains in the United States. (Markton oats and Carleton wheat were named for him, the former, contrary to custom, during his lifetime.) He published extensively on the cereal rusts. One of his first acts was to send out a second circular. This went to farmers in all the States producing substantial amounts of wheat.

Carleton summarized the answers to both circulars (Carleton, 1899). From the States that afterward became the barberry eradication area there were 248 reports of injury to wheat by rust and 520 reports of freedom from rust. The reports did not distinguish between stem rust and leaf rust, but the account makes it clear that most of the rust was leaf rust. The rust attacked mainly late varieties. There were 494 reports of attacks on late varieties to 25 of attacks on early varieties. Evidently the pattern was much the same then as it is now. Rust came late, around ripening time.

Carleton gave few details about subsequent years. But he traveled much. From the absence of any mention of them, it is safe to assume that there were no great devastating epidemics between 1892, when the survey began, and 1899, when Carleton wrote his bulletin.

Carleton anticipated what Waldron was to say. He wrote this about North Dakota: "Wheat does not seem to be commonly damaged in this State, although leaf rust is always present and sometimes abundant." Again: "In [South Dakota], as in North Dakota, it is, as a rule, too dry, cool, and breezy for rusts to do much damage."

15.3. Waldron's Evidence

Our next witness is Waldron. Waldron, a Fellow of the Linnean Society of London, had a hand in breeding wheat for resistance to stem rust almost from the start. There was little resistance in varieties of common wheat (*Triticum vulgare*) at the time of the epidemic of 1916. In 1917 Waldron noticed that a variety of common wheat from Russia, later named Kota, showed resistance. He crossed Kota with Marquis. Out of this cross came Ceres. Ceres was the first popular common wheat variety with resistance. Two hundred bushels were given out in North Dakota in 1926, and Ceres went on to dominate the wheat fields of North Dakota until race 56 of wheat stem rust devastated it in 1935.

Soon after the great epidemic of 1935, Waldron (1935) wrote: "As a matter of fact in 50 years of North Dakota farming there have been only three first class epidemics [of wheat stem rust]. They were in 1904, 1916, and 1935."

What are "first class epidemics" that Waldron speaks of? This chapter centers around that question. Waldron continues: "Other years have shown heavy rust losses, such as 1923 and 1927, but in the 3 years mentioned the epidemics were more severe." Estimated yield losses put Waldron's statement into perspective. It was estimated that in the barberry eradication area stem rust caused a loss of 184 million bushels of wheat in 1916, 33 million in 1923, and 32 million in 1927 (Stakman and Fletcher, 1930). A loss of 33 million bushels of wheat is indeed grievous. But it is not in the same class as a loss of 184 million. There are peaks of devastation that tower high above the others. Waldron's statement was about these peaks. So is this chapter.

15.4. The Evidence of Stakman and Fletcher

Ceres was introduced in a small way to start the era of popular varieties of common wheat with vertical resistance to stem rust. The decade that ended in 1929 was the last in which common spring wheats were not protected by this form of resistance. It is the last decade that can fruitfully be compared with the present era.

Stakman and Fletcher (1930) have left an account of it that gains from having been written at the time. Their account is for the thirteen North-central and Mountain States of the barberry eradication area. Some of the information Stakman and Fletcher gave about barberry eradication was given in Chapter 12. They quoted example after example to show how closely tied stem rust was to barberries: how eradication of barberries largely eliminated stem rust.

To Stakman and Fletcher in 1930 stem rust was mainly a problem of local epidemics around infected barberry bushes. For example, they observed that 1925 was climatically just as favorable to the development of rust as 1916. Barberry eradication began in 1918. By 1925 millions of bushes had been destroyed. "The effectiveness of barberry eradication [they wrote] may be more readily understood when a loss of 61% in Minnesota in 1916 is compared with a loss of about 12% in 1925; a loss of 70% in North Dakota in 1916 with a loss of 5% in 1925; a loss of 64% in South Dakota in 1916 with a loss of 7½% in 1925. . . . There is every reason to suppose that if so many barberries had not been eradicated the epidemic of 1925 would have been almost, if not quite, as destructive as that of 1916."

We are not concerned here with whether the opinion was justified. What matters is that the opinion was held by competent observers who had watched stem rust in the 1920's. That the opinion was held at all shows that stem rust losses in the 1920's could to a great extent be traced to infected barberries. The opinion is entirely in agreement with Stakman and Fletcher's analysis (described in Chapter 12) that showed how damage by stem rust was progressively reduced as barberries were destroyed; the loss computed as a 5-year average fell steadily as eradication proceeded.

Susceptible species of barberry had followed the settlers to the spring wheat area, and multiplied there. If one allows for this, there is no real difference between Carleton's opinion of stem rust in the 1890's and Stakman and Fletcher's of stem rust in the 1920's. And both opinions are entirely consistent with what Waldron said of stem rust in the 50 years before 1935.

15.5. The First-Class Epidemics

In the entire history of the spring wheat area or at any rate, on Waldron's evidence, since 1885, there have been five first-class epidemics: in 1904, 1916, 1935, 1953, and 1954. The term, first class, is used here in the sense in which Waldron used it.

One can take 1930 as the date when spring wheat varieties with vertical resistance to stem rust started to become popular. Admittedly, the date is not sharp. Marquis, before Ceres, was not entirely without vertical resistance; it is resistant to some races. And Ceres itself was not highly resistant even before race 56 became abundant. It is susceptible to other races, and was heavily rusted before 1935 in the rust nursery at Winnipeg (Johnson and Newton, 1941). Nevertheless, it is probably true, and in line with modern literature, to say that Ceres was the first great popular variety bred for resistance to stem rust, and the first popular vertically resistant variety to lose its popularity because a change in race made it susceptible and favored new competitors.

If we accept 1930 as the date when spring wheat varieties with vertical resistance to stem rust became popular, there were two first-class epidemics in the 45 years without vertically resistant varieties, and three in the 32 years with these varieties.

One cannot make a detailed statistical analysis of five epidemics. Many things had changed besides the introduction of vertical resistance. In the 50 years from 1904 to 1954 varieties had become earlier maturing, and more likely to escape rust. And between 1916 and 1935 tens of millions of barberry bushes had been destroyed. But one can at least aver that there is no evidence in history to show that vertical resistance was able to reduce the frequency of first-class epidemics.

15.6. Some Conclusions

This chapter was written to try to correct the commonly held notion about the ravages of stem rust before wheat breeding was directed against it, and to set the achievement of breeding for vertical resistance in better perspective. It is not desired here to try to suggest in great detail why breeding for vertical resistance did not reduce the frequency of first-class epidemics. But two suggestions can quickly be made.

1. For reasons discussed in Sections 13.6 to 13.8, sanitation is not a safe method of control when rt is large, which it is in years of first-class epidemics.

2. Much vertical resistance was lost when the races of the fungus changed. Stabilizing selection—a topic of Chapter 17—failed to act against races 56 and 15B.

Both reasons are good reasons for a change of strategy, even though one concedes that history does not necessarily repeat itself.

CHAPTER 16

Plant Disease in Biological Warfare

SUMMARY

Vertical resistance can be overcome by appropriate races of the pathogen. To protect a great field crop against disease by vertical resistance in time of peace is to invite the use of plant disease as a weapon in time of war. The defense is to use horizontal resistance.

Discussion is centered around wheat stem rust. The dangers from this disease are well known. Biological defense has not been organized, but the horizontal resistance needed for this defense is abundant ready to be used.

The topic is treated quantitatively, and results can be applied to disease in general.

16.1. Epidemics as Explosives

We often call an epidemic explosive. In time of peace the adjective is neatly descriptive. In time of war it could be grimly real in the military sense. An enemy has few explosives to surpass a pathogen that increases at the rate of 40% per day, compounded at every moment, and continues to increase for several months.

Spores are as light as poison gas or smoke. There are 150 billion uredospores of the wheat stem rust fungus to the pound (Stakman and Harrar, 1957): 3×10^{14} to the ton. Many types of spore disperse as easily as smoke. Many are tough and durable. They have only to be dispersed in the proper places at the proper times. Nature sees to the explosion.

16.2. Vertically Resistant Crops as a Target

Vertical resistance makes varieties resistant to some races of a pathogen, but leaves them susceptible to others. An enemy need only introduce the appropriate races, and resistance will vanish. Where vertical resistance is relied upon to protect a crop against disease, enemy action could leave the crop defenseless and open to destruction.

16.3. Wheat Stem Rust as an Example

The danger is not to one continent, but to all except Antarctica; not from one disease, but from several. If we write this chapter around one disease in one continent—wheat stem rust in North America—it is only because we prefer the concrete to the abstract, the familiar to the strange.

Stem rust of wheat is an obvious choice as weapon. It has long been known as the disease which when it is rampant destroys more in less time than any other. In North America it has on occasions spread in a few short months over 20 million acres of wheat and destroyed more than a quarter of a billion bushels of grain.

There are other reasons for the choice.

Great researches, unsurpassed in plant pathology, have uncovered just how stem rust spreads and when it spreads. A long series of publications from Mexico, the United States, and Canada tell how wind, sunshine, temperature, rain, dew, and the humidity of the air govern the dispersal of stem rust over a million square miles of North America. The publications were written in order to build a science and guide the steps to save the wheat fields. They tell an enemy almost all he need know about how and where to attack.

Apart from this wealth of information to guide aggression, there is much else to suggest wheat as a target. Fungicides are not used to protect wheat against stem rust. Up to the present there is no second line of defense behind the vertically resistant varieties. And races of *Puccinia graminis* that can breach this defense are already known. All the work of 50 years, all the pyramiding of resistance genes in sequences of new varieties, could not protect the wheat industry against organized aggression.

16.4. Stem Rust Races that Can Overcome Vertical Resistance

It has been an invariable feature that the release of new wheat varieties vertically resistant to stem rust is followed by the discovery of races of the stem rust fungus that can attack them.

Selkirk, the variety which more than any other has protected the bread wheats of the North American spring-wheat area from stem rust in recent years, and Langdon, which has protected the durums, are susceptible to several races (Miller and Stewart, 1961). In experimental plots in warm and humid weather, Selkirk has been severely damaged by stem rust (Stakman and Christensen, 1960).

There is no real difficulty about producing appropriate races. If a variety is grown on a large enough area in a warm and humid climate, races that can attack it appear sooner or later. The United States Department of Agriculture uses this fact. Varieties of wheat are grown experimentally in Puerto Rica. Conditions there are so favorable to rust that races appear and show what may be expected to happen later on the wheat farms of continental America. The same way of getting races could be used elsewhere as well.

16.5. Weather Adverse to Epidemics

Uredospores of virulent races could be harvested in bulk. They could be harvested at the right time for dissemination. In future they could probably even be stockpiled and stored; storage is now the subject of active research (Loegering *et al.*, 1961). It would be unwise to ignore defense in the hope that an attacker would have difficulty in collecting and delivering inoculum. Many greater difficulties in military matters than this have been overcome.

Progress in weather research adds to the danger. In this age of Tiros satellites, balloons, rockets, and electronic computers we are fast moving toward accurate and early weather forecasts—forecasts that would determine where and when an attack would come.

In most years the weather is not very favorable to natural epidemics; and even before vertically resistant varieties were grown, great epidemics were infrequent. That was the theme of the previous chapter. But there is cold comfort in this thought. One must think of epidemics in terms of time. The difference between a severe epidemic and a mild one is a matter of just a week or two. An unusually severe epidemic is one that reaches high levels of disease a week or two earlier than usual in relation to the time when the fields ripen. This is clear from the analysis of Craigie (1945), which is discussed in Section 16.8. Weather unfavorable to an epidemic could be balanced, partly or wholly, by the earlier dissemination of inoculum. A longer duration of an epidemic can balance a lower infection rate. In the great epidemic year of 1935, little inoculum survived the winter at its source in Texas (Stakman and Harrar, 1957).

Inoculum took time to build up, and the great epidemic developed despite this. The duration of the 1935 epidemic could well be exceeded in an epidemic started artificially.

16.6. Horizontal Resistance for Defense

Crops can be defended; and the defense is horizontal resistance.

Consider the matter quantitatively and take the great rust track from Mexico to Canada as the example. Assume the duration of the epidemic to be 90 days. In 1935 it took about 90 days from the time when uredospores started to blow north from Texas to the time when wheat fields ripened in Manitoba and Saskatchewan. The choice of 90 days as a figure for our calculation is arbitrary; there would be an obvious incentive for the attack to be made further south and earlier. But this does not matter for the argument. Assume the infection rate r to be 0.4 per unit per day. This is somewhat slower than the rate in the experiments of Rowell (1957) and Asai (1960). But this, too, does not matter for the argument. For convenience, assume the whole epidemic, spreading from field to field from south to north, to be one continuous process described by Eq. (3.6). The epidemics of Panama wilt disease of bananas in Jamaica, of sudden death disease of clove trees in Pemba and Zanzibar, and of blight in potatoes in the Netherlands, discussed in Chapter 7, are analogs.

Let X be the quantity of uredospores needed to start an epidemic which ends, after 90 days at an average infection rate $r = 0.4$ per unit per day, at some given level of infection in the fields in the north. If r had some value other than 0.4 per unit per day, what quantity of uredospores would be needed to start an epidemic which reached the same level of infection in fields in the north, also after 90 days? Answers are given in Table 16.1, as multiples of X. An exercise at the end of the chapter illustrates the method of calculation.

If new varieties with better horizontal resistance replaced the old and reduced the average infection rate from 0.4 to 0.374 per unit per day, the quantity of uredospores needed to produce an epidemic of the same final intensity would be $10X$. The horizontal resistance would increase the enemy's military problem tenfold. It would force him to disseminate 10 lb. of uredospores instead of 1 lb.; 10 tons instead of 1 ton.

If r could be reduced to 0.35 per unit per day, the quantity of uredospores necessary for the same final intensity of the epidemic would be increased to $100X$. The enemy's task would be increased one hundredfold; he would have to disseminate 100 tons of uredospores instead of

1 ton. If r could be reduced to 0.25 per unit per day, the quantity of uredospores needed would be increased to $1{,}000{,}000X$. The enemy's task would be increased a million-fold. And so on.

Somewhere along the line all incentive to attack would be lost. Stem rust would cease to be considered a military weapon.

TABLE 16.1

THE QUANTITY OF INOCULUM NEEDED TO PRODUCE ANY GIVEN LEVEL OF INFECTION AFTER 90 DAYS[a]

r per unit per day	Uredospores needed
0.451	$0.01X$
0.426	$0.1X$
0.400	X
0.390	$2.46X$
0.380	$6.05X$
0.374	$10X$
0.349	$100X$
0.323	$1000X$
0.298	$10{,}000X$
0.272	$100{,}000X$
0.246	$1{,}000{,}000X$

[a] The quantity when $r = 0.4$ per unit per day is taken as standard.

Horizontal resistance is only one of the factors that influence the infection rate. The weather and the age of the crop, to mention two other factors, cause r to vary from time to time. We are concerned only with the average value of r for the duration of the epidemic. The average would be higher in seasons favorable to stem rust than in others. Nevertheless, whatever the average value of r might be, reducing that value by 0.0256 per unit per day would increase tenfold the amount of inoculum that is needed.

This figure is for a duration $t_2 - t_1$ of 90 days. (Here t_1 is the date of dissemination of uredospores, and t_2 the date on which fields ripen in the north.) For other durations, the extent to which r must be reduced in order to increase the necessary amount of inoculum tenfold is given in Table 16.2. The shorter the duration of an epidemic, the more r must be reduced.

This is a satisfactory result. It means that using horizontally resistant varieties would more easily stop a great general epidemic than a small local one. It might be difficult to stop an attacker from starting a small local epidemic, but, if the horizontal resistance was high enough, the epidemic would not spread over more than a limited area before the fields ripened and were out of harm's way. The higher the horizontal resistance, the more limited the area would be.

TABLE 16.2

THE EXTENT Δr TO WHICH r MUST BE DECREASED IN ORDER TO INCREASE TENFOLD THE AMOUNT OF INOCULUM NEEDED TO START AN EPIDEMIC OF ANY GIVEN FINAL INTENSITY. Δr IS SHOWN RELATED TO THE DURATION OF THE EPIDEMIC $t_2 - t_1$

$t_2 - t_1$ days	Δr per unit per day[a]
30	0.077
40	0.058
50	0.046
60	0.038
70	0.033
80	0.029
90	0.026
100	0.023
110	0.021
120	0.019

[a] For other values use $\Delta r = 2.30/(t_2 - t_1)$.

16.7. Horizontal Resistance to Stem Rust

This then is the case for horizontal resistance. It protects against epidemics. The greater the epidemic, the better it protects. It is the stuff a grand strategy of defense can be made of.

Horizontal resistance to stem rust in wheat has been called generalized resistance. Stakman and Christensen (1960) have said this of it:

"When wheat varieties with generalized resistance and those without it are grown side by side in the field and inoculated uniformly with races of stem rust to which both are susceptible in the seedling stage, it becomes clearly apparent that the susceptibility to infection differs greatly. The ratio of the number of pustules on the resistant and susceptible varieties may be 1 : 5 or 1 : 10 or even 1 : 20 relatively early in the season. The final amount of rust sometimes is the same, however, if the varieties are inoculated repeatedly over a long period of time. This indicates that there is considerable resistance to entrance of the pathogen; accordingly, it takes more inoculum and more time to produce the same amount of rust on the resistant variety as on a susceptible one, although the type of pustule may be essentially the same.

"There are known morphological characters in some wheat varieties that affect the length of incubation period, the size of pustule, and possibly the amount of damage caused by a given amount of rust. The relative number and size of collenchyma bundles in relation to the amount and distribution of woody sclerenchyma are known to affect the number, size, and speed of development of pustules. At least a few varieties have lignified epidermis, which tends to lengthen the incubation period and reduce the size of pustules, as the epidermis is ruptured less easily than in those varieties which have nonlignified epidermis.

"Clearly, some varieties have combinations of characters that protect them reasonably well against stem rust under a wide range of conditions. Varieties that have only specific resistance against individual rust races are, of course, at the mercy of those races to which they are susceptible. Thus, Kanred wheat, which is immune from a considerable number of races, is likely to be either rust free in the field or completely susceptible, depending on the races present. Certain other varieties, such as Newthatch, not only have specific resistance to a number of physiologic races but also a generalized resistance that protects them to some degree against all races, although the combination of specific and generalized resistance is not sufficient to protect against all races under all environmental conditions." (In our words, Kanred has vertical resistance, but only little horizontal resistance. Newthatch has vertical resistance, and a substantial amount of horizontal resistance as well.)

From Stakman and Christensen's description it will be seen that in some varieties horizontal resistance manifests itself in the same way as horizontal resistance to potato blight does. (See Section 14.16.) Infection is resisted, and sporulation delayed and reduced. These are all factors that reduce r. From an analysis of some results obtained by Hayden, van der Plank (1960) deduced that r was smaller in Lee and Sentry than in Marquis, Mida, Carleton, and Nugget when they were grown under the same conditions.

16.8. Quantitative Interpretation of the Evidence

Consider Stakman and Christensen's statement that the number of pustules could be reduced to 1/5, or 1/10, or even 1/20 by resistance. Table 16.3 gives the results of an analysis that will be discussed later (in Section 20.3). The table shows how resistance of the order mentioned by Stakman and Christensen would affect the amount of uredospores that an enemy would have to disperse. As in Table 16.1 we take the amount needed on the susceptible variety to be X, and the epidemic to continue 90 days before reaching some fixed level of infection.

TABLE 16.3

THE EFFECT OF RESISTANCE TO INFECTION ON THE QUANTITY OF INOCULUM NEEDED TO PRODUCE ANY GIVEN LEVEL OF INFECTION AFTER 90 DAYS[a]

Pustule ratio, resistant : susceptible	Uredospores needed
1 : 1	X
1 : 2	690 X
1 : 5	2.3 million X
1 : 10	710 million X
1 : 20	130 billion X

[a] An infection rate $r = 0.4$ per unit per day is assumed for the susceptible standard.

Even if resistance only halved the proportion of uredospores that formed pustules, it would increase the number of uredospores needed to $690X$. It would increase the enemy's military difficulties 690-fold. Resistance that reduced the proportion to 1/5, the smallest of Stakman and Christensen's ratios, would increase his difficulties 2.3 million-fold. Using the full amount of resistance mentioned by Stakman and Christensen would increase his difficulties 130 billion-fold.

There is resistance to spare.

These results are for the wartime problem. Consider now the peacetime problem of banishing the stem rust fungus into obscurity by means of horizontal resistance. If there is a given source of inoculum, how would resistance delay the culmination of the epidemic? Table 16.4 converts the same information of Stakman and Christensen into the form of delays. (Details of the method are also given in Section 20.3.) If an epidemic takes 90 days in the susceptible standard variety, it takes 110 days in a variety with resistance that halves the proportion of uredospores forming pustules. The resistance delays the epidemic 20 days. Greater resistance, corresponding to ratios of 1 : 5 or 1 : 20, is enough to delay an epidemic from 57 to 172 days. (This last calculation is unreal because it extends the epidemic deep into the winter.)

To see these delays in perspective, consider the data Craigie (1945) gives about stem rust in Manitoba. We can compare 1926, 1928, 1929, 1931, 1932, 1933, and 1934, when stem rust caused little damage, with 1935, when the epidemic was severe. To give averages, in eastern Manitoba the appearance of a light general infection in wheat fields was 5 days later in the years of little damage than in 1935. In western Manitoba it was 12 days later. In the light rust years, harvesting began 2 days earlier than in 1935. In relation to harvesting, a light general infection in 1935 was 7 days earlier in eastern Manitoba and 14 days earlier in western Manitoba than in the light rust years.

TABLE 16.4

THE EFFECT OF RESISTANCE TO INFECTION ON THE TIME TAKEN TO REACH A GIVEN FINAL LEVEL OF INFECTION[a]

Pustule ratio, resistant : susceptible	Time needed (days)	Delay (days)
1 : 1	90	
1 : 2	110	20
1 : 5	147	57
1 : 10	193	103
1 : 20	262	172

[a] A time of 90 days and $r = 0.4$ per unit per day are taken for the standard susceptible variety.

It would have been neater to have had information on a stage later than that of light general infection. But Craigie's evidence suggests that the infection rate after the stage of light general infection was much the same in 1935 as in the other years, so the matter seems unimportant for a preliminary assessment. On a preliminary assessment, a delay of 20 days, for a 1 : 2 ratio, would control rust at the northern end of the rust track, where disease is severest. Ratios of 1 : 5, or more, would give a large margin of safety.

These calculations have a condition. The horizontally resistant varieties must be grown over the whole epidemic area, not just here and there in an occasional field. Herein lies an essential difficulty.

16.9. The Transition

The defense in time of war is horizontal resistance in the varieties that farmers grow in time of peace. It is not enough to have varieties with better horizontal resistance. The varieties must have all the qualities farmers and millers want. Otherwise farmers will not grow them.

In particular the varieties must be free from rust. And this is where the difficulty comes in. Horizontal resistance, used widely, will protect against rust. Horizontal resistance, in a few fields here and there, is less effective and may fail to protect these few fields adequately.

This is the problem of transition: How to protect horizontally resistant varieties while they are still confined to only a few fields? Horizontal resistance acts cumulatively. Horizontally resistant varieties introduced, as they must be introduced, in a small way at first show little of their potential performance. Only when the varieties reach the stage of covering the great majority of fields does horizontal resistance show its full value.

It is a matter of the average value of r. Consider a field of wheat in Manitoba, and suppose a stem rust epidemic starts in Texas. The amount of stem rust in the field in Manitoba depends on the average value of r over the whole track from Texas northward. By track we mean those particular fields through which the fungus passes on its way to the field in Manitoba. Planting a horizontally resistant variety in this field will reduce r in the field. But it will not reduce it over the rest of the track. To get the average value of r as low as possible—to reduce stem rust in the field in Manitoba as much as possible—horizontally resistant varieties must be used over the whole track. Only then does this form of resistance show itself at its best.

Another way of thinking of horizontal resistance is to think of it as a slowing down of the epidemic. Horizontal resistance in all varieties along the track slows the epidemic down all along the track. The epidemic reaches the field in Manitoba late, and loses still more time in the field itself. An epidemic made late enough ceases to be a destructive epidemic at all.

There are alternative solutions to the problem of transition.

The one is to start the transition with varieties having far more horizontal resistance than will be needed later when transition is complete. The resistance must be enough to protect a field on its own, without help from resistance in other fields in the same track.

The other is to couple horizontal with vertical resistance at least in the first few new varieties. For, unlike horizontal resistance, vertical resistance is at its best in new varieties. The next chapter must be consulted about this.

For either alternative, the price will be high. It is the price of past neglect of horizontal resistance. The transition may need a budget of the sort more commonly associated with military defense than agricultural research.

But there is this to be said about horizontal resistance. Expenditure on breeding for it is nonrecurrent. Expenditure on vertical resistance recurs each time the pathogen shifts.

EXERCISE

This is the problem dealt with in Section 16.5: X uredospores start an epidemic. With stem rust increasing at an average rate $r = 0.40$ per unit per day the proportion of infection at the end of 90 days is x_2. If the average rate had been 0.39 per unit per day, how many uredospores would have been needed at the start for the epidemic to reach the same proportion of infection, x_2, in the same time, 90 days?

Use Eq. (3.6), with $t_2 - t_1 = 90$ days. When $r = 0.40$ per unit per day, $x_1 = kX$, where k is a constant; and $1 - x_1$ can be ignored because x_1 is small in relation to the wheat fields as a whole. From Eq. (3.6)

$$0.40 = \frac{2.30}{90} \log_{10} \frac{x_2}{kX(1-x_2)}$$

Rewrite this as

$$\log_{10} \frac{x_2}{1-x_2} - \log_{10} kX = 0.40 \times \frac{90}{2.30}$$

When $r = 0.39$ per unit per day, let the necessary number of uredospores be aX, in which case $x_1 = akX$. Then

$$\log_{10} \frac{x_2}{1-x_2} - \log_{10} akX = 0.39 \times \frac{90}{2.30}$$

By subtraction

$$\log_{10} akX - \log_{10} kX = (0.40-0.39) \times \frac{90}{2.30}$$

and

$$\log_{10} a = \frac{0.01 \times 90}{2.30}$$

$$= 0.391$$

$$a = 2.46$$

There are three points for comment:

1. It does not matter what x_2 is, because the numerical value of x_2 does not enter into the calculation. What matters is that comparisons should all be based on the same final proportion of disease, whatever it may be.

2. The calculation was simplified by assuming the same constant k, whether r was 0.40 or 0.39 per unit per day. That is, it was assumed that it would take the same average number of uredospores to cause a pustule to develop irrespective of the infection rate. But a lower infection rate commonly results from the greater resistance of a variety to infection. For this reason our exercise and Table 16.1 underestimate the benefit of reducing r by using horizontally resistant varieties. Horizontal resistance is of even greater value than we have made it out to be. The error from the simplification is unimportant for present purposes.

3. Calculations are based on a constant final proportion of disease. It is not attempted to say what this means in terms of total loss of crop, i.e., in terms of millions of bushels of wheat. But the use of the final proportion is a realistic way of dealing with the problem, because the greatest destruction has always been in the north: in the Prairie Provinces of Canada, the Dakotas, and Minnesota.

It is not easy to apply the hypothesis, discussed in Section 12.7, that loss of crop is proportional to the area under the rust progress curve, because fields ripen earlier in the south. If an epidemic advances northward with the northward procession of ripening, it is useful to concentrate attention on disease at the end.

CHAPTER 17

The Bases of Vertical Resistance

SUMMARY

Vertical resistance implies two things. More than one genotype of host plant must be involved (unless inoculum comes from the soil). Stabilizing selection must favor simple races of the pathogen on simple varieties of the host, i.e., on varieties with no genes, or only few genes, for vertical resistance.

17.1. Introduction

Vertical resistance is a valuable quality. All sanitation is valuable. If we feel that plant pathologists and plant breeders have given vertical resistance a disproportionately large share of attention, it is not the attention to vertical resistance that we deplore, but the relative lack of attention to horizontal resistance.

Before we go on to other topics, this short chapter is interpolated to examine the bases of vertical resistance: the need for diversity and the need for stabilizing selection in favor of simple races of the pathogen on simple varieties of the host. These topics were introduced in relation to potato blight in Section 14.15.

17.2. The Different Response of Horizontal and Vertical Resistance to Diversity of Varieties

The previous chapter brought to light a point about horizontal resistance. In epidemics of diseases that range from field to field, horizontal resistance gains in potency if all fields, not just a few, have this resistance. The effect of horizontal resistance in a field is increased if the pathogen reaches the field from other fields that also have horizontal resistance. The effect is cumulative. Horizontal resistance gains from uniformity.*

Vertical resistance on the other hand requires, and gains from, diversity. If only one field in a whole epidemic has horizontal resistance, the effect of this resistance is at its lowest. If only one field in the epidemic area has vertical resistance, the effect of this resistance is at its highest.

17.3. The Dependence of Vertical Resistance on Varietal Diversity

One can simplify the argument by considering the best form of vertical resistance: vertical resistance that stops the pathogen from reproducing. A variety can then be vertically resistant only to races coming from a different genotype; to inoculum coming from fields of the same variety as itself it is necessarily susceptible. A potato variety with the gene R_1 is necessarily susceptible to all races of *Phytophthora infestans* coming from other fields with the gene R_1: races (1), (1,4), etc.

When vertical resistance does not entirely stop reproduction, i.e., when resistance is incomplete, it would be possible for a variety to have vertical resistance to inoculum coming from another field of the same variety. But the resistance would be less than if inoculum came from another variety that encouraged a different set of races of the pathogen.

If a variety shows high vertical resistance during an epidemic, it is certain that the inoculum comes from a different variety, species, genus, or family of diseased host plant or it comes from the soil.

17.4. A Reason Why Vertical Resistance Is Commonly Chosen to Protect Varieties

There are various reasons why plant pathologists and plant breeders have so often chosen vertical resistance instead of horizontal resistance to protect new varieties. The inheritance of vertical resistance is

* Uniformly good horizontal resistance throughout the epidemic area does not imply genetic uniformity for resistance. Much horizontal resistance is probably polygenic. It is characteristic of polygenic inheritance that the effects of substitution at different loci are interchangeable. Similar phenotypes can be produced by a diversity of genotypes. For example, horizontal resistance can come from resistance to infection, or delayed sporulation, or sparse sporulation. These factors are presumably governed by different genes. But they all have the same effect: they increase horizontal resistance.

usually less complex, and breeding correspondingly easier; the screening out of susceptible progeny is simpler; and so on.

But one reason is the relation between vertical resistance and diversity. Vertical resistance shines brightest in the breeders' nurseries, on experimental stations, and in the trial plots laid down to test new varieties in comparison with old, because a new variety, if it has new genes for vertical resistance, provides the diversity that vertical resistance needs. It is here that a new variety is chosen. There may be forebodings about what will happen when the variety meets new races of pathogens. But forebodings about the future do not weigh in statistical tests for significance. Horizontal resistance, on the other hand, is at its dimmest in breeders' nurseries, on experiment stations, and in trial plots. It shines brightest where it should, in the broad acres of successful varieties. However, that is not a quality for which a new variety is at present chosen.

17.5. Diversity and Novelty

Much of the hunt for diversity has been a hunt for novelty. In other words, to feed diversity varieties have been changed in rapid succession.

This has been the history of breeding spring wheat varieties in North America. As races became common by directional selection on one group of varieties, new varieties with an added gene for vertical resistance provided the diversity needed for control.

Dependence on vertical resistance has been the main reason for three major replacements of oat varieties in North America in two decades.

Wheat bunt caused by *Tilletia caries* and *Tilletia foetida* is another example. Until quite recently the most widely grown varieties in the Pacific northwest of the United States had no vertical resistance to bunt or were resistant to only a few races (Kendrick and Holton, 1961). In 1949 a variety with considerable vertical resistance was released. By 1953 it was widely grown. The variety was Elmar. Besides being the first vertically resistant variety to become very popular in the northwest, Elmar was the first single variety to cover more than half the wheat acreage in this region. Elmar soon became badly smutted. Indeed it was smutted so soon that one suspects that vertical resistance had brought with it a Vertifolia effect. Be that as it may, the first extensive use of vertical resistance coincided with the quickest change of an important variety in the history of the region. By 1958 Omar had replaced Elmar, now faded to relative unimportance.

The hunt for novelty is nowadays well organized. International variety collections have been established. Interspecific and intergeneric crosses have been made, for example, between *Triticum*, *Aegilops*, and *Agropyron* to enlarge the hunting ground. This is all to the good. But it will only be when we have diversity without novelty that we can stop the tedious and repetitive replacement of varieties, the ceaseless change without apparent progress, for the sake of vertical resistance. There is far more to vertical resistance than slapping a new gene into a new variety every time there is trouble.

How to get vertical resistance without novelty is the theme of Section 17.8.

17.6. The Fitness of Simple Races on Simple Varieties

We define a simple race of pathogen as one that can attack varieties with no genes, or only a few genes, for vertical resistance. Similarly, a simple variety of host plant is one with no genes, or only a few genes, for vertical resistance. Race (0) of *Phytophthora infestans* is simple; it can attack only simple varieties like Katahdin that have no R gene. Race (1, 2, 3, 4, 5, 6) of this fungus is complex; it can attack complex varieties with many R genes, singly or in combination.

We take it as axiomatic that simple races are the fittest to survive on simple varieties. All vertical resistance is based on this. As we pointed out in Chapter 14, if race (1, 2, 3, 4, 5, 6) of the blight fungus had been the fittest to attack a simple variety such as Katahdin, vertical resistance based on the genes R_1, R_2, R_3, R_4, R_5, or R_6 would never have been used. Quite probably, these genes would not even have been identified. So, too, vertical resistance in wheat to the stem rust fungus is based on the resistance genes Sr 1 to Sr 14. If races of the stem rust fungus able to overcome resistance from these genes had been the common races in the old days on varieties such as Little Club or Marquis, wheat breeding for resistance to stem rust would necessarily have been very different.

The matter is evident. One can see clearly that vertical resistance requires that simple races should be the fittest on simple, or fairly simple, varieties. Conversely, where vertical resistance to a disease has not been found, one can assume either that genes for vertical resistance are rare, or absent, in the host plant, or that races of the pathogen complex enough to overcome all available genes for vertical resistance were common even on simple varieties.

This is not to say that the races fittest to survive on simple varieties were invariably the very simplest possible. Race (4) of *P. infestans* is at least as common as race (0) on varieties without R genes. As one would expect, no commercial variety has been bred with resistance based on the gene R_4 alone. All North American races of *Melampsora lini*, which causes rust in flax, attack the flax variety Bison with the gene L^9 (Flor, 1953, 1959), but Bison is resistant to the simplest Australian race (Waterhouse and Watson, 1941). If there originally were simpler races in North America than there are now, they must have become rare after Bison was introduced; the matter cannot be tested because samples of rust were not identified in the pre-Bison era. Conversely, all Australian races of *Puccinia graminis tritici* can attack wheat varieties with the gene Sr 7 (Pugsley, 1959), a gene that confers resistance to race 15B in North America. Even the simplest Australian races of the wheat stem rust fungus have some complexity, which gives special interest to the results in the next section. In North America races 56 and 15B of the stem rust fungus took only the earliest steps toward complexity. Both seem fully fit to survive and are still common. In 1959 about 60% of the isolates of *P. graminis tritici* in Canada were of race 56 and 23% of race 15B. Evidently their complexity is too little to affect survival greatly. The fitness of these two relatively simple races to survive has obvious relevance to the great epidemics of 1935, 1953, and 1954.

In all probability, loss of fitness to survive on simple varieties increases with the complexity of the race. This is fairly clear with potato blight. Race (4) is common. Races (1), (1, 4), and other fairly simple races are not uncommon. But it is rare to find race (1, 2, 3, 4) on a simple variety except at experimental stations or near complex varieties. Races as complex as race (1, 2, 3, 4, 5) do not seem ever to have been isolated from simple varieties, except where they could be traced to a complex source.

It is part and parcel of the argument that if a simple variety is exposed to a mixture of races, stabilizing selection will favor the simple and fairly simple races. This is particularly clear from the results published about potato blight in Mexico and referred to in Chapter 14. Because of the presence there of *Solanum demissum* and other complex hosts, complex races occur, and have probably occurred for ages back. But the common races on simple hosts are simple, or fairly simple.

Holton (1947) observed that complex races of *Tilletia caries* and *Tilletia foetida* are common only in regions where complex varieties are grown. In tests with hybrids from races of *T. caries* he found that, whereas complex wheat varieties promoted the selection of complex hybrids, simple varieties promoted the selection of simple hybrids.

Equally convincing evidence has been given by Flor (1953) for flax rust in the North Central States of the United States. When the relatively simple variety Bison was the common variety, the races of rust that were collected were mostly relatively simple. As more complex varieties were released, they were attacked by more complex races, with genes in the pathogen matching genes in the host variety, one for one. But ever since the testing of races started in 1931 the dominant races have always been the simplest that could survive, i.e., those with the fewest genes needed for attacking the common flax varieties.

It is worth interpolating here that virulence in *Melampsora lini* is recessive, and Flor (1953) suggested that simple races survive best because they are heterozygotes. With *Puccinia graminis tritici* no general rule has been found. The work of Johnson (1954) and Johnson and Newton (1940) in Canada has shown virulence to be recessive on some wheat varieties and dominant on others. Since there is evidence that the simple races of wheat stem rust survive best (see next section) one is inclined at first to think that Flor's suggestion does not cover the wider facts. On second thoughts one remembers that the races studied by Johnson and Newton are all relatively simple. It is not unlikely more recessive genes for virulence in the stem rust fungus will be uncovered as races become more complex, in which case Flor's suggestion would become more attractive. That simple races are the best genotypes to survive is, however, not the only suggestion that can be made. It is equally possible to suggest reasons why simple races are the best phenotypes to survive on simple varieties.

Evidence from Australia for stabilizing selection in favor of simple races of wheat stem rust is convincing. The survival of races has also been studied in North America (Loegering, 1951; Roane *et al.*, 1960), but it is hardly relevant to our particular topic because there was no more complex race concerned than 15B.

17.7. Evidence for Stabilizing Selection of Simple Races of *Puccinia graminis tritici*

Watson and Singh (1952) have discussed the occurrence of four races of wheat stem rust in New South Wales, Australia. The simplest of these races is race 126. It attacks the simple variety Federation, but neither of the varieties Kenya 743 nor Kenya 745. The most complex race is race 222BB which attacks Federation, Kenya 743, and Kenya 745. Races 126B and 222AB are of intermediate complexity, to judge by the reactions recorded in Table 17.1.

TABLE 17.1

ABILITY (+) OR INABILITY (−) OF FOUR RACES OF STEM RUST TO ATTACK FEDERATION WHEAT AND VARIETIES DERIVED FROM KENYA 743 AND KENYA 745[a]

Race	Federation	Kenya 743	Kenya 745
126	+	−	−
126B	+	+	−
222AB	+	−	+
222BB	+	+	+

[a] Data of Watson and Singh (1952).

Initially race 126 was the commonest race. The simplest race was the fittest to survive. Later, after the introduction of new and less simple varieties into cultivation, less simple races appeared. Table 17.2 shows

TABLE 17.2

THE NUMBER OF TIMES FOUR RACES OF STEM RUST WERE ISOLATED AT TWO CENTERS OF THE NEW SOUTH WALES WHEAT BELT IN 1950[a]

	Race:			
Center	126	126B	222AB	222BB
Wagga-Temora	21	9	2	1
Gunnedah	—	—	87	10

[a] Data of Watson and Singh (1952).

that at Gunnedah, where varieties were grown resistant to races 126 and 126B, directional selection made races 222AB and 222BB prevalent. But in the Wagga and Temora districts, where susceptible varieties were still largely grown, race 126 was still the common race. There was evidently stabilizing selection for this simple race on simple varieties. Although the districts are less than 300 miles apart, with continuous wheat cultivation between, the complex race 222BB was unable to establish itself much on the simple varieties. We infer that it was not fit enough to do so.

Note also that race 126 was more abundant than race 126B, and race 222AB than race 222BB.

The results of an experiment of Watson and Singh, recorded in Table 17.3, support this. When mixtures of the four races were cultured on the simple variety Federation, the simple race 126 survived best after

TABLE 17.3

THE EFFECT OF PASSAGE THROUGH THE VARIETY FEDERATION ON THE PERCENTAGE OF VARIOUS RACES OF WHEAT STEM RUST IN NINE DIFFERENT MIXTURES[a]

		No. of passages through Federation			
Mixture	Race	1	3	4	5
1	126	69.0	85.5	86.4	88.8
	126B	31.0	14.5	13.6	11.2
2	126B	71.3	90.4	90.8	86.5
	222BB	28.7	9.6	9.2	13.5
3	126B	61.1	58.6	56.6	47.7
	222AB	38.9	41.4	43.4	52.3
4	126	84.9	96.6	97.8	95.2
	222BB	15.1	3.4	2.2	4.8
5	126	65.7	82.1	77.0	72.6
	222AB	34.3	17.9	23.0	27.4
6	222BB	65.6	4.7	8.3	5.2
	222AB	34.4	95.3	91.7	94.8
7	126	57.9	68.5	63.8	74.2
	126B	33.2	30.4	35.4	25.4
	222BB	8.9	1.1	0.8	1.4
8	126	52.9	68.1	78.1	71.5
	126B	9.4	4.0	3.2	3.7
	222AB	37.7	27.9	18.7	24.8
9	126	66.2	92.1	89.7	87.2
	222BB	19.9	1.1	0.8	0.5
	222AB	13.9	6.0	9.5	12.3

[a] Data of Watson and Singh (1952).

successive passages, and the complex race 222BB worst. The support is not wholly beyond challenge. There was probably interaction between the races (competion is the more usual word) in this experiment; and interaction is probably rare in natural epidemics. (See Section 7.10.) But at least the evidence in Table 17.3 accords fully with the sounder geographic evidence of Table 17.2.

17.8. Vertical Resistance without Novelty

Stabilizing selection in favor of simple races provides a means of finding diversity without novelty. The necessary conditions are that in seasonal epidemics the pathogen should move from field to field, from one variety to another; that varieties in which seasonal epidemics start should be the natural reservoirs of the pathogen from season to season; and that varieties at the start should be simple or nearly simple, and those at the end complex enough to be protected against races that survive on the simple or nearly simple varieties.

Consider potato blight. Van der Zaag (1956, 1959) believes *Phytophthora infestans* moves from early maturing to late maturing varieties. Accept that this is so. Then if early varieties were kept simple, without R genes, and late varieties were made complex enough, with three or four R genes, one could expect stable vertical resistance in the late varieties. Stabilizing selection would keep complex races, able to attack the late varieties, scarce in the early varieties. The natural flow of blight from early to late varieties would thus be stemmed.

The illustration brings out two points.

First, the natural path of the epidemic must be known. If blighted early varieties are not the chief source of infection for late varieties—if inoculum for late varieties overwinters in tubers of late varieties—the whole suggestion falls away.

Second, there would have to be some authority to control the use of genes by breeders. If complex early varieties were bred and grown, the suggestion loses its purpose.

Potato blight is used here just as an illustration. Drastic measures are probably not needed in most countries; a little extra horizontal resistance against blight would probably be enough. But the illustration brings out the point that the haphazard use of genes for vertical resistance, in varieties where the epidemic starts as well as in varieties where the epidemic ends, would throw away the chance of getting diversity without novelty.

17.9. Natural Stability in Vertical Resistance

The use of vertical resistance does not always cause varieties to change in quick succession. There are many examples of vertical resistance accompanying stable genotypes. This can well occur, particularly if inoculum survives outside the crop itself—in grasses or weeds, for example, or in the soil.

Fusarium oxysporum f. *lycopersici* causes a wilt disease of the tomato. Vertical resistance to it was found by Bohn and Tucker (1939, 1940) in *Lycopersicon pimpinellifolium* accession 160. This gives resistance to race 1 of the fungus. Almost immediately, in 1939, a second race, race 2, was isolated, and its discovery announced by Alexander and Tucker (1945). This race attacks *L. pimpinellifolium* accession 160. Nevertheless commercial tomato varieties with resistance to race 1 but not to race 2 have been widely and successfully grown in most tomato growing areas without succumbing to race 2. The vertical resistance has been stable.

F. oxysporum f. *lycopersici* is very mutable. The early appearance of race 2 is evidence for that. Other evidence is found in the results of Gerdemann and Finley (1951). For 5 years they inoculated a field which had never grown tomatoes before with three single-spore isolates of race 1. Another field was inoculated for 2 years. Tomatoes of different varieties were then planted in these fields. Varieties susceptible to race 1 went down to this race. In addition, some plants resistant to race 1 wilted. From 12 of these plants the pathogen was isolated and proved to be race 2. This race had evidently arisen from race 1.

On the evidence of Alexander and Tucker and of Gerdemann and Finley it seems likely that in the United States race 2 has arisen thousands, probably millions, of times by mutation or by some other process of change. If one accepts this, then one must attribute the success and stability of vertical resistance against race 1 to stabilizing selection. That is, directional selection favors race 2 in fields in which tomato plants with resistance to race 1 are growing. But stabilizing selection favors the survival of race 1 in the soil itself after the crop is off. This is a comforting notion. It implies that varieties resistant only to race 1 have been successful in many soils, not because new races have not appeared, but because selection in the soil is against these races. This in turn implies that broadened parasitism impairs the race's ability to survive in a saprophytic or dormant state. There is nothing improbable about this.

This suggestion does not mean that race 2 is totally unable to survive outside the infected host plant, i.e., independently of its parasitism. The suggestion is that race 2 is less able than race 1 to survive independently. We are dealing with relative ability to survive. The presence of race 2 on six farms in a county in Florida (Stall, 1961a, b) does not invalidate the suggestion. It would invalidate the suggestion if it were later found that race 2 could exist independently as easily as race 1 in all soils of the tomato wilt area of North America.

When the pathogen comes from soil, weeds, or other source outside the crop itself, stable vertical resistance does not necessarily imply immutability in the pathogen. It could equally mean that though the pathogen can vary by mutation or in other ways, the variations are curbed by stabilizing selection.

CHAPTER 18

General Resistance against Disease

SUMMARY

General resistance is defined as an extension of horizontal resistance. It is resistance inherited as a unit not only against all races of a species of pathogen, but against different species, genera, or even larger groups. Evidence for general resistance against a wide range of different viruses has been found in a tobacco variety; and there is evidence that general resistance is common in nature.

18.1. General Resistance

Horizontal resistance is against all races of a pathogen, e.g., against all races of *Puccinia graminis*. Its logical extension is resistance against several different pathogens, e.g., against *P. graminis*, *Puccinia recondita*, and *Puccinia striiformis*, the causes of stem rust, leaf rust, and stripe rust in wheat, as well as against all races within each of these species. Far fetched, you may think. Why? Troutman and Fulton (1958) and Holmes (1959, 1960, 1961) have recently found a tobacco variety resistant to tobacco mosaic virus, cucumber mosaic virus, turnip mosaic virus, potato mottle virus (potato virus X), tomato ring spot virus, tobacco ring spot virus, tobacco streak virus, tobacco necrosis virus, severe etch virus, and aspermy virus. Resistance genes to several different diseases are of course quite common in a variety. But the resistance found by Troutman and Fulton and by Holmes seems to be governed by a single genetic mechanism and to operate through some physiologic process effective against all viruses that have been tried. Moreover, on the evidence of Troutman and Fulton, the resistance against cucumber mosaic virus is effective both with mechanical inoculation and with inoculation by aphids. If a tobacco variety can be resistant against such a heterogeneous conglomeration of viruses, why should not a wheat variety be resistant to three species of rust in a single genus?

Resistance in a single species to many viruses was discussed earlier by van der Plank (1949), on evidence from the natural distribution of viruses. He called it general resistance.

This chapter is about general resistance, which we define as resistance in a host against different species, genera, or even wider groups of pathogens. We discuss first the work of Troutman and Fulton and of Holmes on viruses; and then the wider implication for the control of disease of all sorts.

18.2. Results of Troutman and Fulton and of Holmes

Troutman and Fulton (1958) tested 400 accessions of *Nicotiana tabacum* and other species of *Nicotiana* for resistance to cucumber mosaic virus. They selected the variety T.I.245, of Mexican origin, as the most resistant of all that they tested. The resistance of this variety is a resistance to infection. When leaves of this variety are mechanically inoculated with diluted sap containing cucumber mosaic virus fewer lesions are formed than on leaves of more susceptible varieties. (See Fig. 18.1.)

Similar tests with tobacco mosaic virus, tobacco necrosis virus, tobacco streak virus, and turnip mosaic virus showed that the variety T.I.245 has resistance against them as well. The resistance against these viruses (to judge by a comparison between T.I.245 and a susceptible variety) is of much the same order as the resistance to cucumber mosaic virus. It expresses itself in the same way, as a low lesion count on leaves that have been mechanically inoculated with sap.

Field tests with cucumber mosaic virus indicated that the resistance of T.I. 245 was expressed against the aphid-borne virus as well as against the mechanically transmitted virus. In the field fewer plants of T.I.245 became systemically infected than plants of the control variety. There was an exception in 1 year when infection was so severe that all plants of both varieties were infected; but even in that year infection was slower in T.I.245.

Holmes (1959, 1960) found that in addition to the five viruses tested by Troutman and Fulton, T.I.245 was resistant to tobacco ring spot virus, potato mottle virus, severe etch virus, tomato aspermy virus, and tomato ring spot virus. Holmes (1961) states that no virus has been tested to which T.I.245 is not more resistant than other control varieties.

Holmes (1960, 1961) crossed T.I.245 with an unrelated tobacco variety. In the second filial generation he found some plants of unusually high resistance to tobacco mosaic virus. He self-pollinated these and selected a line that bred true for increased resistance.

The whole process of screening and selection was carried out with tobacco mosaic virus alone. But when the line selected for high resistance to tobacco mosaic virus was tested against cucumber mosaic virus, turnip mosaic virus, potato mottle virus, tomato ring spot virus, tobacco ring spot virus, and tobacco streak virus, it was found to be resistant to them as well. (No tests were made with the other three viruses which T.I.245 is known to resist: tobacco necrosis virus, severe etch virus, or aspermy virus.)

Selection for resistance against tobacco mosaic virus alone was also a selection for resistance against six other viruses as well. The data of Holmes (1961) on the seven viruses are given in Table 18.1. The table compares Holmes' newly selected line with a control variety. As Holmes pointed out, this concomitant inheritance of resistance to many virus diseases is evidence that the resistance depends on the same genetic mechanism.*

TABLE 18.1

A COMPARISON BETWEEN THE TOBACCO VARIETY WHICH HOLMES SELECTED FOR RESISTANCE TO TOBACCO MOSAIC VIRUS AND A CONTROL TOBACCO VARIETY[a]

Virus in inoculum	Lesions per plant[b]	
	Selected variety	Control variety
Tobacco mosaic	89.7	355.5
Tobacco ring spot	5.5	55.8
Tomato ring spot	2.5	9.0
Potato mottle	14.0	151.0
Tobacco streak	0.4	2.3
Turnip mosaic	1.0	20.8
Cucumber mosaic	0.8	8.2

[a] Data of Holmes (1961)
[b] Average number of lesions per plant.

The resistance against at least four of the viruses (the others were not tested) seems also to depend on the same physiologic mechanism. The

* The actual method of inheritance is not wholly clear. Troutman and Fulton concluded that the resistance to cucumber mosaic virus was polygenic. Holmes (1960) got similar experimental results with tobacco mosaic virus to what they got with cucumber mosaic virus, and concluded that probably two or more genes, segregating independently, confer the resistance.

four are cucumber mosaic virus, tobacco mosaic virus, tobacco streak virus, and tobacco necrosis virus. Troutman and Fulton found that keeping the tobacco plants of T.I.245 in the dark for 24 hr. before inoculating them greatly reduced their resistance. This effect of a pre-inoculation dark period was markedly greater on T.I.245 than on a control variety.

The evidence of Troutman and Fulton and of Holmes is then that the tobacco variety T.I.245 has resistance to several viruses, determined by the same genes and operating in the same way.

18.3. The Commonness of Resistance

The evidence for general resistance against viruses that van der Plank (1949) used was the great resistance of perennial plants that have achieved ecological dominance in their communities. We need not consider this now, because the evidence of Troutman and Fulton and of Holmes is neater. But let us consider evidence for general resistance not just against viruses but against fungi and bacteria as well.

Resistance is the rule in plants, and susceptibility the exception. For example, the potato is attacked by a large number of fungi, bacteria, and viruses. But there are vastly more fungi, bacteria, and viruses that do not attack it. The potato is resistant to all the rust fungi, for example. Let us consider them.

It is possible that during its evolution the potato was protected by a mass of genes for resistance, each one protecting against separate species or even races of rust. But the potato left home and traveled to most parts of the world. It still remained resistant to the rusts, even to the new contacts it made. It is highly improbable that the potato started its travels with separate genes for resistance to all the rusts it was still to meet. The easier explanation is that there is something about a potato that makes it inhospitable to rust fungi. That is to say, it has general resistance to rust fungi. This does not necessarily imply that one single genetic mechanism gives protection against the whole group. But if there are more mechanisms than one, they must nevertheless be general enough to give wide protection.

The suggestion that there may be more mechanisms than one against a wide group of pathogens such as the rust fungi must be made for, say, wheat. Wheat is susceptible to three species of *Puccinia*. But it is resistant to all the great number of other species and other genera and has remained resistant to them during its travels around the world.

18.4. The Need to Stress Affinities

The work of Troutman and Fulton and of Holmes may be one of the great landmarks in the progress of plant pathology. Biffen (1905) showed that resistance to stripe rust in wheat was inherited as a simple Mendelian character. This gave a Mendelian basis to vertical resistance. Indirectly, but nevertheless inevitably, the stress fell on difference: differences within species, within subspecies, within races. The work of Troutman and Fulton and of Holmes could lead to stress on affinity, not difference, between pathogens—to trying to find how species or even genera agree in their parasitism rather than to find how biotypes differ from one another.

The distinction between vertical resistance, on the one hand, and horizontal resistance, which includes general resistance, on the other, brings in far more than a distinction in definition and mathematical approach. It brings in a fundamental distinction in strategy. The strategy of vertical resistance is to match each change in the pathogen—each new race—with a change in variety of the host plant. The strategy of horizontal resistance is to stop the pathogen from gaining any advantage from a change of race—to make changes of race irrelevant to control. Concentration on vertical resistance has brought the complaint that pathogens are shifty enemies. The purpose of horizontal resistance is to give the answer: Let them shift.

General resistance is the widest possible horizontal resistance. A strategy of general resistance would be the grandest strategy of all.

18.5. The Search for General Resistance

It will be asked why general resistance against pathogens has not been described more often. The answer has probably been given in the previous section. One does not look both ways at once. A generation schooled to admire (and rightly admire, for the work is part of the finest fabric of plant pathology) the detection of 20,000 biotypes of *Ustilago maydis* is not schooled to seek affinities between species or between genera in their parasitic behavior. General resistance has not been found more often probably because it has not been looked for.

The literature does not lack hints that general resistance occurs. In dent corn, for example, there appears to be a positive correlation between resistance against leaf blight caused by *Helminthosporium turcicum* and resistance against leaf blight caused by *Xanthomonas stewartii* (Ullstrup, 1961). The one pathogen is a fungus dispersed by spores, the other a bacterium transmitted by insects. We are reminded by this to keep an open mind and not to assume *a priori* that general resistance is likely to be found only against pathogens that are closely related or similarly dispersed.

The test for general resistance is concomitant inheritance of resistance against two or more species. Statistical analysis can assess whether concomitant inheritance is likely to be due to general resistance and not just to linkage between genes.

CHAPTER 19

The Choice of Type of Resistance

SUMMARY

Vertical resistance is the usual choice of breeders nowadays, because it is relatively easy to use. But vertical resistance should in general be preferred only when there has been a record of varietal stability or when disease multiplies slowly. Horizontal resistance should be preferred when there has been a record of varietal change because of the disease, when *rt* is high, when there is danger of a Vertifolia effect, when there is danger of the disease being used as a weapon in war, or when it is desired to make fungicides more effective.

19.1. The Aims of Plant Breeding

All resistance to plant disease is valuable, except vertical resistance when it brings with it a Vertifolia effect. But this does not mean that all possible resistance should be bred into new varieties.

In the final analysis plant breeding has only two aims: to achieve the highest yield, and to achieve the highest quality. All else is subsidiary.

Disease reduces yield and quality. So breeding for resistance against disease is necessary. But breeding for resistance can conflict with the primary aims. Sometimes resistance is linked with poor yield or quality. Undesirable linkages of this sort have often been broken, and resistance incorporated without apparent detriment to yield and quality. But in a general sense breeding for resistance always conflicts with the primary aims. Other things being equal, the larger the breeding project, the greater is the chance of improving yield and quality. Breeding for resistance in effect reduces the size of the breeding project. If nine-tenths of the lines have to be discarded for susceptibility, the project is reduced by anything down to one-tenth, the actual reduction depending largely on how soon the susceptible lines can be discarded.

There is therefore an optimal amount of attention that can be given to disease in plant breeding—neither too little nor too much.

With many crops the need for resistance has been the major need for plant breeding. Resistance has dominated the project. Stevens and Scott (1950) found that in the central corn belt of the United States oat varieties had a useful life of 5 years, because of disease. They estimated that a new oat variety would be needed every 4 or 5 years to counter changes in the races of stem rust and crown rust. Obviously, oat breeders were obsessed with the need for resistance.

But even here one must be cautious about the interpretation. What was happening to oat varieties in the corn belt in 1950 can be taken to show the importance of resistance in plant breeding. More properly it shows the importance of the right sort of resistance in plant breeding. The violent changes of variety were the direct consequence of using vertical resistance. Had horizontal resistance been used to control stem rust and crown rust from the start, it is possible that the oat breeders in 1950 would have been happily carrying on with the primary tasks of improving yield and quality, with no more than balanced attention to disease.

19.2. The Difficulty about Generalizing

This chapter tries to give a guide to the choice of strategy in breeding against plant disease. But generalization is inevitably difficult. For example, sanitation theory suggests that relying on vertical resistance to control stem rust in wheat is more hazardous in Manitoba than in California. The relative value of vertical and horizontal resistance necessarily varies with the circumstances. No hard and fast universal choice can be made.

19.3. The Usual Preference for Vertical Resistance

There is almost no need to recommend vertical resistance. It is the sort of resistance breeders usually choose when they can.

Vertical resistance is relatively easy to work with. Differences between susceptible and resistant plants are commonly clear, and can be seen early, often in young seedlings and on a single leaf. Unwanted lines are quickly discarded. Resistance is often inherited in a known Mendelian way, and the breeding project can be planned ahead on a satisfactory genetic basis. Even in the ordinary technique of field plot experiments the dice are loaded in favor of vertical resistance, as we shall see in Chapter 23.

One can, therefore, understand that in default of a powerful recommendation for horizontal resistance it is vertical resistance that will be used.

A recent book on the principles of plant breeding illustrates this. It has three chapters on breeding for resistance against disease. So strong is current prejudice in favor of vertical resistance that the chapters are written as though vertical resistance were the only form of resistance useful for plant breeders. There is little in the chapters to suggest that horizontal resistance even exists.

We need not therefore waste much time praising vertical resistance, even though it often merits praise.

19.4. The Case for Vertical Resistance

There are two situations in which vertical resistance is the indicated method of control.

1. Vertical resistance can be, and probably should be, used when it has a good record of varietal stability to back it up. Good records are common. We discussed one example in Chapter 17: fusarium wilt of tomato.

The principle here is that, though pathogens are on the whole shifty enemies, there are many exceptions for one reason or another.

2. Vertical resistance is satisfactory when the disease multiplies and spreads slowly or for only a short time. Wart disease of potatoes caused by *Synchytrium endobioticum* is an example of disease that spreads slowly. It has been controlled very satisfactorily by planting vertically resistant varieties. New races have appeared, but the spread of the disease is so slow that there have been no explosive epidemics.

The pathogen shifts too slowly to get out of control.

A case for using vertical resistance against "simple interest" disease was made in Chapter 13.

19.5. The Possibility of Improving Vertical Resistance

Much of the history of vertical resistance has been of new destructive races of pathogens and rapid changes of varieties. One cannot assume that history will always repeat itself. There have been several suggestions for improvement, which can be regarded with hope or skepticism according to one's temperament.

1. More and more genes for vertical resistance are being added to varieties as they become available. This is the policy against stem rust in wheat in the United States and Canada. There is a chance that one can eventually exhaust the pathogen's ability for adapting itself to attack new varieties of wheat.

2. Virulence in the pathogen tends to increase step by step to match a step-by-step increase in the host's resistance. If one could get the host two or more steps ahead, the pathogen might not be able to catch up easily. In Australia, Watson and Singh (1952) proposed adding genes for resistance against wheat stem rust in pairs.

3. Multilines were suggested by Jensen (1952) for oats and are being developed in Mexico by Borlaug (1959) for wheat. By backcrossing, Borlaug developed lines of wheat that are similar in height, maturity, etc., but differ in genes for vertical resistance against stem rust. The lines are mixed together to give a composite variety that is uniform in agronomic and milling quality, but diverse in resistance. Mixtures of lines can be changed as races change in prevalence.

The three methods contradict one another to some extent. Method 3 contradicts method 2 by giving the pathogen the maximum opportunity for stepwise increases of complexity. Methods 1 and 3 have divergent aims. Method 1 aims at getting all available resistance genes into a variety. Method 3 aims at using these same genes but dispersing them so that no one variety (line) has them all.

To these suggestions we would add a fourth, applicable only to epidemics that follow a regular course from start to finish. The suggestion is that genes for vertical resistance should be added to varieties where the epidemic finishes but not to varieties where it starts. This suggestion, designed to make use of stabilizing selection in favor of simple races on the simple varieties where the epidemic starts, was discussed in Section 17.8 and need not be discussed further here.

19.6. The Choice of Horizontal Resistance when There Has Been a Record of Varietal Change

It is characteristic of vertical resistance that varieties of the host lose their resistance when the pathogen changes to a race that can attack them. A quick change of varieties because they lost resistance shows that protection was based on vertical resistance, or, to put it more correctly, it shows that horizontal resistance was inadequate.

Puccinia graminis tritici is a shifty enemy. In retrospect it seems unfortunate that when the breeding of new wheat varieties against stem rust began in earnest after the 1904 and 1916 epidemics in North America, it was vertical resistance that was used, and not horizontal resistance. If it was unfortunate, it was also understandable. As was pointed out earlier, vertical resistance has an immediate appeal to breeders which horizontal resistance has not. Indeed, the fact that it was chosen shows that vertical resistance was thought to be the better type. It is easy now to be wise after the event. From the work of Stakman and Piemeisel (1917a,b) and Levine and Stakman (1918), it was known at the time that races of the stem rust fungus occurred. But the full extent of the stem rust fungus' shiftiness was not realized. What is less understandable is that later, as its shiftiness became glaringly apparent, knowledge of this shiftiness was used to intensify and support breeding for vertical resistance—the opposite strategy from what one would have expected; the strategy that allows the enemy to reap the maximum advantage from shifting.

The real misfortune about most wheat varieties is not their vertical resistance, but their lack of enough horizontal resistance. Yet, as we saw in Chapter 16, there is plenty of horizontal resistance, if only it were used. Had it been used enough, stem rust in wheat might have been as unimportant now as rust in corn.

Consider what horizontal resistance has done to rust in corn (*Zea mays*). Rust is caused by *Puccinia sorghi*. Corn has considerable field resistance to it. We interpret field resistance to mean horizontal resistance, for the same reason as we interpreted field resistance in potatoes to blight to mean horizontal resistance—there is no evidence of change in the resistance of varieties relative to one another. (See Section 14.4.) The inheritance of this field resistance against corn rust is probably polygenic (Stevenson and Jones, 1953), in contrast with that of vertical resistance to the rust which is monogenic (Mains, 1931; Russell and Hooker, 1959; Hooker and Russell, 1962).

The field resistance to rust is so great that many strains of corn survive natural and artificial epidemics almost unscathed. They survive without vertical resistance or with vertical resistance to only a few of the common races. The results of Hooker and le Roux (1957) bring this out. Of 85 corn strains from Iowa and Wisconsin tested at Wisconsin, 16 had high horizontal resistance as shown by high field resistance. Only 1 had vertical resistance, and that was against no more than 2 out of the 15 isolates of *P. sorghi* used. In Iowa, 160 strains of corn had high resistance in the field; of these, only 35 had vertical resistance, and this resistance was not against all the isolates of *P. sorghi*.

Vertical resistance clearly plays relatively little part in protecting corn against rust. Indeed, in tests of over 300 corn strains from various countries, Hooker and le Roux found only one variety, from Peru, that had vertical resistance to all the races of the corn rust fungus they used. Most had none at all.

Corn farmers are fortunate that the resistance of their crop to rust is horizontal. The resistance has done what all good resistance should do—it has kept the disease out of the news.

19.7. The Choice of Horizontal Resistance when *rt* Is High

Some epidemics start from small amounts of initial inoculum, but end with a destructively high level of disease. The infection rate is high or multiplication of disease carries on over a long time. The smaller the amount of initial inoculum that can cause a destructive epidemic, the more it is indicated that horizontal resistance should be used.

Reasons against using vertical resistance, such as sanitation, when *rt* is high were given in Sections 13.6 to 13.8.

19.8. The Choice of Horizontal Resistance when There Is Danger of a Vertifolia Effect

If new varieties are selected while they are being protected against disease by vertical resistance, it is ordinarily not possible to ensure that enough horizontal resistance is present as well. As we saw in Chapter 14 even the relatively small vertical resistance against potato blight conferred by the gene R_1 allowed much horizontal resistance to be dispersed and lost in late maturing German potato varieties. The greater vertical resistance of Vertifolia, by virtue of its genes R_3 and R_4, allowed even more to be lost.

One must be cautious about using vertical resistance when there is a danger of a Vertifolia effect.

The danger of a Vertifolia effect depends on how much horizontal resistance there is at the start. If there has been no previous selection for horizontal resistance, if there has been just a random assortment of genes in relation to the disease, there can be no Vertifolia effect. But if horizontal resistance was previously selected for, it is likely to be dissipated progressively in new varieties bred under protection of vertical resistance.

An essential feature of the Vertifolia effect is that horizontal resistance was the usual form of resistance in the past. Selection for vertical resistance is mostly a recent practice. When, at the start of the era of breeding for vertical resistance, breeders took over commercial varieties and used them as parental material, they had in their care whatever horizontal resistance farmers had gathered through the ages.

The old potato variety Epicure has vertical resistance to potato viruses A and X. There are other examples.

But most vertical resistance is recent. When, for example, wheat breeding against stem rust was started in the United States by Federal and State Departments of Agriculture under the spur of the 1904 and 1916 epidemics, bread wheat was not protected by vertical resistance. It is a matter of history that a long search had to be made before bread wheats were found with "resistance," i.e., vertical resistance. Vertical resistance against stem rust in bread wheat seems to have been even more uncommon 50 years ago than vertical resistance against rust in corn is now.

It is not difficult to see why vertical resistance was not used much until modern times. Individual breeders without the backing of government or other institutions, farmers choosing varieties for their own use, and amateurs could not isolate and maintain pure cultures of different races. This is the foundation of breeding for vertical resistance. They had not the glasshouses and necessary paraphernalia for breeding for vertical resistance; they observed in the field, and field resistance tends to be horizontal resistance. Nor was individual effort unbacked by public funds likely to cope with the varietal somersaults that breeding for vertical resistance often leads to.

How much horizontal resistance was gathered by farmers and inherited by professional plant breeders and how much has since been dissipated through lack of selection while new varieties were bred under protection of vertical resistance, we shall not know until appropriate tests are made.

We need not concern ourselves much with the past except to guide us in the future. One point is clear. Where breeding for vertical resistance was not undertaken before, vertically resistant varieties should be introduced only with the greatest caution.

Consider corn rust again. Corn has been grown in conditions favorable to rust for millennia. A great mass of horizontal resistance has been accumulated: enough to get rust rated as a relatively unimportant disease. Work was started in the United States in 1951 with the stated object of finding vertical resistance against corn rust for breeding purposes. There is no harm in vertical resistance, provided that positive steps are taken to conserve the great horizontal resistance that corn now has against rust. But blindly to follow the path against corn rust that was followed against the rusts of wheat and oats half a century earlier could be disastrous.

What can happen when horizontal resistance is dissipated (though the reason was not a Vertifolia effect) is shown by the history of tropical rust of corn in Africa. Tropical rust caused by *Puccinia polysora* is much like common rust caused by *Puccinia sorghi*, except that it needs higher temperatures. It, too, is mainly controlled by horizontal resistance, which in the Western Hemisphere is so effective that *P. polysora* was not identified on corn until just over 20 years ago (Cummins, 1941), such is the insignificance of the disease. In 1949, or shortly before, *P. polysora* crossed the Atlantic Ocean, found corn in Africa with its resistance dissipated after more than four centuries out of contact, and started a great epidemic. Corn was killed before it could ripen, and the epidemic swept across tropical Africa. The sequel is also relevant to what has been said: it shows farmers as selectors. By 1954 the epidemic began to abate, largely because farmers grew the most resistant varieties they could (Cammack, 1961), and resistance accumulated.

19.9. The Choice of Horizontal Resistance when There Is Danger of Biological Warfare

Vertical resistance is resistance to some races of a pathogen but not to others. One of the others could be introduced by enemy action. This was the subject of Chapter 16. Wheat stem rust was used as an example from the many that could have been chosen.

Horizontal resistance is the indicated defense. Sense of responsibility will see that it is used.

19.10. The Choice of Horizontal Resistance when Fungicides Are Used

Horizontal resistance improves the performance of fungicides. This is discussed in Chapter 21. There has, for example, recently been research on using nickel salts and dithiocarbamate fungicides for controlling rusts of cereals and sunflower. Against some of the diseases, especially leaf rust of wheat, the use of fungicides seems to be on the brink of success. The deciding factor may well turn out to be the horizontal resistance of the host.

It is just as important and relevant to the use of fungicides to increase the resistance of the host as to search for better chemicals. Ideally, fungicide research and breeding for resistance should both be part of the same research project. The need for coordination may not seem pressing when the crop is of a type that gives a high gross return of money per acre. But coordination is essential when it is hoped to control foliage disease of cereal crops by fungicides.

CHAPTER 20

The Quantitative Effect of Horizontal Resistance

SUMMARY

For this chapter it is necessary to return to the topics of Chapters 5 to 8.

Horizontal resistance manifests itself, according to circumstances, as a decrease in the basic infection rate R or R_c, an increase in the latent period p, or a decrease in the infectious period i as a result of removals. A decrease of R or i has its greatest effect relatively when pR is small; an increase of p has its greatest effect relatively when pR is great.

These conclusions hold for pathogens that multiply in the sense that Chapter 4 uses the verb, i.e., for the "compound interest" type of disease. For "simple interest" diseases only R is relevant to the horizontal resistance of a crop during a single season. On the whole, a decrease of R is less able to decrease damage by "simple interest" than by "compound interest" epidemics.

20.1. The Components of Horizontal Resistance

In Chapter 14 we discussed how horizontal resistance to potato blight manifests itself:

1. Plants resist infection. When plants are treated with the same number of spores, fewer lesions are formed on plants of a resistant variety.

2. Sporulation is less abundant in lesions on a resistant variety than on a susceptible one.

3. From the time of inoculation it takes longer for sporulation to start on a resistant variety than on a susceptible one.

4. Infected tissue ceases to be infectious sooner. It is sooner removed, to use a general word. In blight lesions on potato leaves an infertile zone follows behind the sporulating zone and removes infected tissue from further part in the epidemic.

Resistance in forms 1 and 2 reduces R, the basic infection rate. In form 3 it increases the latent period p. In form 4 it reduces i, the period during which tissue is infectious. Some equations relating the logarithmic and apparent infection rates, r_l and r, to R, p, and i were given in Chapters 5, 6, and 8.

This chapter analyzes some effects of changes in R, p, and i brought about by resistance.

20.2. Changing the Basic Infection Rate

Consider first the effect of a change in R on r_l, when other things are equal. To keep the analysis simple we ignore removals. This sort of analysis applies fairly well to the cereal rusts.

Tropical corn rust caused by *Puccinia polysora* has been studied in Africa (Anonymous, 1958; Cammack, 1961). Pustules develop, and under favorable conditions start releasing uredospores 9 days after inoculation, i.e., $p = 9$ days. They continue releasing spores for another 18 to 20 days, i.e., $i = 18$ to 20 days. With such a high value of i, one can, for our present purpose, neglect removals and apply Eq. (5.7) instead of Eq. (8.5).

Each pustule liberates from 1500 to 2000 spores a day under favorable conditions if the variety is susceptible, and from 600 to 1150 spores if it is resistant. As a figure for numerical calculation we shall assume that resistance reduces spore production from 1750 to 875; it halves it.

In tests with single-spore inoculations, 15% of the spores established lesions on susceptible varieties, but only 1.9% on resistant varieties. We assume these figures for calculation.

On these figures combined, the ratio of R for susceptible varieties to R for resistant varieties is as 1750×15 is to 875×1.9. That is, R for susceptible varieties is 16 times as great as for resistant varieties.

If r_l was 0.5 per unit per day in the susceptible variety, it would be 0.26 per unit per day in the resistant variety. The following are the details of the calculation: Because $p = 9$ days, $pr_l = 4.5$ per unit in the susceptible variety. From Table 5 in the Appendix, this corresponds with $pR = 405.1$ per unit. For the resistant variety $pR = 405.1/16 = 25.32$ per unit. This corresponds with $pr_l = 2.37$ per unit. Hence $r_l = 2.37/9 = 0.26$ per unit per day in the resistant variety.

If r_l was 0.4 per unit per day in the susceptible variety, it would be 0.18 per unit per day in the resistant; if 0.3 in the susceptible, 0.11 in the resistant; if 0.2 in the susceptible, 0.05 in the resistant.

The effect of resistance thus varies with the speed of the epidemic, which in turn varies with the weather, the age of the plants, etc. Resistance that reduces R to 1/16 of the value it has in a susceptible variety would (approximately) halve r_l if this were 0.5 per unit per day in the susceptible variety, but would reduce r_l to one-quarter if this were 0.2 per unit per day in the susceptible variety.

For another example, consider the statement of Stakman and Christensen, quoted in Chapter 16, that in wheat varieties with generalized (horizontal) resistance to stem rust the number of pustules may be 1/5, or 1/10, or even 1/20 of those in susceptible varieties. For calculation take the middle figure, and assume that susceptible varieties form 10 times as many pustules as resistant varieties form from the same amount of inoculum. If the resistant varieties have no other advantage, R for the susceptible varieties is 10 times as high as for resistant varieties. As a fair figure for "rust weather" take $p = 8$ days. Then resistance would reduce r_l from 0.5 per unit per day in the susceptible varieties to 0.278 per unit per day in the resistant; from 0.4, to 0.199; and from 0.3, to 0.124.

The reduction is fairly uniform. Written correct to two digits, it is 0.22, 0.20, and 0.18 per unit per day when r_l for the susceptible variety is 0.5, 0.4, and 0.3 per unit per day, respectively. For calculation we shall use 0.4 per unit per day for the susceptible variety. It is quite close to the value calculated from Asai's (1960) data.

20.3. A Return to Tables 16.3 and 16.4

Table 16.3, in Section 16.8, was concerned with how resistance increases the number of uredospores needed to start an epidemic which reaches a given intensity after 90 days.

It makes it simpler if we answer the problem in two steps.

First, consider pustules, not uredospores, as the initial inoculum. By the method in the exercise at the end of Chapter 16, reducing r by 0.0256 per unit per day increases tenfold the initial number of pustules needed if the epidemic is to reach some fixed level of infection after 90 days.

One can apply the results in the preceding section on the assumption that r retains its logarithmic value r_l. This assumption was discussed in detail in earlier chapters. For wheat stem rust it has the experimental backing of the data in Figs. 4.3 and 4.4., which should be interpreted with allowance for the "simple interest" phase of increase analyzed in Exercise 2 at the end of Chapter 5.

Resistance that reduces the proportion of spores forming pustules to one-tenth lowers r from 0.4 to 0.199, a reduction of 0.201 per unit per day.

This increases the numbers of pustules needed from X to $10^{7.85} X$, which is 71 million X. (Here $7.85 = 0.201/0.0256$.)

The second step is to change from pustules to uredospores. By hypothesis, it takes 10 times as many uredospores to form a pustule on the resistant variety as on the susceptible variety. So one must multiply the answer in the first step by 10. The final answer is then 710 million X uredospores.

In drawing up Tables 16.1 and 16.2 it was not possible to take the second step, because the nature of the horizontal resistance was not specified. Therefore, as was pointed out in comment 2 on the exercise at the end of Chapter 16, Tables 16.1 and 16.2 underestimate the value of resistance.

Other answers for Table 16.3 follow the same pattern as the answer, 710 million X.

Table 16.3 is concerned with the extent to which the initial number of uredospores would have to be increased if resistant replaced susceptible varieties, in order that the epidemic would reach any given level of disease 90 days after inoculation. That is the wartime problem. The peacetime question is different. With the same initial number of uredospores, by how long would resistance delay the reaching of that final level of disease?

It is again easier to do the calculation in two steps.

First, assume the initial and final levels of disease, x_1 and x_2, to be fixed. Hence by Eqs. (3.5) or (3.6), the product $r(t_2 - t_1)$ is fixed. From the previous section, resistance that reduces the proportion of spores forming pustules to one-tenth lowers r_l from 0.4 to 0.199 per unit per day. Hence (again on the assumption that r retains its logarithmic value) the duration of the epidemic would be extended from 90 days to $(90 \times 0.4)/0.199 = 181$ days. The first step puts the delay at $181 - 90 = 91$ days.

As the second step, allow for the fact that the same initial of number of uredospores causes only one-tenth the initial number of pustules in the resistant variety. That is, x_1 for the resistant variety is one-tenth of x_1 for the susceptible variety. So we have to add the time disease takes to increase tenfold in the resistant variety. Because x_1 is small, this time is $(2.30/0.199) \log_{10} 10 = 12$ days.

The total delay is $91 + 12 = 103$ days. This is the value entered in Table 16.4.

20.4. Relative Insensitivity of the Apparent Infection Rate to Change when the Basic Rate Is High

It was noticed, in Section 20.2, with tropical rust of corn that reducing R to $R/16$ reduces r_l relatively less when R is large.

This can be seen in the four graphs of Fig. 20.1, calculated from

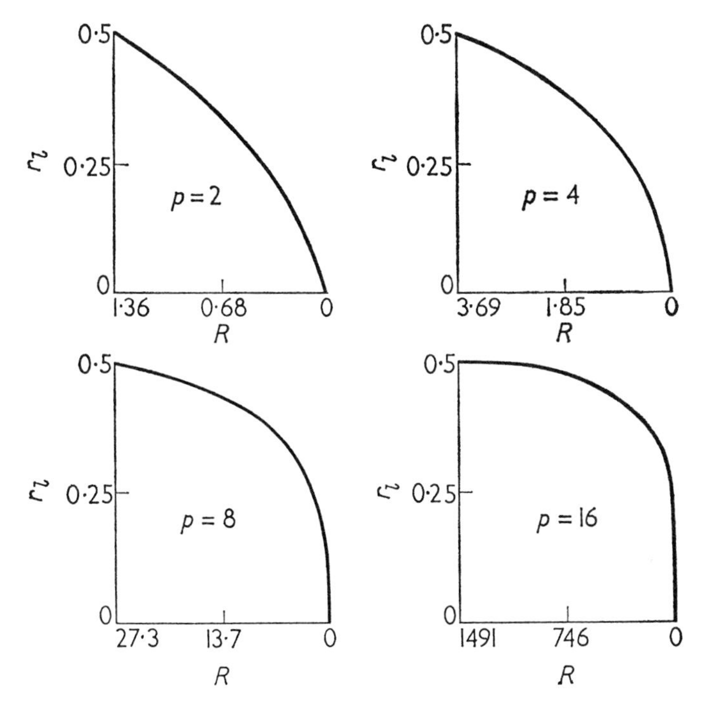

FIG. 20.1. The effect of R on r_l. The four graphs are for four different values of p.

Eq. (5.7). Because of the latent period p, the relation between r_l and R is not linear. When pR is great, r_l changes slowly as R changes. When pR is small, r_l changes quickly.

Resistance is only one of the factors that affect R. Weather, for example, is another. But whenever pR is great, whatever the reason may be, r_l is less sensitive to changes in resistance that affect R.

20.5. The Effect of a Change of the Latent Period on the Apparent Infection Rate

An increase in p has been studied in relation to potato blight by various workers mentioned in Section 14.16.

As a rule, enough resistance has been found in varieties to increase p by about 1 day, from, say, 4 days to 5 days. In "blight weather" very large differences have not been observed.

Table 20.1 gives some results of Lapwood (1961b) for the time taken

TABLE 20.1

VARIETAL DIFFERENCES IN THE LATENT PERIOD OF POTATO BLIGHT[a]

	Days after inoculation:				
Variety	3	4	5	6	7
Up-to-date	0	12	23	24	—
King Edward	0	8	22	24	—
Majestic	0	14	21	24	—
Ackersegen	0	0	18	23	24
Ås	0	1	13	20	23
Ontario	0	3	18	22	24
Average susceptible	0	11	22	24	—
Average resistant	0	1	16	22	24

[a] Data of Lapwood (1961b) for droplet inoculations. Figures are the number of discs out of 24 that were sporing.

for blight sporangia to form on discs cut from inoculated potato leaves. The varieties Up-to-Date, King Edward, and Majestic are varieties with little or relatively little horizontal resistance. Ackersegen, Ås, and Ontario have greater resistance.

It is not easy to assess how much p can be increased by resistance alone. Great differences in p in some diseases have been reported for other reasons. For example, with stripe rust of wheat caused by *Puccinia striiformis* p varies from 11 days at optimal temperatures to nearly 120 days in winter (Zadoks, 1961). But great differences between varieties grown side by side in the same conditions do not seem to have been reported.

With stem rust of wheat caused by *Puccinia graminis*, Hart (1931) found that the epidermal membrane of the variety Webster is so strong that the fungus often cannot rupture it. The collenchyma is then crushed by the mass of unreleased uredospores. Subepidermal pustules are common in Webster, both in the leaf sheath and in the peduncle. Very often at least 60% of the pustules of the culm never break the epidermis. Spores are formed in abundance, but they remain within

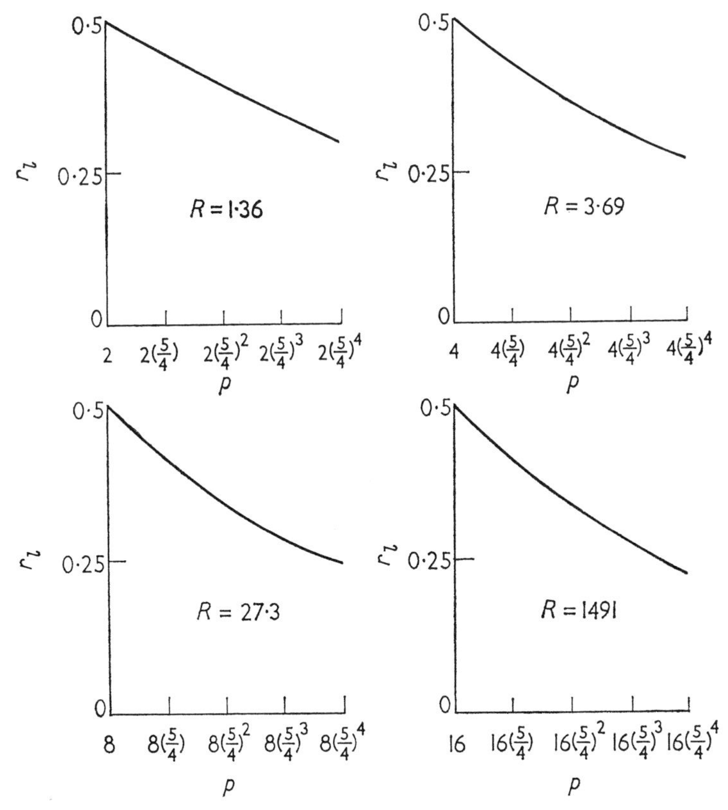

FIG. 20.2. The effect of p on r_l. The four graphs are for different values of R.

the host. Insofar as a tough membrane retards the release of spores, it increases p. But when the pustules remain subepidermal, one must assign this to a reduction of R. That is, other things being equal, R is reduced 60% in Webster if 60% of the pustules fail to break the epidermis.

Figure 20.2 shows how a change of p affects r_l. The four graphs are for different values of R. The effect of p is greatest when R is great. That is,

r_l is most sensitive to a change in p when pR is great. This is in contrast to changes of R, because, as we have seen, r_l is least sensitive to a change in R when pR is great. Resistance obtained from increasing p is at its best when it is needed most.

It may seem anomalous that if pR is high enough to cause r_l to be high, r_l may be reduced by making pR higher still. The anomaly comes from a change in the unit of time. R is measured in conventional units: per unit per day or per year. But pR makes p the unit of time; pR measures R per unit per p. Similarly, pr_l measures r_l per unit per p.

20.6. The Effect of Removals on the Apparent Infection Rate

It will be remembered from Chapter 8 that Lapwood marked the progress of the sporing zone of lesions of potato blight. In susceptible varieties in wet weather the zone was continuous from day to day. The zone was 1 day's growth wide; $i = 1$ day. Up-to-Date was one of the susceptible varieties he used. In Table 20.2 it is compared with Arran Viking, a moderately resistant variety. For the data of 1958, a "blight

TABLE 20.2

WIDTH OF THE SPORING ZONE OF POTATO BLIGHT LESIONS IN TWO VARIETIES AT THREE DIFFERENT HEIGHTS IN THE CANOPY. THE WEATHER WAS PARTICULARLY WET IN 1958[a]

Year and position in the canopy	Width of sporing zone (mm.) in: Up-to-Date	Arran Viking
1957		
Upper	2.7	1.6
Middle	3.9	2.6
Lower	4.4	3.0
1958		
Upper	4.6	3.5
Middle	5.0	4.2
Lower	5.5	4.5

[a] Data of Lapwood (1961b). Figures are the means of 26 days in 1957 and 12 days in 1958.

year", the zone was about one-fifth less for Arran Viking than for Up-to-Date. The rate of advance of the lesion through the leaf tissue was about

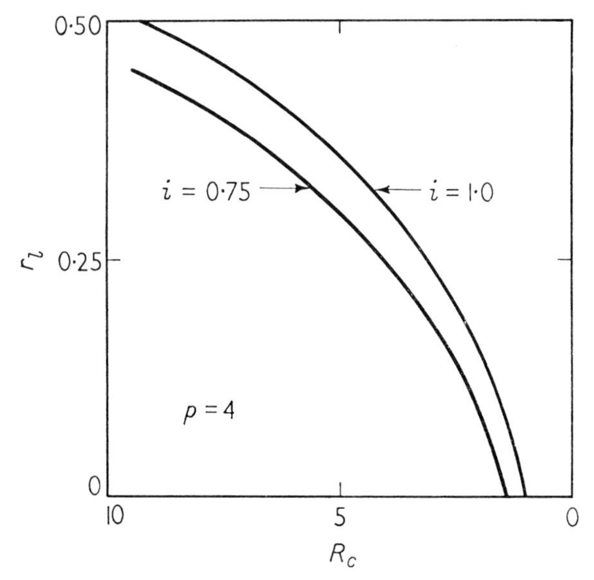

FIG. 20.3. The relation between R_c, r_l, and i, determined by Eq. (8.5) with $p = 4$.

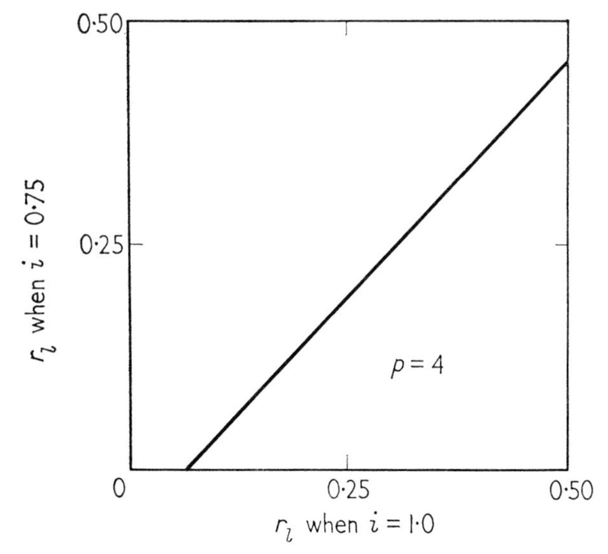

FIG. 20.4. The relation between r_l when $i = 0.75$ and r_l when $i = 1.0$, with $p = 4$.

the same in the two varieties: when $i = 1$ day for Up-to-Date, then $i = 0.8$ day for Arran Viking. There are varieties more resistant than Arran Viking; and, assessing the data of Lapwood (1961c) as best we can, we take $i = 0.75$ day for a resistant variety in "blight weather" to compare with $i = 1$ day for a susceptible variety.

Figure 20.3 shows how reducing R_c reduces r_l, when $i = 1.0$ and when $i = 0.75$. The calculations use Eq. (8.5). Figure 20.4 shows the relation between r_l when $i = 0.75$ and r_l when $i = 1.0$. For both figures $p = 4$. With days as units of time, these figures are apt for potato blight.

The effect of changing i from 1.0 to 0.75 day is at its highest relatively when r_l is small. Changing i to this extent reduces r_l from 0.5 to 0.447 or from 0.1 to 0.037 per unit per day. The first of these reductions of r_l is by about one-ninth, the second by about two-thirds.

20.7. The Effect of Resistance

To discuss resistance it is useful to adopt the classification of disease in Chapter 4. There are "compound interest" diseases in which the pathogen "multiplies": it passes from plant to plant or from lesion to lesion during the life of its host. For them we plot $\log [x/(1-x)]$ against time. Then there are "simple interest" diseases in which the pathogen does not "multiply": it does not move in appreciable amounts from plant to plant or from lesion to lesion during the life of its host. For them we plot $\log [1/(1-x)]$ against time.

Against the compound interest type of disease, horizontal resistance affects x_0, R, p, and i. The combined effect on R, p, and i is calculated as an effect on r. For notes on the effect of horizontal resistance on x_0, see Section 10.3.

Against the simple interest type of disease, horizontal resistance is simpler. It affects R. Neither p nor i is relevant to behavior during a single season or lifetime of the host plants. Even R is restricted in relevance. Only resistance to infection matters.

To plan a strategy one wants to know what effect resistance has. With disease of the compound interest type it is convenient to calculate the effect of resistance as a delay in reaching any particular level of infection, i.e., as an increase in the duration of the epidemic. According to Eq. (3.5) and (3.6), the duration $t_2 - t_1$ is inversely proportional to r. This was the method used earlier in this chapter, in Section 20.3, to calculate data for Table 16.4.

With the simple interest disease the effect can be calculated directly as the effect on R defined by Eq. (4.2). If there is no disease at the beginning of the season (no disease then is usual), the equation means that at any time the effect of resistance on log $[1/(1-x)]$ is proportional to its effect on R. At any time log $[1/(1-x)]$ is proportional to R.

One can determine whether any particular amount of resistance is adequate only in relation to particular problems. In general, horizontal resistance goes further with the compound interest type of disease than with the simple interest type. Consider, e.g., resistance that reduces R to one-fifth. With wheat stem rust, in the circumstances postulated for Tables 16.3 and 16.4, this is enough for a vast effect on the final amount of disease. With a simple interest disease the resistance would reduce x from, say, 0.2 to 0.044, a fairly satisfactory reduction, or from 0.99 to 0.60, an inadequate reduction. Resistance that reduces R to one-fifth could stop a severe compound interest epidemic, but not a severe simple interest epidemic, from being destructive. More about this was said in Section 13.8.

20.8. The General Simplification

In this chapter we have simplified the discussion, e.g., we have used Eq. (5.7) instead of Eq. (6.2). In particular, we have taken r to be constant. In actual fact the infection rate of an epidemic moving from field to field or from State to State will vary from time to time. But the use of some particular constant value of r—a high value to represent a fast rate and a low value to represent a slow rate—is legitimate heuristic procedure. The complicated computations needed to follow the course of some particular epidemic in detail will be justified only when there are accurate experimental data available on which to base the computations and when there is some particular precise purpose for which the computations are needed. There is no reason now to believe that, when such a purpose arises and when the data are available, difficulties in computation will be insurmountable.

CHAPTER 21

Control of Disease by Fungicides

SUMMARY

In chemical control of plant disease the action of the chemical on the pathogen—fungicidal action in the limited sense of the term—is only part of the story. From spraying experiments one can estimate the proportion of spores prevented by the fungicide from infecting the host; against potato blight in the field four, five, or six sprays with copper fungicides, zineb, or maneb during the season destroy an average of about 3 spores out of 4. The effect on the infection rate of any given degree of destruction, say, of 3 spores out of 4, depends on the weather or other environmental factors, and on the horizontal resistance of the host.

Against diseases of fruit, vegetable, and other crops that give high gross returns of money per acre, fungicidal inefficiency can be overcome by spraying frequently. But much of our food still comes from fields unprotected against foliage diseases by fungicides. High costs of production cannot be absorbed; and the use of chemicals will depend greatly on how far fungicidal action can be enhanced by putting horizontal resistance into the host plants. To control foliage diseases of cereal and other low-priced crops by fungicides, the strategy must be to stress the breeding of new varieties for horizontal resistance as much as to search for new and cheaper chemicals and for new and better techniques of applying them.

21.1. The Fungicide Square

In a numerical example in Section 8.7 it was calculated that if a fungicide destroyed 6/7 of the spores of *Phytophthora infestans* before they could establish infection this would be enough to reduce r_l from 0.44 to 0 per unit per day, i.e., to stop the epidemic altogether. To give similar control of wheat stem rust the fungicide would have to destroy 359/360 of the spores of *Puccinia graminis*. The reason is that, for the same value of r_l, potato blight and wheat stem rust differ greatly in the basic infection rate, the latent period, and the infectious period. But for the moment we are more concerned with the general principle that very different degrees of fungicidal activity are needed to control different diseases: that fungicidal activity, in the broad sense of the term, is far more than the activity of a chemical against a fungus.

Consider the matter in another way. Suppose a fungicide destroys 9/10 of the spores of *P. graminis* before they establish infection. If $p = 10$ days and there are no removals, the fungicide would reduce r_l from 0.4 to 0.23 per unit per day. This is equivalent to reducing R from 21.8 to 2.18 per unit per day, in accordance with our hypothesis that the fungicide destroys 9/10 of the spores.

Now suppose the wheat variety is changed to one with more horizontal resistance. Suppose the resistance reduces by 9/10 the proportion of spores that germinate and establish lesions. This would reduce r_l from 0.4 to 0.23 per unit per day, corresponding, as before, to a change in R from 21.8 to 2.18 per unit per day. These figures are for varieties unprotected by fungicides.

In effect, there is no difference between the action of the fungicide on a variety without resistance and the action of resistance on a variety without fungicidal protection.

Consider the susceptible variety without fungicidal protection. Suppose the weather turns drier, and this reduces by 9/10 the proportion of spores that germinate and establish lesions. This would reduce r_l from 0.4 to 0.23 per unit per day, as before by reducing R from 21.8 to 2.18 per unit per day.

In effect, drier weather would behave like a fungicide or resistance.

Now suppose that the resistant variety is treated with a fungicide. Suppose that the fungicide and resistance each reduce by 9/10 the proportion of spores that germinate and establish lesions. Together (if one assumes no interaction between fungicide and resistant host) they reduce the proportion by 99/100. This would reduce r_l from 0.4 to 0.089 per unit per day, corresponding with a reduction of R from 21.8 to 0.218 per unit per day.

So, too, a combination of the fungicide and drier weather (to the extent used in the previous examples) would reduce r_l in the susceptible variety from 0.4 to 0.089 per unit per day. (One assumes here that there is no interaction between weather and fungicide, e.g., that rain does not affect the fungicide deposit.) So, too, a combination of drier weather and a resistant variety unprotected by a fungicide would reduce r_l from 0.4 to 0.089 per unit per day.

In these examples one cannot distinguish between the action of fungicide, resistance, and dry weather. This is well recognized in practice. We say that the weather was dry enough during the season to reduce the need for some sprays. Or we say that a variety needs less spraying because it is more resistant.

When we think of a disease we should think of the disease triangle: host, pathogen, environment. When we think of fungicidal action, we should think of the fungicide square: host, pathogen, environment, fungicide. This somewhat arbitrarily separates the fungicide from the environment, but it is convenient to do so.

We talk of a square, not a quadrilateral, because there is nothing to show that one side is greater than another. It depends on the perspective. The fungicide specialist thinks the fungicide side the greatest. The plant breeder thinks the host side the greatest. Each specialist sees the side nearest him the greatest.

If fungicides were highly efficient, there would be no useful purpose in thinking of a fungicide square. If, e.g., a fungicide destroyed 99.999% of the spores before they could establish infection, one could concentrate on fungicides and ignore other means of control.

But fungicides are not always highly efficient. An analysis of experimental results often leaves one with the impression of inefficiency rather than efficiency.

The greatest mass of appropriate experimental data are for copper fungicides, zineb, and maneb against *Phytophthora infestans* on potato foliage.

21.2. Laboratory and Glasshouse Experiments with *Phytophthora infestans*

McCallan and Wellman (1943) sprayed tomato plants in a glasshouse with fungicides and then inoculated them with spores of *P. infestans*. After a few days they counted the number of lesions that had formed. They found that standard copper fungicides destroyed up to about 99% of the spores before they could initiate lesions. The amount of fungicide per leaf area was about as large as would be used in the field; and a combination of air pressure from the spray gun, compound turntable, and air currents produced by hood draft and gravity tended to give uniform coverage and deposit on the leaves. One can, therefore, take a 99% kill as about the highest one could expect from copper fungicides. McCallan's (1958) results confirm this.

Björling and Sellgren (1957) sprayed detached potato leaves with Bordeaux mixture or zineb and, when the fungicide deposit was dry, inoculated them with spores of *P. infestans*. After a few days they counted the number of lesions.

If one groups all their tests with Bordeaux mixture together, there were 12,732 lesions on the unsprayed leaves to compare with 1193 lesions on the sprayed leaves. Bordeaux mixture reduced infection to roughly one-tenth. As an average this figure probably flatters the fungicide, because the best results were obtained with droplet sizes smaller than those produced by most commercial spraying machines, and because most of the doses of fungicide were higher than would be given in the field. (The highest dose was equivalent to from 33 to 108 kg. of copper, as Cu, per hectare of field, i.e., from 29 to 97 lb. per acre.) Nevertheless the best treatments, which included redistribution of the fungicide deposit by artificial rain, seem to agree with those of McCallan and Wellman.

Björling and Sellgren's results with zineb cannot be used for direct calculation. But the control of infection was poor, even with good doses of the fungicide. Indirect calculation suggests that the kill with zineb was less than 9 spores out of 10.

21.3. The Performance of Fungicides against Potato Blight in the Field

Figure 8.5 reproduced Hooker's (1956) results in Iowa on the progress of blight epidemics in plots of potatoes sprayed with various fungicides. Spraying stopped the epidemic for a while. We calculated in Chapter 8 that to stop the epidemic the fungicide had to destroy 6 spores out of 7 before they could initiate infection. Maneb was somewhat better than zineb or copper.

There are many other data that suggest that Hooker's results are typical of what one can expect in a good spraying experiment, with adequate sprays applied frequently. The greatest amount of evidence comes from the Netherlands, where the progress of potato blight in experimental plots and fields is recorded year by year as a matter of routine, and published. It is unnecessary to reproduce all the evidence here, but a few selected examples will be discussed.

Figure 21.1 summarizes data (Anonymous, 1954) for the progress of blight in the variety Bintje in the sand areas of the Netherlands in 1953. The data for the unsprayed fields are the combined data for 29 fields; those for the sprayed fields are for 31 fields. The sprayed fields are divided into those sprayed 1 or 2 times with fungicides and those sprayed from 3 to 5 times. The sprays were mostly with copper oxychloride. The Netherlands records were converted into percentages by the conversion graph of Cox and Large (1960). Records of less than 0.1% of disease are excluded. $\text{Log}_e\,[x/(1-x)]$ is plotted against time in days, day 0 being June 29.

The regression coefficient of $\log_e [x/(1-x)]$ on time in days is r per unit per day. (See Exercises 8 and 9 at the end of Chapter 3.) The regression coefficients, estimated in the usual way, give $r = 0.42$ per unit per day for untreated fields and $r = 0.35$ and 0.082 per unit per day for fields treated 1 or 2 times and from 3 to 5 times, respectively.

We assume these values of r to apply to the phase of logarithmic increase. Then, if $p = 4$ days and $i = 1$ day, Eq. (8.5) gives $R_c = 6.57$ per unit per day for the untreated fields, and 4.81 per unit per day for the fields treated once or twice. Spraying fields once or twice reduced

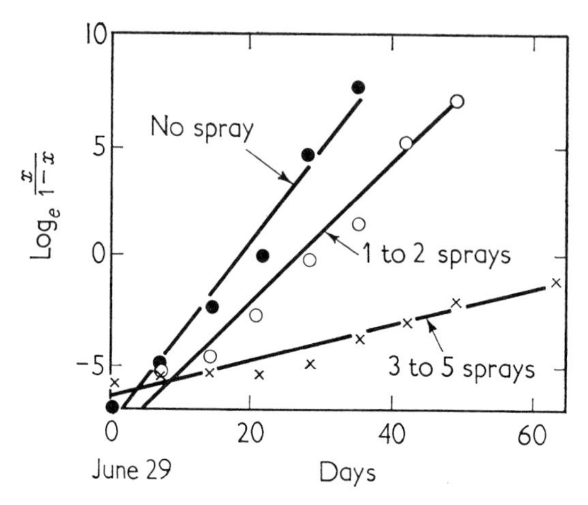

FIG. 21.1. The progress of blight in 29 unsprayed and 31 sprayed fields of the susceptible potato variety Bintje. Data from the Netherlands (Anonymous, 1954).

R_c from 6.57 to 4.81, a reduction of 27%. That is, spraying once or twice reduced the number of spores that were able to initiate infection by 27%. Spraying destroyed about 1 spore in 4 before it could start an infection, and let 3 spores out of 4 through. Clearly one or two sprays were inadequate to do much good at the time the records were taken. This is also evident directly from the graph without any detailed analysis.

Spraying from 3 to 5 times reduced R_c to 1.45 per unit per day. It destroyed about four-fifths of the spores before they could start infections, and let about 1 spore in 5 through.

The data in Fig. 21.1 have the special merit that they were derived from whole fields. When one is dealing with small experimental plots in a conventional design, one is never certain how many spores travel from unsprayed to sprayed plots next to one another. This is the topic of

Chapter 23. But normally one has no choice, and must use data from plots. Some data from plots follow.

Figures 21.2 and 21.3 illustrate good control by fungicides. There are many examples of poor fungicidal action, because the plots were sprayed too early or too infrequently. But they are not relevant to our inquiry, which is to find out what can be expected from the frequent and efficient use of fungicides.

From Figs. 21.2 and 21.3 all records of less than 0.1% or more than 90% disease have been excluded.

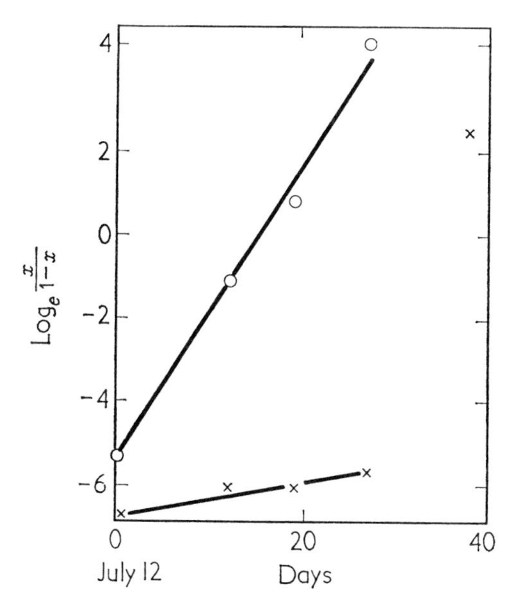

FIG. 21.2. The progress of blight in sprayed and unsprayed plots of the susceptible potato variety Bintje. The fungicide was copper oxychloride. Data of de Lint and Meijers (1959).

Figure 21.2 is drawn from the data of de Lint and Meijers (1959) for a spray trial with the potato variety Bintje. The fungicide was copper oxychloride. For the first 27 days after July 12 control was good; but then blight in the sprayed plots increased fast. This ultimate increase is discussed in Section 23.7. We are concerned now with the earlier period of good control.

From results analyzed in the same way as before, with $p = 4$ and $i = 1$ day, spraying reduced R_c from 4.81 to 1.20 per unit per day. It destroyed 3 spores out of 4 before they could initiate infection.

The writer analyzed the published results of seventeen potato blight trials, which met this double requirement: the sprays were applied at least as frequently as 4 times during the season, and spraying reduced infection significantly at the 1% level of significance. Copper fungicides were used in all the trials, and on an average destroyed 3 spores out of 4 before they could initiate infection. Zineb was used in many of the trials and maneb in some. Results with these fungicides did not differ significantly from those with copper.

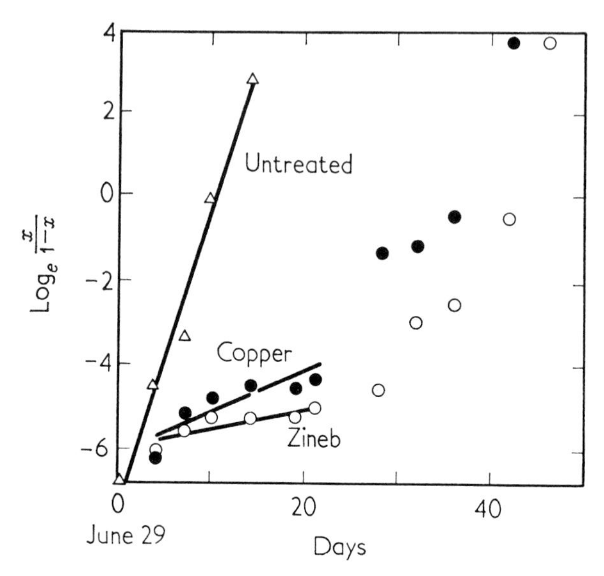

FIG. 21.3. The progress of blight in sprayed and unsprayed plots of the potato variety Bintje. Data from the Netherlands (Anonymous, 1954).

Of the seventeen trials, the trial (Anonymous, 1954) on which Fig. 21.3 is based gave the greatest apparent degree of control. Copper oxychloride apparently destroyed 14/15, and zineb 17/18, of the spores before they could initiate infection. But one must be cautious about how one interprets a selected example. Analysis shows that the destruction was not significantly greater than 5 spores out of 6, either by copper oxychloride or zineb. This is for the 5% level of significance.*

One may achieve a 99% kill of spores in a laboratory or glasshouse experiment, but the figure has no relation to results in the field. Why so many spores escape the fungicide barrier in the field is not known with certainty. Presumably the escape of 1 spore in 4 means that one-quarter

* In the unsprayed plots, $r = 0.70$ per unit per day, which is one of the highest values met. But r was not significantly greater than 0.50 per unit per day.

of the spores reach parts of the surface not adequately covered with fungicide. This would tally with Large and Taylor's (1953) finding that in sprayed fields only from 40 to 70% of the upper surface, and still less of the lower surface, of leaves is covered by copper in amounts detectable by prints on paper impregnated with rubeanic acid. Hirst and Stedman (1962) found even less coverage.

The poor showing of fungicides in the field must also be related to the speed with which the fungus infects, another reminder of the fungicide square. *Phytophthora infestans* does not dawdle on the leaf's surface, waiting to be killed. In "blight weather" it takes only from 2½ to 4½ hr. from the time a sporangium settles on the leaf to the time the germ tubes of the zoospores establish infection within the leaf (Cox and Large, 1960). One arrives at a slightly longer, but not very different, estimate from the results of Lapwood (1962) and Mooi (1962).

21.4. Citrus Black Spot and Apple Scab

It is not our purpose to suggest that the proportion of spores killed is always small for all diseases.

The control of citrus black spot caused by *Guignardia citricarpa* is an example of efficient control.

The disease has been studied by Kiely (1950), Wager (1953), Kotze (1963) and others. Young fruits, up to 4 months old, are infected by ascospores released from infected fallen leaves. Spots show many months later. No complicated analysis of data seems necessary, and we take the number of spots to be proportional to the number of spores that escape being killed.

Spraying starts when half the petals fall. A second and third spray are given at 6-week intervals. During the 4 months in which the fungicide protects, the young fruits, especially those of grapefruit, swell to many times their original size. Yet the three copper sprays (Bordeaux mixture, copper oxychloride, or other fixed coppers) at the dilute strength of 0.0625 to 0.075% Cu destroy up to 98% of the spores.

The fungus infects slowly; and one surmises either that the fungus is unusually sensitive to copper, or that it stays long enough on the fruit's surface to be caught by fungicide redistributed by rain and dew and long enough to absorb enough copper to kill itself.

Another disease that contrasts with potato blight is apple scab, caused by *Venturia inaequalis*. Two differences can be mentioned.

Against potato blight, at least in many countries, fungicides are used simply to delay an epidemic, without trying to hold disease down right until the end of the season; against apple scab, fungicides are used to keep fruit healthy right until the end.

With potato blight, the infectious period i of leaf tissue is short. With apple scab, lesions continue to form spores for three weeks or more if the weather is suitable; i is large. Because i is large, disease can be controlled only if R_c is kept small.

Some fungicides are very effective against scab, if they are applied often enough at strengths great enough. Two of them are known to act in at least three ways. Captan and dodine form a protective barrier against the infection of healthy tissue. On infected tissue they reduce the number of spores formed in lesions (Albert and Lewis, 1962). And the spores that do get formed germinate poorly even when they are transferred to a suitable medium, free from fungicide (Albert and Lewis, 1962; Heuberger and Jones, 1962). In all three ways, these fungicides reduce R_c.

21.5. The Calculated Effect of Horizontal Resistance or Adverse Climate on the Degree of Fungicidal Activity Needed

Table 21.1 shows what percentage of spores must be destroyed by a fungicide to stop an epidemic from continuing to develop. By destroyed we mean that they must be prevented from germinating and starting a

TABLE 21.1

THE CALCULATED PERCENTAGE OF SPORES A FUNGICIDE MUST DESTROY IN ORDER TO STOP AN EPIDEMIC[a]

r_l per unit per day	Destruction needed (%) when:	
	$i = 1.0$ day	$i = 0.75$ day
0.5	89	—
0.4	83	—
0.3	74	65
0.2	59	46
0.1	36	15

[a] $p = 4$ days

lesion. The table is designed for potato blight. We take $p = 4$ days throughout, and $i = 1.0$ day. For lower values of r_l, such as would occur in resistant varieties or less humid weather, values are also given for $i = 0.75$ day. This conforms with Lapwood's (1961b) finding that the width of the sporing zone of blight lesions is less in resistant varieties or less humid weather. We assume an epidemic to stop when $iR_c < 1$. (See Chapter 8.)

The table shows that a fungicide must destroy 8 spores out of 9 to stop a fast epidemic that mounts at the rate $r_l = 0.5$ per unit per day in unsprayed fields. But it need destroy only 1 spore in 3, if $i = 1$ day, or 1 spore in 7, if $i = 0.75$ day, when $r_l = 0.1$ per unit per day in unsprayed fields.

On the evidence at present available, there is no need to distinguish here between the effect of horizontal resistance in the host and adverse weather.

21.6. Other Calculated Effects

Table 21.1 is meant to represent fungicidal activity on potato blight. In Fig. 21.4, curve A represents activity against potato blight, and curves

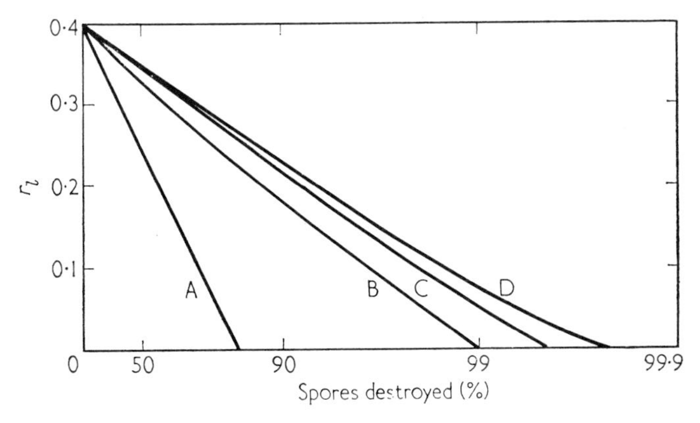

FIG. 21.4. The effect on r_l of the percentage of spores destroyed by a fungicide. Without fungicidal treatment, $r_l = 0.4$ per unit per day. For curve A, $p = 4$ and $i = 1$ day; for curve B, $p = 8$ and $i = 10$ days; for curve C, $p = 10$ and $i = 10$ days; and for curve D, $p = 10$ and $i = 20$ days.

B, C, and D activity against cereal rusts. The percentage of spores destroyed is on a logarithmic scale, which gives a fairly simple relation between r_l and spore destruction. (The abscissae are proportional to minus the logarithm of the proportion of surviving spores.)

The curves bring out the essential difference between potato blight and the cereal rusts. For any given percentage of spores destroyed, blight is more effectively controlled than rust. This is not a direct comparison of fungicidal activity, because it says nothing about the relative difficulty of destroying spores of the different fungi. Figure 21.4 measures factors other than the direct action of the chemical on the fungus.

Curves B, C, and D show the effect of varying p and i. It appears from the literature that p for the cereal rusts can vary from 8 days, or somewhat less in "rust weather", to 10 days and more when conditions are not optimal for infection. For i, no precise general figure can be given. (See Chapter 8 for evidence.) The three curves B, C, and D reflect this variation and probable differences among the cereal rusts. They are meant to apply to "rust weather".

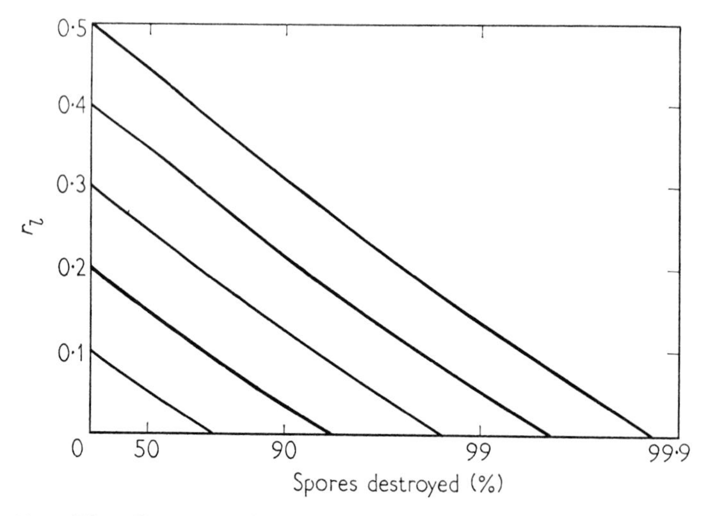

FIG. 21.5. The effect on r_l of the percentage of spores destroyed by a fungicide. For all curves, $p = 10$ and $i = 10$ days. The five curves represent conditions in which, without fungicidal treatment, r_l is 0.5, 0.4, 0.3, 0.2, and 0.1 per unit per day, respectively.

A high value of p associated with some given value of r_l makes fungicidal control more difficult. All forms of horizontal resistance that reduce r make control by fungicides easier. But, of all the forms, increasing p is the least compatible with the use of fungicides. This links with the fact, discussed in the previous chapter, that horizontal resistance from an increase of p is most effective when pR is great.

The lower the value of i, the more effective are fungicides. We are concerned here only with the removal of infectious tissue in the lesions. One cannot generalize with certainty about other forms of removal without a detailed mathematical analysis, but it seems likely that most forms of removals will enhance control by fungicides.

Figure 21.5 has the same purpose as Table 21.1. The curves show how difficult it is to control disease by fungicides when r_l is great, either because the host is very susceptible or because the weather favors the disease.

21.7. Variable Results with Fungicides and the Need for Recording *r*

Variable results with fungicides are often reported, especially when it is touch and go whether there will be control. The recent results with nickel salts for the control of cereal rusts are examples.

Consider leaf rust of wheat caused by *Puccinia recondita.* Rowell (1959) records that in 8 years of observations of Marquis wheat at St. Paul, Minnesota, the time taken for leaf rust to increase from 10 to 50% varied from 4 to 27 days. This means that r varied from 0.55 to 0.08 per unit per day. In the year of fastest increase, r was faster than for the top curve of Fig. 21.5. Adequate control of leaf rust by fungicides in such a year would be difficult. In the year of slowest increase, r was slower than for the bottom curve of Fig. 21.5. If a fungicide is to have any prospect at all of commercial use against leaf rust, it should be very effective in such a year.

Results of fungicide trials against cereal rusts (and to a varying extent other diseases) lose much of their meaning unless one can get an idea of the infection rate. A general correlation with severity of infection in the control plots is too vague. Disease can be severe even with a slow infection rate if inoculum arrives early and abundantly. It can be mild even with a fast rate if inoculum arrives late and in small amounts.

21.8. The Timing of Applications of Protectant Fungicides

It is commonly believed, and often emphatically asserted, that spraying with protectants must start before inoculum arrives. The belief is founded on experience and is supported by theory.

First, there is a measure of inherent stability about the infection rate. It is usually harder to stop an epidemic when once it has started at a fast rate. The rate determines the proportion of infected tissue that is still in the latent stage, the proportion that is infectious, and the proportion that has passed the infectious stage and is removed from the epidemic. (Some figures were given in Section 7.3 on the age of infected tissue at different infection rates.) With a fast rate the proportion that is removed is small. A protectant applied for the first time after an epidemic has started at a fast rate has more inoculum to cope with than a protectant applied when the rate has previously been reduced by earlier applications. It takes time to slow the epidemic down and to establish an age structure of the lesions compatible with a slow rate. In Hooker's (1956) experiment with potato blight, discussed in Section 8.7, it took about 2 weeks after spraying started on July 18 to bring the rate down. But if spraying starts before the inoculum first arrives, the rate is curbed from the beginning.

Second, if spraying starts before the inoculum first arrives, this usually means that more applications of the protectant will be given during the season. The rate is curbed over a longer period, and disease is less likely to reach a harmful level by a given date.

But when the rate is likely to be low even in unsprayed fields (when, e.g., the variety has some horizontal resistance) and it is desired to apply the protectant only a few times during the season, a late start may be warranted to avoid wasting materials by applying them long before the inoculum arrives. De Lint and Meijers (1959) tested the timing of fungicidal sprays against potato blight. They assumed that a farmer would spray a moderately resistant variety such as Voran only once, and found that he would get the best result if he waited until he saw blight in his fields before he put on this one spray (of copper oxychloride).

21.9. Eradicant Fungicides in Relation to the Infection Rate

Tense defines fungicides usefully. An eradicant acts against infections of the past; a protectant acts against infections of the future, against disease still to come.

It is convenient to think of an epidemic as starting anew after eradication. The effect of the eradicant is then to reduce the initial inoculum x_0 of this new epidemic. The problem becomes one of sanitation, and one can apply the principles set out in Chapters 11 and 13.

The efficiency of eradicants, just as that of protectants, is heightened by a slow infection rate. By Eq. (11.1) or (11.2), halving r doubles the effect of any given degree of eradication, when we measure effect as a delay in the progress of an epidemic. Halving r doubles the time the pathogen needs to make up for what it lost during eradication.

It does not matter what reduces r. Horizontal resistance of the host and weather adverse to disease, if they affect r equally, equally increase the benefit from a given degree of eradication. The fungicide square is as real for eradicants as for protectants.

The spread of disease—the movement of spores, for example—affects the benefits of eradication even more than those of protection. There is little purpose in eradication if inoculum floods back from sources nearby. Eradicants, by themselves, are most likely to be useful when inoculum moves slowly or when there are few sources nearby. Eradication of soil pathogens (by fumigation, etc.) because they usually move slowly is an example of the first alternative; eradication of apple scab because all neighboring orchards are well tended is an example of the other. But the swift spread of pathogens like *Puccinia* spp. may prove to be a major obstacle to efficient eradication.

Against the invasion of inoculum from without, as well as against the increase of inoculum from within, a slow infection rate is a safeguard. Eradicants applied to a horizontally resistant variety are likely to be more effective than applied to a susceptible variety, whether or not the field is sheltered from reinfection from outside.

21.10. The Fungicide Frontier

Great progress has been made with fungicides. As late as 25 years ago there were few fungicides except Bordeaux mixture and other copper compounds, sulfur and lime sulfur, formalin, and mercurial seed dressings. Now the range of fungicides is wide.

There has not been a corresponding widening of frontiers. Half a century ago we were spraying apples against scab, using copper and sulfur against downy and powdery mildew in grapevines, protecting potatoes against blight, and guarding many fruit, vegetable, and ornamental crops against various diseases. We still protect these crops. With the new chemicals, we protect them better now than before. And we can now protect some crops against diseases that were not easily controlled by the chemicals we had before. Nevertheless the bulk of our food still comes, directly and indirectly, from fields and pastures with yields commonly reduced by foliage diseases against which no fungicides are used. Here lies the fungicide frontier that has not changed much for many years.

Essentially the difficulty of extending this frontier is one of cost. Chemicals that control rust in wheat, for example, have long been known; sulfur or zineb is effective if applied often. The problem is to cheapen the treatment.

Progress has recently been made against the cereal rusts. Peturson *et al.* (1958) controlled leaf rust in the field with nickel compounds, which seem both to eradicate and to protect. (These workers also usefully state the increase of this rust in untreated plots: from 4 to 84% in 21 days, which estimates $r = 0.23$ per unit per day.) More attention has lately been given to dithiocarbamates and nickel sulfate. With four applications per season, Forsyth and Peturson (1960) got good control of leaf rust but not of stem rust with zineb and nickel sulfate. Others have, however, had less promising results.

21.11. The Forgotten Factor

New, cheaper, and better fungicides and therapeutants will be found. With better weather forecasts and better knowledge of the effect of climate on disease, spraying and dusting with fungicides will be better timed. All this will improve chemical control of disease.

But we need not wait for this. Better chemical control will come from better horizontal resistance in the host plants. This follows from the fungicide square.

There is nothing new or profound in this statement. It is general experience that it is easier to control disease with fungicides in a moderately resistant variety than in a very susceptible variety. Pathologists take this relation between fungicides and resistance for granted, as self-evident. It has often just been forgotten to make use of it in projects for chemical control.

One can understand why the relation has been forgotten. Fungicides had to be adapted to the host plants. Few farmers would rip out apple orchards in order to replace them with varieties better suited to the available fungicides. But the notion that the fungicide must fit the host is no unalterable law; and if we wish to extend the fungicide frontier we must start thinking of fitting the host to the fungicide.

We need not (except in matters of phytotoxicity) think of any specific fungicide. Horizontal resistance in the host will help all fungicides: sprays or dusts, protectants or eradicants, superficial or internal, local of systemic.

If the strategy of control of, say, the foliage diseases of wheat is to use fungicides, then, along with the search for better chemicals and better timing of their applications, new varieties are needed with horizontal resistance not just against one or other rust, but against all relevant diseases.

It is possible that in the process of breeding for resistance the need for fungicides will fall away. No farmer will complain about this.

CHAPTER 22

How Disease Spreads as It Increases

SUMMARY

At first, early in an epidemic, disease is mostly confined in clear foci close to the initial inoculum. Later, spread becomes general, and the pathogen moves from field to field. At this time gradients flatten and the distance between given levels of disease increases. These changes accompany an increase in the amount of disease. The quicker the increase, the sooner the pathogen begins to spread widely. The pattern of spread is much the same for different diseases and is related to the whole disease triangle rather than just to how individual propagules disperse.

22.1. Increase and Spread of Disease

As disease increases it also spreads over a wider area. It seems easier to relate measures of control, except through quarantine and exclusion, to increase rather than to spread. For this reason we have been concerned primarily with increase. But this and the next chapter deal mainly with spread.

Chapter 7 discussed foci: areas of disease more concentrated than the average. Disease that "multiplies"—the "compound interest" type of disease—does not increase uniformly within a field or large area. It increases and spreads in foci of various sizes and conspicuousness. In Chapter 7 we were concerned to show that this method of increase does not invalidate our equations. In the present chapter we discuss foci as units of spread of disease.

Most pathogens seem to disperse in several ways. To repeat examples from Chapter 7, it is believed that *Puccinia striiformis*, the cause of stripe rust in wheat, spreads by the rubbing together of infected and healthy leaves, by water dispersal, and through the air. *Puccinia graminis* and *Phytophthora infestans* are dispersed by water and through the air.

Cacao swollen shoot virus is spread by crawling mealy bugs and by mealy bugs blown about by wind. All methods of dispersal enlarge existing foci, but only some of them are apt for taking the pathogen over long distances to start new foci.

The general pattern of spread seems to be much the same for different diseases. At first, when the epidemic S is only just beginning, disease seems to be mainly confined to focal outbreaks. As the epidemic develops, individual foci become less conspicuous. The epidemic seems to pass from a stage of focal outbreaks to one of general disease, though the stage of general disease is probably no more than one of numerous small foci, closely packed, and less conspicuous.

22.2. The Spread of Wheat Stem Rust

Lambert (1929) described the epidemic process for stem rust in wheat. At first there are single pustules on leaves or culms. In the second "generation" (presumably p later) there are several fresh pustules near the old pustule, mostly on the same plant but with possibly one or two on adjacent plants. In the third generation each plant within a 2-ft. radius has several pustules, usually on the side of the culm facing the old infection. In the fourth generation the foci start coalescing, but are usually still discernible. Thereafter the epidemic becomes fairly general throughout the field, with foci no longer clearly discernible.

Schmitt *et al.* (1959) studied the spread of stem rust from artificial foci in winter wheat in Maryland. On April 20, rusted plants were transplanted in a field. On May 2, 13 days later, some pustules were found, mostly within 2 ft. of the transplanted focus. On May 23, about $2p$ later, nearly all obvious infection was confined within a radius of 20 to 40 ft. from the focus. Thereafter disease multiplied fast far from the focus.

Bromfield *et al.* (1959) and Underwood *et al.* (1959) also studied how stem rust spreads in winter wheat in Maryland. A plot 80 by 90 ft., the initial focus, was inoculated with uredospores on April 23. The plants were then from 8 to 10 in. tall and tillering. The first pustules were therefore mainly on the four lower leaves of the plants. They erupted on May 3 and 4. The next pustules appeared on May 17, p later. Many were on stem tissues and higher leaf blades. The plants were then from 32 to 42 in. tall and just past flowering. Toward the end of May (from $2p$ to $3p$ after the first pustules erupted) pustules appeared in plots 1800 ft. from the initial focus, and increased fast. Although this dispersal over a long distance first became obvious toward the end of May, evidence was found that it had, in fact, started as early as May 8 and 9, from the first pustules on the lower leaves of the plants. Both abundance of spores and their release higher on the plant favor the spread of disease, but neither is essential for spread over long distances. There is no abrupt change. Dispersal over short distances to enlarge existing foci and over long distances to establish new foci occurs simultaneously.

22.3. The Spread of Potato Blight

Potato blight follows much the same focal pattern. At first blight is found more or less restricted to foci surrounding the initial sources of inoculum. Then the epidemic becomes general, and the fungus spreads from field to field (Hirst, 1955; Hirst and Stedman, 1960a).

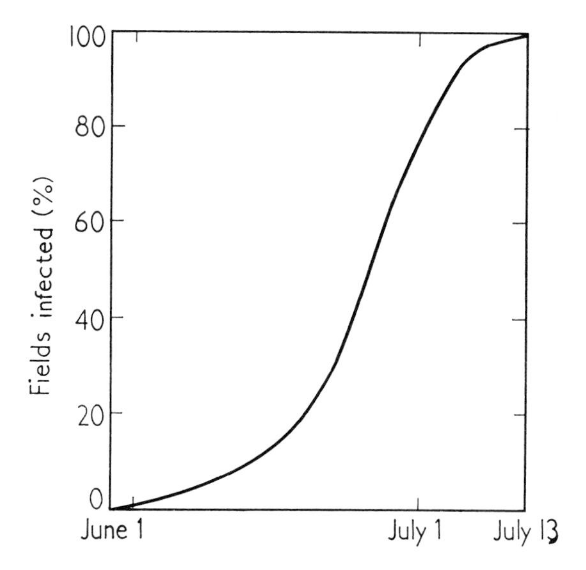

FIG. 22.1. The percentage of potato fields in which blight had been found at different dates. The data are for the variety Bintje in the sand areas of the Netherlands in 1953 (Anonymous, 1954).

When once the fungus starts to spread from field to field it spreads quickly. Figure 22.1 shows the percentage of fields infected at different dates. A field was considered to be infected as soon as a trace of blight was first found in it. The data are for the susceptible variety Bintje in the Netherlands (Anonymous, 1954). Similar data for other varieties are also available (Anonymous, 1954; de Lint and Meijers, 1956). A few fields were infected by the end of May. By the end of June, 75% of the fields, and by the middle of July all the fields were infected. It is not known in how many fields infected seed had introduced initial foci. But, on van der Zaag's (1956) observation that there is one primary focus in 1 km.^{2} of potato fields, most of the fields in Fig. 22.1 must have started healthily. The steep curve thus indicates swift spread of infection from field to field.

Hirst and Stedman (1960a) inquired why there was such a dramatic change from the early phase of a seasonal epidemic with blight confined

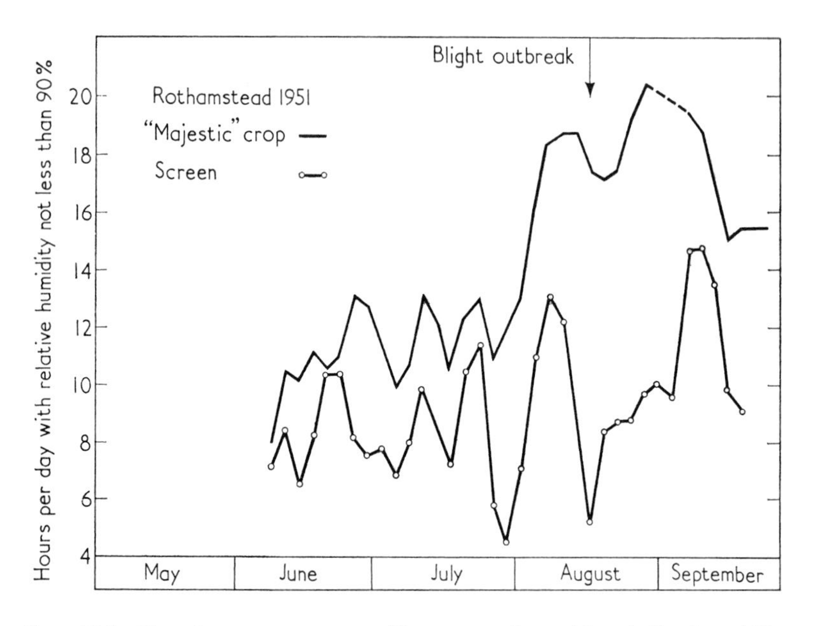

FIG. 22.2. Running mean curves of hours per day with relative humidity not less than 90%. The records in the crop of Majestic potatoes were taken level with the tops of the potato ridges. The screen records were taken in standard louvered screens 4 ft. above ground. From Hirst (1958), by kind permission.

in foci to the later phase when blight spreads rapidly and the epidemic becomes general. They found no climatic changes reflected in readings from instruments in a standard screen to account for it and decided that the change came in the ecoclimate within the crop itself. (This and the rival, or supplementary, theory of Grainger were discussed in Chapter 12.)

Figure 22.2, from Hirst (1958), shows how an ecoclimate favorable to blight develops in the variety Majestic in England. At first, when the field is young, the plants small, and much soil exposed to the sun and wind, the hours of high humidity within the foliage of the crop itself are not much greater than in a standard screen above the crop. But early in August the rows close. The foliage becomes dense and creates within itself its own humid ecoclimate largely independent of the climate above the crop. Later, when blight has destroyed 50% of the foliage and opened it again to sun and wind, the climate within the crop again becomes more like the climate above it.

Hirst and Stedman found that the change from focal outbreaks to rapid spread from field to field occurs at the time when the foliage first closes up and creates a humid ecoclimate within itself. The humid ecoclimate provides the conditions that moisture-loving *Phytophthora infestans* needs to reproduce fast.

This is the anomaly: the fast spread of *P. infestans* from field to field comes at a time when an unusually large proportion of the sporangia are enclosed in close foliage and cannot get away to move from field to field. For by all laws of diffusion the same close growth that encloses molecules of water vapor to make a humid ecoclimate will also enclose, and enclose much more efficiently, large sporangia released within this ecoclimate.

One might generalize and say that if spores are to move freely in air they must form in air that moves freely. If disease increases fast as a result of a humid ecoclimate within the crop very different from the climate above the crop, this disease necessarily festers locally to an unusually great extent.

There is another factor that applies to potato blight, though it may not apply to all diseases. Hirst (1958, 1959) has found that, though ripe sporangia are easily detached by water, they remain attached to the sporangiophores so long as the air remains humid, even though the current of humid air is fast. To detach the sporangia, the air must become drier. Thus it is that sporangia in the air above potato fields are trapped in greatest numbers during sunny spells in the forenoon. Close foliage which keeps the air within it humid and stops the sun from reaching the sporing zone also stops sporangia from being easily detached. Close foliage is likely to hinder both the release of sporangia into the air below the canopy and the movement of the sporangia through the canopy to the winds above.

If these deductions are correct it follows that a high proportion of sporangia get away into the air moving over the crop only before the foliage closes or after blight opens it up again. Before the foliage closes, the amount of disease and therefore the number of sporangia are usually relatively small. Rapid spread of disease from field to field is thus probably brought about by sporangia escaping after blight has opened the foliage up.

22.4. Wheat Stem Rust and Potato Blight Contrasted Again

It is most useful for purposes of studying how disease spreads that wheat stem rust and potato blight differ in many ways but are alike in others.

Uredospores of *Puccinia graminis* seem to be able to get away and move freely in the air. Movement of spores over a distance can be correlated with wind, temperature, and humidity above the crop (Bromfield *et al.*, 1959; Underwood *et al.*, 1959). Sporangia of *Phytophthora infestans*, on the evidence presented in the previous section, get away less freely, at least until the crop is mostly destroyed.

Dispersal of uredospores of *Puccinia graminis* conforms with Gregory's (1945) equation for dispersal in air of normal turbulence (Smith, 1961). Dispersal of sporangia of *Phytophthora infestans* conforms with the equation for dispersal in air of low turbulence (Gregory, 1961).

We use Gregory's equations in preference to those of others because we believe they fit the experimental evidence of observed gradients better. But even Gregory's equation for low atmospheric turbulence, which predicts a steep gradient of disease away from a point source, does not quite match the actual steepness of gradients of potato blight that have been observed in the field. It seems that, at a little distance from the source, the amount of blight varies inversely as the fourth, fifth, or even sixth power of the distance (van der Plank, 1960, Fig. 4). A comparison of the theories of Gregory and Schrödter and how they predict steepness of gradient are given by Schrödter (1960) and Gregory (1961).

On the basis of Gregory's equations, with *Puccinia graminis* and *Phytophthora infestans* taken to disperse at normal and low turbulence respectively, the proportion of air-borne spores that travel 1 km. or more is 10 million times greater for *Puccinia graminis* than for *Phytophthora infestans*. Of spores released near the ground and carried by air of normal turbulence, about 4% travel 1 km. or more. Of those carried by air of low turbulence, less than 4 in a billion travel 1 km. or more. The reason, on Gregory's theory, is that propagules moving in air of low turbulence are rapidly deposited on the ground and vegetation, and thereby removed from the air-borne spore cloud.

It must be interpolated that rapid depletion of the spore cloud does not mean that *Phytophthora infestans* cannot under any circumstances travel far in one move. Billions of sporangia are produced. Admittedly, the observed range of dispersal of blight directly from known sources is usually small. But it has already been remarked that pathogens commonly have several ways of dispersing; and dispersal on a low turbulence pattern over short distances does not preclude occasional long-distance dispersal on some other pattern. (Conversely, evidence that *P. infestans* can spread 50 km. must not be used to justify some theory of dispersal over short distances.)

The evidence of van der Zaag (1956) is convincing that *P. infestans* spread 11 km. over the sea to the island Rottumeroog, off the mainland of the Netherlands. Hyre (1950) trapped sporangia 14 km. from any known source. Harrison (1947) found blight in isolated seedbeds and fields of tomatoes in Florida, from 48 to 64 km. from known plantings of tomatoes and potatoes. There is evidence (Robinson, verbal communication) that when *P. infestans* first came to Kenya during the Second World War it crossed the great Rift Valley, some 50 km. wide, in a few weeks. The last example is noteworthy because practically no sporangia would be deposited between one side of the Rift Valley and the other; there would be no loss by deposition.

To return to a comparison of wheat stem rust and potato blight, although these diseases differ so much in some details they spread along much the same pattern. Focal outbreaks change to general epidemics. This change is also the pattern of many other diseases quite unlike either wheat stem rust or potato blight. This makes one believe that the pattern is not connected with one or other detail of spore dispersal, but is general for epidemic disease. The pattern follows from the increase of disease itself.

22.5. The Flattening of Gradients

Cammack (1958) artificially infected corn plants in pots with *Puccinia polysora*, the cause of tropical rust. When these plants were heavily rusted, he brought sixteen of them into a healthy corn field as a compact source of infection.

The latent period of corn rust is from 9 to 10 days in favorable weather. On the tenth day after bringing the infected plants into the field he observed how the disease had spread from them. His results are given in the lower curve of Fig. 22.3, which shows the average number of pustules per plant at varying distances from the initial source of inoculum. The curve falls sharply with the distance from the source: the gradient is steep.

On the thirtieth day (about $2p$ later) the field was again sampled, with results shown by the upper curve in Fig. 22.3. The gradient had practically disappeared over the line of observation.

A steep gradient, shown by the lower curve in Fig. 22.3, represents a small focus, clearly defined from the rest of the field in which it occurs. A flat gradient, shown by the upper curve in Fig. 22.3, represents a large focus, only vaguely defined, if it is defined at all, and grading almost imperceptibly into the rest of the field.

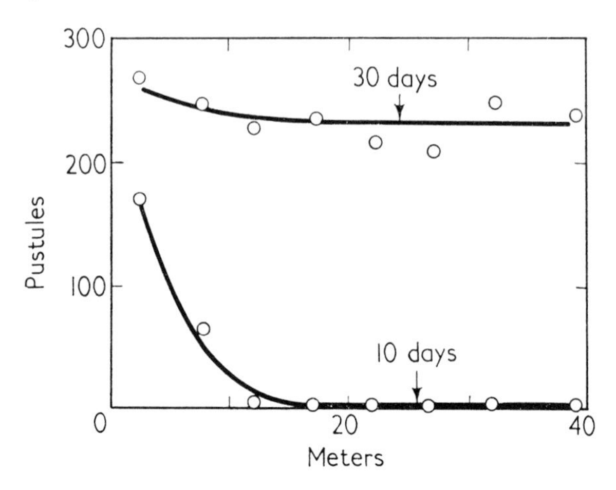

FIG. 22.3. The dispersal of *Puccinia polysora* from a point source in corn. The average number of pustules per plant is plotted against the distance from the source, 10 and 30 days after the source was introduced. The data are for dispersal in a line along the direction of the prevailing wind. Data of Cammack (1958).

The trend toward the flattening of gradients as disease multiplies seems to be quite general. The speed of the flattening in the example just discussed was possibly faster than usual. This may be because Cammack used a large initial source of inoculum and a very susceptible variety of corn in conditions favorable to rust. In Lambert's (1929) observations, already quoted, on wheat stem rust it took about $4p$ for the foci to disappear, i.e., for the gradients to flatten. Gradients of potato blight around initial point sources also flatten (Waggoner, 1952).

22.6. Increasing Disease and Increasing Scale of Distance of Spread

The lower curve in Fig. 22.4 reproduces a part of the lower curve in Fig. 22.3. The curve is for the spread of *Puccinia polysora* from sixteen heavily infected corn plants brought into a healthy corn field and grouped together.

Now suppose sixty-four instead of sixteen infected plants had been brought in and grouped together as before. With 4 times as much inoculum, 4 times as many pustules would be expected on the plants in the field. (We are dealing in Fig. 22.4 with low levels of infection, and can ignore competition between pustules for space.) The expected curve is the upper curve in Fig. 22.4, which at any distance shows 4 times as

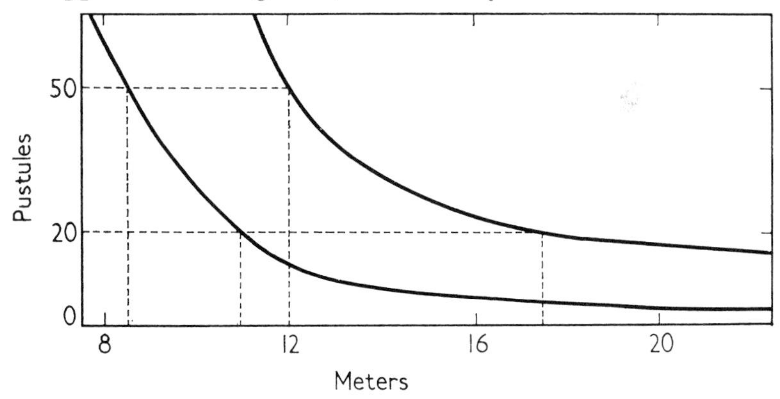

FIG. 22.4. The dispersal of *Puccinia polysora* from a point source in corn, 10 days after the source was introduced. The lower curve reproduces part of the lower curve of Fig. 22.3. The upper curve is for a hypothetical point source 4 times as strong.

many pustules as the lower curve. Both curves in Fig. 22.4 are for the tenth day (about p) after the inoculum was brought in, and therefore show direct spread, without secondary multiplication, from this inoculum.

Consider the infection levels of 50 and 20 pustules per plant. The lower curve shows there were 50 pustules per plant at 8.5 m. from the initial inoculum, and 20 pustules at 10.8 m. from it. Infection dropped from 50 to 20 pustules per plant in $10.8-8.5=2.3$ m. The upper curve shows that infection dropped from 50 to 20 pustules per plant in $17.5-12.0=5.5$ m.

We define scale of distance of spread as the distance between any two given levels of infection. In our example the two levels are 50 and 20 pustules per plant, and the distances 2.3 and 5.5 m. in the conditions represented by the lower and upper curves, respectively.

A large scale of distance of spread means that disease is spreading far. Take, for example, the distance between 95 and 5% infection. This is the distance between destruction virtually complete and destruction just starting. With stem rust there are often hundreds of miles between 95 and 5% infection. The disease is then spreading far. Around a focus of swollen shoot disease of cacao, disease may fall from 95 to 5% within a few yards. (See Fig. 7.1.) Disease is closely confined and spreads slowly.

To return to Fig. 22.4, the mere fact of more disease (in our hypothetical example, because a more powerful source of inoculum was brought into the field) ordinarily makes for a larger scale of distance of spread. It makes disease spread further.

To summarize this and the preceding section, as disease increases in amount in a field or country, it spreads over greater distance in the field or country. Increase and spread are two phenomena of the same process. Increase promotes spread in two ways: by flattening the gradient—the topic of the preceding section—and by increasing the scale of the distance of spread—the topic of this section. These two ways are distinct. (This can be confirmed by plotting the logarithm of the number of pustules instead of the number itself. In Fig. 22.3, the two curves would retain different slopes despite a change to logarithms. In Fig. 22.4, the two curves would take on the same slope, at a given distance from the source, if logarithms were used.)

22.7. The Behavior of Populations of Pathogens and of Individual Propagules; the Disease Triangle

The change from disease mostly confined in clear foci and spreading slowly to disease occurring generally and spreading fast from field to field does not imply a change in the way individual propagules move. The change comes automatically when the whole population of the pathogen increases.

It can even happen that the change to a general epidemic comes at a time when individual propagules move on an average less far than usual. There is the example of potato blight, which, it seems, becomes epidemic when the foliage closes and hinders propagules from moving.

The spread of disease is not just a matter of propagules moving. The whole disease triangle—host, pathogen, environment—is involved. The susceptibility of the host is as much a factor in spread as is the way propagules move. The temperature and humidity that determine whether a spore can infect after it has settled are as much factors as the atmospheric conditions of, say, turbulent wind that brought the spore to settle in the first place. One must beware of studying the spread of disease simply as a study of spore flight or of the dispersal of propagules generally.

One despairs at present of linking the spread of disease directly with the disease triangle. Instead, it seems less difficult to link spread of disease with multiplication of disease, and multiplication of disease with the disease triangle. The more we study multiplication the more we shall understand spread.

CHAPTER 23

The Cryptic Error in Field Experiments

SUMMARY

This chapter deals with field experiments laid out to collect information that can be applied to farming. Plots in the experiment are meant to represent farmers' fields (or orchards) receiving the same treatment as the plots receive. But plots represent fields only when the plots within an experimental area do not interfere with one another: only when spores do not move freely between differently treated plots. The representational error—the error of taking plots to represent fields when they do not—is often great. The more freely spores move the less signs they leave to show they have moved; the greatest representational errors are cryptic. Ordinary guard (border) rows cannot contain pathogens that move freely. If one wishes to use simple conventional layouts for experiments with freely moving pathogens without incurring large representational errors, one must restrict the experiment to treatments that do not differ greatly from one another in the amount of disease they allow to develop. Alternatively, if one has to compare treatments that differ greatly, special layouts must be used. Some are suggested.

23.1. Errors of Representation in Results from Plots in a Field Experiment

Consider the simplest form of field experiment. Plots treated in some way are compared with plots not treated in that way. If yield is the criterion, the mean yield of the treated plots is compared with the mean yield of the untreated plots. The comparison is subject to experimental error; and by standard methods one estimates the least difference between the means that is statistically significant.

But what do the plots represent? In agriculture one takes them to represent farmers' fields or orchards. One experiments with plots in order to learn what should be done in fields or orchards: how to fertilize them, what varieties to grow, when to spray them with fungicide, insecticide, or herbicide. One experiments with plots not for the sake of learning about plots but for the sake of applying any results obtained from the plots to general agricultural practice. To pretend otherwise would be insincere.

If one grants the premise that plots are meant to represent fields, then one must grant the possibility of error if the plots do not accurately represent the fields they are meant to. The error is distinct from the ordinary statistical errors that arise from variability among plots receiving the same treatment.

The error when plots differ from the fields they are meant to represent will be called the "representational" or the "cryptic error".

A representational error arises when (to give a simple example) disease moves from unsprayed to sprayed plots. Suppose fungicides are being tested against disease in an area where all farmers spray their fields regularly and efficiently. In the experiment we suppose some plots to be sprayed with fungicide and others left unsprayed. In the unsprayed plots disease develops and multiplies fast. Soon the plants in these plots are heavily infected and release masses of spores, many of which blow on to the sprayed plots. Because the sprayed plots receive masses of spores from outside their borders, the plants in them become more heavily infected than they would otherwise have been.

The sprayed plots in this experiment differ from sprayed fields. The sprayed plots are heavily contaminated from outside; the sprayed fields are not, because (by hypothesis) all farmers in the area spray regularly and efficiently. The fungicide has a harder task to perform in the sprayed plots than in the sprayed fields; and anyone who advises farmers about spraying on information gained from the sprayed plots underestimates the potency of the fungicide.

It is an accepted principle that plots must not interfere with one another. In agricultural field trials, guard rows are left between plots in order to prevent mutual interference. In Cox's (1958) "Planning of Experiments", p. 21, for example, a plant disease problem is used for illustration. In an experiment in which some plots are inoculated with virus-carrying aphids and other plots left untreated, one must not only leave substantial space between treated and untreated plots but also check as far as possible that disease does not move from one plot to another.

To accept the principle that plots should not interfere with one another is not to solve the problem of how to stop them from interfering. This chapter discusses in three ways the problem that arises out of interference.

1. Some results are analyzed (in Sections 23.2 to 23.7) to see how large the error from interference is in ordinary field experiments in plant

pathology. It appears that the error is often many times greater than the least significant difference at the 5% level, which reflects the probable statistical error from variation among plots with the same treatment. If how to reduce a statistical error is worth studying—and the error has been studied in many books and countless papers in statistical and agricultural journals—how to reduce the error from interference is equally worth studying. And the error is worthy of a name of its own. Admittedly, a representational error is probably of great importance mainly in plant pathology and entomology. But that does not make it any less real.

2. Analysis (in Sections 23.8 to 23.11) shows that the representational error may be large without giving any obvious sign of its presence. Indeed, the larger the error the less its presence is likely to be observed. Representational errors are likely to be largest when they are cryptic. Advice that one must check as far as possible that disease does not move from one plot to another misses the point that the freest movements leave little trace that they have occurred.

3. The last part of this chapter deals mainly with how to reduce representational errors. It is usually accepted that plots should be separated from one another by guard strips; but the idea that guard strips in conventional layouts greatly reduce interference will be shown to be based on a false premise. Especially when propagules move in air of normal turbulence, guard strips do not have the efficiency they are commonly thought to have. An admonition that one should leave substantial space between treated and untreated plots is all very well. But what is substantial space when one is experimenting with pathogens that travel for miles, as many pathogens do?

Two general methods for reducing representational error are suggested. The treatments in any one experiment are limited to those that do not differ greatly from each other in the amount of disease they allow to develop. (See Section 23.21.) Experiments are laid out in such a way that the spores that leave the plot of their origin also mostly escape from the whole experimental area, or in such a way that the effect of the spores leaving the plot of their origin is adequately diluted. (See Section 23.22.)

A few paragraphs back we grouped plant pathology and entomology together. The problem of a great representational error is common to both. There are, of course, differences in detail: insects commonly move instinctively, plant pathogens mechanistically. But the movement of insects and pathogens is alike enough for Wolfenbarger (1946, 1959), in his comprehensive reviews, to deal uniformly with the dispersion of insects, viruses, bacteria, and fungus spores.

An analysis of the problem of representational errors in entomological experiments has been made by Joyce (1956), Joyce and Roberts (1959), and Roberts (1960). They studied the "interplot effect" in experiments with cotton plants in which some plants were sprayed with DDT and others were not.

23.2. An Experiment with Tomato Fruit Diseases

Christ (1957) tested directly how inoculum from unsprayed plots affects sprayed plots. The fungicide was copper oxychloride, used to control infection of tomato fruits by *Alternaria solani* and *Xanthomonas vesicatoria*.

Four experiments were laid out at 1/4 to 1/2 mile from each other and from other tomatoes. The plots were three rows of ten plants each, and records were made on fruits from the middle rows. There were four replicates for each treatment. In two of these experiments, half the plots were sprayed with copper oxychloride every 2 weeks and half were left unsprayed. In the other two experiments, half the plots were sprayed with copper oxychloride and the other half, instead of being left unsprayed, were sprayed with Bordeaux mixture.

In the unsprayed plots, 28.0% of the fruit was infected. In the plots sprayed with copper oxychloride next to the unsprayed plots, 19.3% was infected. In the plots sprayed with copper oxychloride next to plots sprayed with Bordeaux mixture, 7.8% was infected. (In the Bordeaux plots themselves 6.2% was infected.) The difference between the means, 19.3 and 7.8%, was significant at the 0.1% level.

An experimenter using an equal number of sprayed and unsprayed plots might report that spraying every 2 weeks with copper oxychloride reduces the amount of diseased fruit from 28.0 to 19.3%. But for a farmer who uses the same fungicide in the same way but sprays his fields from end to end without leaving unsprayed patches, it would probably be nearer the mark to estimate that spraying would reduce the amount from 28.0 to 7.8%.

In the account just given, and in the next few sections, we neglect the effect of loss of inoculum from the unsprayed plots on disease in these plots themselves. If the unsprayed plots had 28.0% of infected fruits, unsprayed fields would have had somewhat more than 28.0%, because fields, being larger, lose relatively less inoculum by dispersal. Using plots to represent fields doubly underestimates the benefit of spraying to the farmer. The cryptic error in using unsprayed plots to represent unsprayed fields is discussed in Section 23.13, and we shall say no more about it now.

23.3. An Experiment with Leaf Rust of Wheat

Wheat varieties with vertical resistance to leaf rust caused by *Puccinia recondita* are hypersensitive to the fungus. The leaves react by forming a small sterile necrotic area around the point of entry of the fungus instead of a normal sporing pustule. In the field the number of these necrotic areas is few. But in experimental plots resistant, hypersensitive varieties may be so bombarded with spores from susceptible varieties grown next to them that large numbers of necrotic areas develop and reduce the yield.

Samborski and Peturson (1960) grew four varieties of wheat in experimental plots: Thatcher, which is very susceptible to leaf rust; Selkirk, which is moderately resistant; and two highly resistant hybrids, Thatcher^7 × Frontana and Selkirk^8 × Exchange. There were twelve plots of each variety, each plot being four rows wide. The center two rows were harvested, the outside rows being left as guards. Half the plots of each variety were sprayed 11 times with zineb. This kept all the varieties, even susceptible Thatcher, practically free from rust. The other plots became naturally infected with leaf rust at an early stage of growth.

Results are given in Table 23.1. With sprayed plots taken to represent

TABLE 23.1
THE EFFECT OF LEAF RUST ON FOUR WHEAT VARIETIES IN EXPERIMENTAL PLOTS[a]

Variety	Treatment	Necrotic or chlorotic area of leaf (%)	Average yield per acre (bu.)	Average weight per 1000 kernels (gm.)
Thatcher	Sprayed	—	33.6	29.9
	Unsprayed		14.0	22.7
Thatcher^7	Sprayed	trace	34.8	30.4
× Frontana	Unsprayed	60	24.8	24.5
Selkirk	Sprayed	trace	33.6	37.8
	Unsprayed	40	24.0	32.0
Selkirk^8	Sprayed	trace	29.8	37.1
× Exchange	Unsprayed	30	26.2	31.6
		LSD 5%	5.9	1.14

[a] Data of Samborski and Peturson (1960).

healthy plots, leaf rust reduced the yield of Thatcher from 33.6 to 14.0 bushels per acre and the weight of 1000 kernels from 29.9 to 22.7 gm. (Rust caused fewer kernels to be set.) So far as we know, the reduction in a farmer's yield would have been of much the same order, if he had sown at the same time and under the same conditions. But with Thatcher[7] × Frontana (to cite one of the three resistant varieties), the yield in a farmer's (unsprayed) field would have corresponded to that of healthy (sprayed) experimental plots, i.e., 34.8 bushels per acre. The drop in yield to 24.8 bushels per acre in the unsprayed plots is the result of a bombardment of spores and consequent necrosis of leaf tissue that would not normally occur in a field.

Samborski and Peturson put this clearly. "Within a large area [i.e., in farmers' fields] planted to resistant varieties, the spore population is kept at a low level and a total amount of dead tissue resulting from infection is also low. However, when resistant varieties are grown in experimental plots, the proximity of susceptible varieties provides a constant and heavy supply of inoculum and considerable reduction of leaf tissue of resistant varieties may occur .... It should be emphasized that this is an agronomically abnormal situation and the results do not give a true appraisal of the practical performance of the resistant varieties under field conditions."

23.4. An Experiment with Stem Rust of Wheat and Rye

Kingsolver *et al.* (1959) arranged plots in a 5×5 Latin square. The plots were 20×20 ft. Between the plots a 20-ft. strip was left uninoculated as a guard. There were separate Latin squares for wheat and rye.

Four-fifths of the plots were inoculated with spores of *Puccinia graminis* when the wheat was jointing and the rye heading. The other plots were left uninoculated. After about $2p$, when the inoculated plots had from 1 to 8% of rust, the fungus began to spread uniformly over the entire experimental area, and the 20-ft. uninoculated guard strips did not stop it from entering the uninoculated plots. But fields 400 m. upwind from the Latin squares and sown at the same time as them remained practically free from rust until the plants ripened.

Results are given in Table 23.2. To discuss only the figures for wheat, the field upwind that escaped rust yielded 43.1 bushels per acre; the uninoculated plots in the Latin square that soon became infected yielded 11.6 bushels per acre; and the inoculated plots yielded still less. The least significant difference at the 5% level was estimated at 3.2 bushels per acre.

Now suppose that a chemical manufacturer produced a perfect eradicant, able instantly to destroy all stem rust in the plants it is sprayed on. If all farmers used it on their fields, they could practically eradicate stem rust for the season and expect a yield (on a figure from the previous paragraph) of 43.1 bushels per acre.

TABLE 23.2

THE EFFECT OF STEM RUST ON WHEAT AND RYE IN 20 × 20 FT. PLOTS SEPARATED BY 20-FT. GUARD STRIPS[a]

Rate of inoculation (gm.)	Bushels per acre	
	Wheat	Rye
0.0	11.6	24.4
0.1	8.6	24.6
1.0	7.5	23.9
10.0	4.1	17.8
100.0	0.9	11.8
Check fields upwind[b]	43.1	39.8
LSD 5%	3.2	5.0

[a] Data of Kingsolver *et al.* (1959).
[b] Fields upwind from the inoculated area. Rust never exceeded trace severity. For further details about the wheat plots see Fig. 12.4.

This perfect eradicant, we suppose, goes to an experiment station to be tested along with other materials. An experiment is designed with 20 × 20 ft. plots in a Latin square. Between the plots 20-ft. guard strips are left. Showers of uredospores arrive, and disease begins to mount in the plots. When the level of rust is about 1 to 8% the experimenter acts. He sprays the perfect eradicant onto a fifth of the plots and destroys all rust in them. For good measure he sprays all the guard strips as well, and destroys all rust in them too. At harvest he records the yield in the plots treated with the perfect eradicant: 11.6 bushels per acre, with a least significant difference of 3.2 bushels per acre. The experiment station reports back to the manufacturer that his eradicant is interesting —it can destroy all existing pustules—but the yield of treated plots was disappointing because of rust infection. More probably, the station reports back that the eradicant is interesting, but needs to be applied 2 or 3 times during the season to act properly.

With current, conventional designs of experiments it is possible for an experimenter to strain at a least significant difference of 3.2 bushels per acre and swallow a cryptic error of $43.1-11.6=31.5$ bushels per acre. With current designs it is possible for a perfect eradicant to be judged mediocre. Admittedly, the particular experiment we are discussing puts the cryptic error rather high. The design is not bad as conventional designs go; indeed it is very much better than most. But disease started unusually early for stem rust; and an estimated cryptic error 10 times as large as the least significant difference is unlikely except in a season of severe rust.

23.5. The Deposition of Spores

Gregory (1961) quotes experiments of Stepanov and of Gregory, Longhurst, and Sreeramulu. They released spores from a point near ground-level and then measured deposition on the ground at several distances along lines radiating from the point of dispersion. About 10% of spores of *Tilletia caries* were deposited in a sector of annulus 105° wide and from 5 to 20 m. from the point of dispersion. About 18% of *Lycopodium* spores were deposited in a sector of annulus 120° wide and from 2.5 to 10 m. from the point of dispersion.

In many experiments there are times, before disease has developed far in sprayed plots, when unsprayed plots have from 50 to 500 times as much disease as the sprayed plots. For example, in the experiment recorded in Fig. 21.3 about 95% of the potato foliage of the unsprayed plots was destroyed by blight when only 1% of the foliage of the sprayed plots was destroyed. If we assume that other spores spread much the same as *Tilletia* or *Lycopodium* spores, a little arithmetic applied to ordinary designs shows that sprayed plots can receive up to 20 spores from unsprayed plots for each spore they themselves produce, the actual figure depending on the arrangement of the plots. If 20 spores come from outside for each spore produced within a plot, R is increased 21 times in the plot. If the field a farmer sprays is not contaminated with spores from outside, this 21-fold increase of R is the source of a large cryptic error that can be estimated by methods discussed in earlier chapters.

23.6. Some Calculations for Potato Blight

The equations of Gregory (1945) for spore dispersal in air of low turbulence fit observed potato blight gradients well. In order to assess representational errors quantitatively, van der Plank (1961a) applied the equations to the results of an experiment by Large (1945). The plots in this experiment were 20×5 yd. and were in three randomized blocks of fourteen plots each. Of the fourteen plots in each block, ten were sprayed and four left unsprayed. It was calculated that the Bordeaux-sprayed plots in this experiment, at the time when they had 0.1% infection, were receiving an average of 20.7 spores from the unsprayed plots for every spore they themselves produced. The effect of this contamination from outside was estimated in Exercise 2 at the end of Chapter 8. It was estimated that if there had been no contamination by spores from unsprayed plots, the Bordeaux mixture would have altogether stopped blight from increasing, i.e., $r_l = 0$. But because of contamination blight was, in fact, increasing at the rate $r_l = 0.34$ per unit per day. The difference between 0 and 0.34 per unit per day estimates the error in using sprayed plots to represent sprayed fields, if these fields are not contaminated from outside.

This estimate of the representational error uses Gregory's equations for dispersal of spores in air. Hirst (1958) estimates that about half of the new lesions of potato blight are formed by spores dispersed in water. (The point we make would not be materially altered even if the estimated number of migrating spores was reduced to half.) But for the purpose of this section one can, if one wishes, regard the equations as satisfactory empirically, because they fit observations fairly well, and not inquire closely into the agents of dispersal.

23.7. Large's Observations on Potato Blight

From a study of blight progress curves Large (1945) concludes that when about 1% of the foliage of sprayed plots is destroyed the fungicide ceases to act, and disease increases as if unhindered by the fungicide.

Figures 21.2 and 21.3 illustrate this. With 1% disease, $\log_e [x/(1-x)] = -4.6$. Somewhere near this point, with $\log_e [x/(1-x)]$ between -6 and -4, the infection rate in sprayed plots begins to increase.

This increase is not the result of fungicides decomposing or being washed off the foliage. This is clear from much internal evidence that need not be discussed here. It is also clear from Fig. 21.3. In the experiment that Fig. 21.3 describes, the effect of zineb lasted rather better than that of copper oxychloride, although zineb is less stable and more easily removed by leaching.

It would be unreasonable to expect Large's observation to hold without exception for every fungicide experiment with potato blight. Sometimes, if too little fungicide is applied or if it is applied too early, the fungicide appears to lose control before there is 1% disease. Occasionally, as Hooker's results in Fig. 8.5 show, the fungicide continues to control blight even after 1% of the foliage is destroyed. But taken as an approximate rule Large's observation holds for scores of spraying trials that have been made against blight in various countries.

As it stands, the observation cannot be explained. When 1% of the foliage is destroyed, about 1 leaflet in 50 is infected. Interpreted directly, the observation means that when 1 leaflet in 50 is diseased, the disease inactivates the fungicide on the other 49 leaflets. One would have to postulate that a diseased leaflet produces a substance that inactivates copper fungicides, zineb, maneb, and organic tin compounds (for the observation holds for all of them) on 49 healthy leaflets.

We believe that Large looked at the wrong plots. He should have looked at the unsprayed plots. When disease is about 1% in the sprayed plots it is often more than 50% in the unsprayed plots (i.e., $\log [x/(1-x)] > 0$ in the unsprayed plots, as in Figs. 21.2 and 21.3). When blight reaches 1% in the sprayed plots, it reaches the stage of opening up the foliage in the unsprayed plots; and this (we deduced from Hirst and Stedman's findings discussed in the previous chapter) allows sporangia to get clear of the foliage and move easily from plot to plot.

On this suggestion, unsprayed plots interfere relatively little with sprayed plots while the foliage in the unsprayed plots is still dense and encloses most of the sporangia. But once blight destroys most of the foliage in unsprayed plots, sporangia are free to spread in mass to the sprayed plots. The fungicide then seems to lose its potency, whereas in reality it simply has more spores to cope with.

Other details support the suggestion. For example, full-strength Bordeaux (with 0.25% Cu), half-strength Bordeaux, cuprous oxide, and copper oxychloride all gave similar control of blight despite great differences in copper residues on the leaves (Large *et al.*, 1946). The suggestion of massive interference, timed by the destruction of foliage in unsprayed plots, easily explains this. In detail, Large's results show that when blight in the unsprayed plots was high, blight increased in the sprayed plots apparently unchecked by copper residues and at a rate not much modified by the type of copper fungicide used.

The argument in this section is that we cannot suggest any reason other than massive mutual interference among plots to explain the common phenomenon that Large first observed. Argument by exclusion is weak on its own, but usefully adds to other evidence.

23.8. The Inadequacy of Guard Rows in Conventional Designs

The evidence we have presented suggests that the representational error is often large. An objection can be raised. If the error is large, why has it not received much attention?

If one asks an experimenter why he is sure disease from his unsprayed plots is not interfering with his sprayed plots, he answers that he has left guard rows between the plots and that disease is not much worse in sprayed plots where they border on unsprayed plots.*

On the matter of guard rows we need spend little time. It was noticed in Section 23.4 that the wheat stem rust fungus made short work of 20-ft. guard strips. What is a 20-ft. guard strip to a fungus that moves annually from Texas to Saskatchewan in a couple of months? The evidence of the previous chapter was that pathogens spread fast and far as disease mounts. When a fungus invades all the fields of a country as quickly as *Phytophthora infestans* does (see Fig. 22.1), a few guard rows cannot fence it in and stop it moving from plot to plot.

The use of guard rows in order to keep plots apart and prevent pathogens from moving from one plot to another is based on the premise that pathogens settle near the borders of the plots they move out of. How false this premise is, especially for pathogens that move in normally turbulent air, will be seen in Section 23.15.

We do not condemn the use of guard rows. Guard rows are convenient and useful in several ways, and should be present in most field experiments. What we dispute is the belief that the sort of guard rows found in conventional experiments greatly reduce representational errors, if the pathogen moves freely. It needs more than guard rows to justify an experimental layout, when one deals with disease.

23.9. The Gradient Fallacy

The argument runs that if disease spreads from an unsprayed to a sprayed plot, there would be more disease in that part of the sprayed plot nearest to the unsprayed plot: in the sprayed plot the amount of disease would fall in a gradient away from the unsprayed plot. Gradients are indeed often noticed in an experimental area. But conspicuous

* The problem is not just one of fungicide experiments; but it is convenient to discuss the general problem in terms of fungicide experiments. Conclusions can easily be adapted to variety trials and other experiments.

gradients need not occur; and we regard as wholly fallacious the argument that absence of a conspicuous gradient means the absence of considerable spread of disease, for three reasons.

1. In a replicated experiment there is more than one unsprayed plot to act as a source of inoculum. The presence of several sources confuses the gradients. This is seen on a giant scale in Fig. 12.1. In between the groups of barberry bushes it is not always clear how gradients run.

2. A plot is an area source, not a point source, of inoculum. Gradients are less steep from an area source than from a point source, and the notion of a sharp drop in the amount of disease as one moves away from an infected plot is incorrect. Spores that move out of a plot tend to keep moving, at least until they are out of the experimental area. This is especially true of spores carried in air of normal turbulence. In ordinary designs there is little scope for steep gradients within the confines of the experimental area.

A quantitative analysis of the movement of spores out of plots and experimental areas is made in Section 23.15.

3. As disease multiplies, gradients flatten. Figure 22.3 shows this for tropical rust of corn. At the start one might expect some sort of a gradient in the sprayed plots when the amount of infection in them is small. But as disease, initially derived from unsprayed plots, begins to multiply in the sprayed plots, the gradient gradually fades as it flattens.

Absence of a conspicuous gradient in sprayed plots can just as well mean that disease had spread and multiplied fast as that disease had not spread at all.*

23.10. The Gradient Fallacy Illustrated

You have stopped for the evening at a camping site. Other campers are there besides you. Each tends his own fire. It has rained. The

* In Chapter 7 it was said that disease spreads in gradients, and foci are the units in which disease multiplies. Here, in Chapter 23, it is said that gradients tend to disappear, and need not be clear. There is no contradiction.

The basic gradient is from a point source. Daughter pustules form when uredospores are released from a single parent rust pustule. The parent pustule is the point source. The concentration of daughter pustules falls off in a line from the parent pustule; they fall off in a gradient. This gradient is the subject of Gregory's equations. The parent pustule and daughter pustules together form a focus. In the next generation each daughter pustule becomes a parent. There is now no longer one point source and one focus, but many. Each daughter lesion is a new parent source and the center of a new focus. With each generation the process continues, each generation blurring more and more the individual gradients and individual foci until there is seemingly uniform distribution of disease. As Chapter 22 shows, seeming uniformity is achieved in two or more generations.

firewood is wet and the fires smoky. Space is cramped, and the average distance between fires is only 2 or 3 times the diameter of a fire. You look for a place to sit between the fires where the smoke is less. But the wind is gusty and veering. As soon as you find a place where the smoke seems less, the wind veers and you find yourself in the smoke again.

The gradient fallacy assures you that if you can find nowhere to sit between the fires that is reasonably free from smoke there is no smoke between the fires.

23.11. The Cryptic Error

Epidemic foliage diseases such as the cereal rusts or potato blight are known to spread far and fast. When therefore the designer of a conventional experiment states, on the strength of the presence of guard rows and the absence of marked gradients, that a cereal rust fungus or the potato blight fungus has not spread from plot to plot during an epidemic, he implies that an ordinary process of nature conveniently ceased in the experimental area. No one has yet dared to try to show the implication to be true.

Spores and other propagules move, on the balance, more from points of high disease to points of low disease than in the reverse direction. The more freely they move across plots and across the experimental area, the more they smooth out disease inequalities and the less evidence there is of gradients to show that they have moved. The greatest errors of representation are likely to show themselves least. The errors are likely to be greatest when they are cryptic.

23.12. Where the Onus of Proof Lies

It is now universally accepted that an experimenter must show his conclusions to be statistically significant. One should not distinguish in principle between one source of error and another. The experimenter must also show that there is no representational error in his conclusions.

The onus is not on the critic to prove that in an experimental design the movement of disease stops experimental plots from truly representing farmers' fields. Unless the experimenter states that he is interested only in plots as such and that his experiments are not intended to have any bearing on farming, he must himself prove there are no great errors of representation.

23.13. The Error when Unsprayed Plots Represent Unsprayed Fields

When (to use fungicide experiments again for illustration) spores blow out of unsprayed plots and settle on sprayed plots, they increase disease in the sprayed plots. This is the source of representational error we have been considering. Sprayed plots are likely to have more disease than sprayed fields. But there is another source of error. The unsprayed plots, because they are small, lose a larger proportion of spores than unsprayed fields. (See Table 23.3.) For this reason, unsprayed plots are likely to have less disease than unsprayed fields, other things being equal.

The two sources of error supplement one another. Both cause the efficiency of fungicides on the farm to be underestimated.

The loss of spores from a plot and a field was the topic of Exercise 1 at the end of Chapter 8. From a plot, 20×5 yd., the proportion of sporangia of *Phytophthora infestans* lost was estimated as 12%, and from a square 2-acre field as 2%. In the plot $r_l = 0.460$ per unit per day. It was calculated that in the field r_l would be 0.486 per unit per day. In other words, $0.486 - 0.460 = 0.026$ per unit per day is the estimated error in r_l of representing an unsprayed field by an unsprayed 20×5 yd. plot.

The experiment discussed here was the same as that discussed in Section 23.6. In the sprayed plots $r_l = 0.34$ per unit per day; and it was estimated that in sprayed fields disease would not have increased at all.

We can therefore compare the two sources of error in this particular experiment. Using unsprayed plots to represent unsprayed fields caused r_l to be estimated as 0.46 instead of 0.486 per unit per day. Using sprayed plots to represent sprayed fields caused r_l to be estimated as 0.34 instead of 0 per unit per day.

It is not claimed that the one source of error—that of using unsprayed plots to represent unsprayed fields—is always small relative to the other. But it is often relatively small; and we shall concentrate on the other source of error—that of using sprayed plots to represent sprayed fields, or its equivalent in experiments other than with fungicides.

23.14. The Loss of Air-Borne Spores from Plots and Fields

When Gregory (1945) published his equations for the dispersal of air-borne spores, he brought together data from the literature to see how observations on disease in the field compared with what was expected from theory. With two pathogens, *Botrytis tulipae* on tulip and *Cercosporella herpotrichoides* on wheat, observation does not agree with expectation; and Gregory suggests reasons for the disagreement. But with other pathogens the agreement is fair.

We shall therefore use the equations here. Insofar as observation and expectation disagree, the observed gradients (other than with the two fungi mentioned in the last paragraph) tend to be flatter near the source and steeper at a distance from the source than the equations predict. This means that by using Gregory's equations we tend to underestimate representational errors.

TABLE 23.3

THE CALCULATED PERCENTAGE OF AIR-BORNE SPORES LOST FROM SQUARE PLOTS AND FIELDS

Side of square (m.)	Area of square (m.^{2})	Percentage spores lost	
		Normal turbulence	Low turbulence
1	1	30.8	38.7
3.162	10	25.6	23.9
10	10^2	20.8	12.3
31.62	10^3	16.7	5.16
100	10^4	12.9	1.83
316.2	10^5	9.39	0.619
1000	10^6	6.41	0.214
3162	10^7	4.33	0.097

Table 23.3 gives estimates of the losses of spores from square plots and fields. The method of estimation was suggested by van der Plank (1961b). It is assumed that wind does not change its direction while moving across the plot or field; and the figures in the table are weighted averages for winds from different points of the compass. We discuss square plots because estimates lose accuracy when the length of a plot is much greater than the width.

According to the table a small plot, 1 m.^{2}, loses 30.8% of its spores by wind of normal turbulence. That is, wind of normal turbulence blows out 30.8% of the spores produced within the plot. Wind of low turbulence blows out 38.7%. The corresponding figures for a large field, 1 km.^{2}, are 6.41 and 0.214% for normal and low turbulence, respectively.

Increasing the size of the square reduces the proportional loss of spores, and this reduction is much more marked with air of low than with air of normal turbulence.

23.15. The Movement of Spores between Plots within an Experimental Area

From Table 23.3, a square with a side of 1 m. loses 30.8% of its spores in air of normal turbulence; a square with a side of 10 m. loses 20.8%. Consider an experiment in which 100 small square plots each with a side of 1 m. are arranged in a square experimental area with a side of 10 m. For every 100 spores produced within the experimental area 30.8 leave the plot of origin and 20.8 leave the whole experimental area. That is, of every 100 spores produced within the experimental area 10.0 eventually settle somewhere in the 99 plots other than the plot of origin. With air of normal turbulence, 10% of the spores produced within square plots with a side of 1 m. interfere with other plots within a square experimental area of 100 plots.

This figure and others calculated in the same way are entered in Table 23.4.

TABLE 23.4

THE CALCULATED PERCENTAGE OF SPORES PRODUCED WITHIN PLOTS THAT ESCAPE FROM THE PLOTS BUT SETTLE AGAIN WITHIN THE EXPERIMENTAL AREA (THE PLOTS AND THE AREA ARE SQUARE)

Side of plot (m.)	9-Plot area		100-Plot area	
	Normal turb.	Low turb.	Normal turb.	Low turb.
1	5.0	14.1	10.0	26.4
3.162	4.6	11.1	8.9	18.7
10	3.9	6.8	7.9	10.5
31.62	3.6	3.1	7.3	4.5
100	3.3	1.1	6.5	1.6

In air of normal turbulence the majority of spores that leave a plot also leave the experimental area. This is true even when the area has as many as 100 plots. Spores that move as far as the edge of a plot tend to keep moving, and many get clear of the whole experimental area.

The assumption behind the use of guard rows in conventional layouts —that most spores settle near the plots they move out of—is not warranted either by the theory of spore dispersal or by everyday observations that epidemic diseases spread far.

The turbulence of the air matters greatly in estimates of representational errors. The data in Table 23.4 show that, when small plots are used, greater representational errors can be expected with pathogens that disperse at times of low turbulence.

How spores move greatly influences the representational errors. They move further when atmospheric turbulence is normal and not low, and paradoxically this reduces representational errors if plots are small. For small plots to interfere greatly with one another, spores should move far but not too far; in air of normal turbulence they move too far for maximal interference. In air of normal turbulence many spores move clear of the whole experimental area in which they arise.

23.16. The Effect of the Size of Plot

The turbulence of the air in which spores disperse greatly affects the choice of size of plot in experiments.

The data in Table 23.4 show that much can be gained by increasing plot size if pathogens disperse in air of low turbulence. Increasing the side of square plots from 1 to 10 m. reduces mutual interference by more than one-half; increasing it from 10 to 100 m., reduces mutual interference by more than five-sixths. This statement holds for both a small 9-plot experimental area or a large 100-plot area.

But when air is normally turbulent while spores disperse, increasing the side of square plots from 1 to 100 m. does not even halve mutual interference among plots. This statement, too, holds for both a 9-plot experimental area and a 100-plot area. If the spores disperse in normally turbulent air one cannot greatly reduce representational errors by juggling with plot size.

It follows that mere difference in size between a plot and the field it is meant to represent can easily be overrated as a source of representational error, at any rate when spores disperse in air of normal turbulence. Putting differently treated plots in a compact experimental area is a worse source of error.

Largeness of plot by itself must not be taken as evidence of safe design.

23.17. The Effect of Shape of Plot

It has not yet been possible properly to analyze how shape of plot affects representational errors. But it seems safe to conclude two points:

for a given area, a square plot is safer than an elongated plot; and elongation has less effect when dispersal is at normal atmospheric turbulence than when it is at low turbulence.

The conclusion, that square plots are safer than elongated plots of the same area, does not necessarily hold when it is possible to orientate plots in relation to a prevailing disease-carrying wind.

23.18. Interference between Plots

For a numerical example, consider 10×10 m. plots in a square 100-plot area. Table 23.4 states that in normally turbulent air 7.9% of the spores of a plot escape and settle on the other 99 plots. Let us suppose that in an experiment 1 untreated plot is left for every 4 plots treated with fungicide. That is, there are 20 untreated plots in the experimental area. For every 100 spores produced within an untreated plot, $(7.9 \times 20)/99 = 1.6$ spores from untreated plots land on each treated plot. If untreated plots have 100 times as much disease as treated plots (which often happens), 1.6 spores from untreated plots fall on treated plots for each spore the treated plot itself produces. Spores moving between plots make R in treated plots 2.6 times as great as it would be without spore movement. All these figures are averages.

Calculations of this sort can be repeated without end. Mutual interference among plots depends on the turbulence of the air, the size and number of the plots, and the difference in disease between plots treated differently. The last variable is often the greatest of them all. If one keeps the range of disease small between the worst and the best treatments, one automatically avoids great representational errors. This is the principle of control in Section 23.21.

In these calculations we have ignored spore movements of minor effect, as from treated to untreated plots.

23.19. Different Types of Experiment

We discuss representational errors mainly to learn how to avoid or reduce them. Three types of experiment will be considered first in relation to fungicide treatment. It is assumed that the pathogen disperses freely.

1. There are experiments in which a representational error is permissible, although it is a nuisance and lowers efficiency. Experiments

carried out to screen fungicides or rank them in order of effectiveness against diseases are examples. They are discussed in the next section.

2. Representational errors arise mainly when some plots in an experiment have much more disease than others. In order to avoid great errors and at the same time still retain the convenience of conventional layouts, one must confine treatments in any one experiment to those that do not differ much from one another in their effect on disease. This is discussed in Section 23.21.

3. When the plots in an experiment must unavoidably differ greatly in disease, special layouts are needed to reduce interference. Special layouts are often needed to compare, for example, treated and untreated plots. They are discussed in Section 23.22.

In regard to experiments 2 and 3, one can suggest no way in which the convenience of conventional layouts can be combined with the luxury of putting all sorts of different treatments into one experiment. It seems one must pay one price or the other when one experiments with pathogens that disperse freely and far.

23.20. The Screening and Ranking of Fungicides in the Field

Every year new fungicides and new formulations arrive for testing. The inefficient must be screened out, the good ranked with other fungicides.

To select more or less at random a short example of good screening and ranking out of the hundreds available, consider Schroeder's (1959) results at Geneva, New York, with early blight of tomatoes caused by *Alternaria solani*. He sprayed 6 times during the season with 100 gallons of spray per acre, and ranked his materials in order of disease control as shown in the accompanying table.

Manzate, 4 lb. per acre	1
Dithane M-22, 4 lb. per acre	1
Dyrene, 4 lb. per acre	1
Phaltan 50-W, 4 lb. per acre	2
Orthocide 50-W, 4 lb. per acre	3
Bayer Cu. 4848, 4 lb. per acre	4

Schroeder rightly omits unsprayed check (control) plots. In a tomato experiment the real basis of comparison is with the standard fungicide maneb, represented in Schroeder's experiment by the two formulations,

Manzate and Dithane M-22. To prove that another fungicide is worth using on tomatoes is to prove that it is as good as, or better than, maneb. The real check plots are those treated with maneb; and plots untreated with any fungicide at all are both irrelevant to the comparison and a danger to other plots in the experiment.

But in the usual run of screening and ranking one must unavoidably often include treatments little better, or no better, than no treatment at all. There is nothing much that one can do about this, and one must accept that plots will interfere with one another.

There are usually two rather different aims in experiments of the type we are discussing. The first is to rank treatments in order of preference. The other is to estimate the effect of treatment, for example, on yield. Of the two, the ranking of treatments is less subject to representational error if all fungicides are protectants, unmixed with eradicants.

23.21. Quantitative Comparison of Fungicide Treatments

Many crops are treated with fungicides by farmers as a matter of routine. It is known that fungicides are needed. When an experiment is conducted, it is not to prove that a fungicide is better than no fungicide, but to compare one fungicide treatment with another.

Often, treatments must be compared accurately. To select some treatment implies that one has weighed costs and labor against benefits likely to be received. To give a simple example, one may know that with some particular fungicide spraying every 5 days is more effective than spraying every 10 days. But is it worth spraying every 5 days? Would the gains justify it? To answer the question one must compare the two schedules accurately.

To compare treatments accurately, when plots lie side by side with no more than a few guard (border) rows between, the effect of spores moving between the plots must be reduced. To reduce the effect of spores moving between plots, the plots must not differ greatly from one another in the level of disease in them. To keep plots from differing greatly from one another in the level of disease in them, plots untreated with a fungicide must be omitted.

It is as simple as that. We assume, of course, that we are dealing with the sort of pathogen that spreads fast; slow moving soil pathogens, for example, are not being discussed.

From a survey of published descriptions of experiments the writer is convinced that the worst representational errors could be avoided simply by a change of attitude toward check (control) plots. Too often it is taken for granted that the only proper check plots are plots untreated with fungicide. But we are discussing here crops that farmers normally treat with fungicide as a matter of course: apples in a scabby area, for example, or potatoes in Aroostook County, Maine. The relevant check treatment then is what the farmers commonly use against the disease, or what the local experiment station recommends to farmers. This is the check treatment that new treatments must improve on to be acceptable. To improve on unsprayed plots is (in the present context) just *not* relevant at all.

It will be argued that an experiment needs untreated plots to gauge how conditions favor disease during the season. The results of a fungicide trial, it will be argued, can be interpreted properly only in relation to the disease conditions of the season. The argument is sound. But it is not an argument for having untreated plots as an integral part of the experimental area. One needs to know the season's rainfall in order to interpret the results of a fungicide experiment properly. But most of us would agree to use records taken in a rain gauge a few hundred yards from the experimental area.

If an untreated plot is put outside the experimental area as a disease gauge, the safety of the arrangement increases quickly with the distance. The further away the untreated plot is, the narrower is the sector in which wind blows spores to the experimental area. This is in contrast with what happens when untreated plots are integrated within a conventional experimental area. The untreated plots are replicated and occur among treated plots in such a way that wind from almost any quarter will blow spores from untreated to treated plots.

If there is a prevailing disease-carrying wind, putting the disease gauge downwind from the experimental area is an obvious choice.

We are concerned with trying to reduce the range of disease in plots within the experimental area. Plots treated inefficiently, e.g., with a fungicide not very effective against disease, are almost as dangerous as plots receiving no fungicide treatment at all. They too must be excluded from the experimental area in order to reduce representational errors. If one takes the check treatment to be a standard treatment used by farmers or recommended by experiment stations, this should remind one that the object of the experiment is to improve on the check. Candidate treatments that have no hope of improving on existing treatments should not be included. Include only treatments that have previously been screened; and keep accurate quantitative comparisons of treatments separate from routine screening trials.

It is taken for granted that only approved statistical designs will be used.

23.22. Experiments with Plots That Differ Greatly in Disease

The more the plots in an experiment differ from each other in the amount of disease they have, the more one must reduce the effect of movement of propagules between plots, if one is to avoid a large representational error. The most difficult experiments are those comparing plots untreated with fungicides with treated plots. They are the experiments needed to inquire whether a fungicide can profitably be used: whether it is better to treat than not to treat.

We suggest two general ways of laying out an experiment. There are probably many more. But we believe that all ways will have this in common: they will need plenty of space. Experiments with plots that differ greatly in disease cannot be tucked away in some small corner of space on an experiment station.

One accepts, on evidence given earlier, that narrow guard strips between plots in a conventional compact experimental area are unlikely adequately to reduce the number of spores moving from plot to plot. But two general methods can be used: escape and dilution. One can place treated plots where they escape as many spores from untreated plots as possible. And one can dilute the effect of spores from untreated plots as much as possible.

In order that treated plots should escape most spores that come from untreated plots the plots can be put in a line, with large guard plots between treated and untreated plots. Only wind blowing up or down the line will disperse air-borne spores up or down the line.

Guard plots between the experimental plots are doubly useful: they not only separate the plots but also reduce the points of the compass from which wind can blow spores from one experimental plot to another. It was estimated (van der Plank, 1961a) that, with 10×10 yd. plots in a line and conditions of low atmospheric turbulence, only 0.1% of the spores produced in an untreated plot cross a guard plot (also 10×10 yd.) and fall in a treated plot next to it. If the untreated and treated plots are separated by two guard plots in the line (all plots being 10×10 yd.), only 0.03% of the spores from the untreated plot fall in the treated plot. These figures are averages for winds from all points of the compass. If there is a prevailing disease-carrying wind, putting the line across the direction of the wind reduces the transfer of spores still more.

Within a line several standard designs are possible, including randomized blocks and Latin squares. (With Latin squares, columns represent segments of the line, and rows represent positions within the segments.)

Details would have to be adapted to the particular experiment. Guard plots could be of some neutral crop, e.g., brassicas between experimental potato plots. Or they could be of the same crop as that in the experiment, but well sprayed.

The other method, of dilution, uses small unsprayed plots in large sprayed fields. The simplest pattern would use one unsprayed plot to each sprayed field. The position of the unsprayed plot would have to be randomly selected, and the number of fields adequate for proper statistical analysis.

Countless variations are possible. Instead of comparing an unsprayed plot with the sprayed remainder of the field, one can pair the unsprayed plot with a plot measured off in the sprayed remainder. Fields large enough can each have several unsprayed plots. There is no point in probing all possible variations here.

Tables 23.3 and 23.4 give an idea of how much dilution is needed when spores are dispersed in air. Consider air of normal turbulence. Suppose there is an unsprayed plot, 10×10 m., randomly placed in a 100×100 m. field. The remainder of the field is sprayed, and it is proposed to compare the plot with the remainder. The area of the field is 100 times that of the plot; the sprayed area is 99 times the unsprayed area. Table 23.4 states that for every 100 spores produced in the unsprayed plot 7.9 fall in the sprayed remainder of the field. If the percentage of foliage diseased in the unsprayed plot is a times as great as in the sprayed remainder of the field, then for every 100 spores produced within the sprayed remainder $7.9a/99$ arrive from the unsprayed plot. If $a = 100$, about 8 spores come from the unsprayed plot for every 100 spores produced within the sprayed remainder. If the fungicide is a protectant, R in the sprayed remainder is about 1.08 times as large as it would have been if there were no contamination from the unsprayed plot. The representational error is not very large.

As a rough and ready rule for testing protectant fungicides against pathogens that disperse in air of normal turbulence one might say that the area of the unsprayed plots should not exceed $1/a$ of the area of the fields they are in. If one expects the percentage of foliage diseased to be 50 times as high in the unsprayed plots as in the sprayed remainders of the fields, not more than 1/50 of the fields should be left unsprayed for experimental purposes. We suggest a should be measured at its greatest. When, for example, the purpose of spraying is simply to delay the epidemic without trying to keep disease under control all the time until the crop is ripe, it would be pointless to estimate a at ripeness.

The same rule can be used to estimate how large a field should be. Suppose it is desired to carry out the whole experiment in a single field. Eight unsprayed plots, each 10×10 m., are to be situated randomly in the field, and compared with marked areas in the sprayed remainder of the field. It is expected that the percentage of foliage diseased will be 50 times as high in the unsprayed plots as in the remainder of the field. How large should the field be? It should not be less than $50 \times 8 \times 10 \times 10$ m.$^2 = 4$ hectares $= 10$ acres roughly.

The rule for protectants is not exact. There are many variables, including the position of the plots in the fields, the sizes and shapes of the plots and fields, and the time of treatment. Nor can one lay down the standard of accuracy to be obtained. But the rule is possibly good enough to go on with.

No satisfactory general rule can be given for eradicants. For one thing, the representational error depends greatly on the time of eradication in relation to the progress of the epidemic.

Any rule that applies to unsprayed plots applies equally to plots given inefficient fungicidal treatment. When one sets out to estimate whether it is better to treat than not to treat, there is much to be said for confining treatments to the best available. Experiments of the sort described in this section are too expensive to be mixed up with screening tests. One assumes that they will not be undertaken until screening tests have revealed promising treatments, and that they will then be undertaken only with the promising treatments.

23.23. Variety Trials

Variety trials to assess horizontal resistance do not differ from fungicide trials in principle. One must just think of plots of susceptible varieties instead of unsprayed plots, and plots of resistant varieties instead of sprayed plots. If one wishes to use the usual conventional ways of laying out experiments, one must be prepared to limit the range of disease in plots: one should use separate experiments to compare fairly resistant varieties with fairly resistant varieties, moderately susceptible varieties with moderately susceptible varieties, and very susceptible varieties with very susceptible varieties. If one wishes to compare very susceptible varieties with moderately resistant varieties, one must be prepared to use methods suggested in the preceding section.

There are times when representational errors are large even in comparisons between vertically resistant varieties. It was noted in Section 23.3 that when the wheat variety Selkirk, which is moderately resistant to leaf rust, and two hybrids, which are highly resistant, were grown in experimental plots they were so lashed with spores from susceptible Thatcher that the yield was considerably reduced. The small necrotic spots, which are all the vertically resistant varieties show on infection, were numerous enough to damage the plant, which is something that does not happen conspicuously in farmers' fields.

But this is unusual. More often the individual necrotic areas are too small to do great harm; and vertically resistant varieties are not greatly affected by growing next to susceptible varieties. Indeed, it was a theme of Chapter 17 that vertical resistance is at its best when varieties are diverse.

It is horizontally resistant varieties that suffer most from proximity to susceptible varieties. Representational errors are normally much greater when horizontal resistance is tested than when vertical resistance is tested. The errors cause horizontal resistance to be unfairly undervalued when it is compared with vertical resistance.

23.24. The Conservative Error Theory

When spores from unsprayed plots contaminate sprayed plots, they cause differences between treatments to be undervalued. There is a school that believes this to be an advantage. Undervaluation, they believe, means that results from the experiment are conservative: the results are on the safe side. A farmer will be agreeably surprised, the argument runs, if he gets better results from using fungicides than the experiment station said he would.

Two examples suffice to show the argument to be dangerous nonsense.

With proper precautions fertilizers do not spread much from plot to plot, and in a fertilizer experiment great pains are taken to ensure that results are accurate. On what logic is it assumed that an agricultural chemist needs true estimates about fertilizers, whereas a plant pathologist can be trusted only with information "on the safe side" about fungicides?

Underestimates are not conservative estimates. They are wrong estimates. In Section 23.21 the example was given of having to decide whether, with a given strength of a given fungicide, it is worth spraying once every 5 days instead of once every 10 days. By underestimating the difference between treatments one might decide in favor of spraying once every 10 days, whereas the real difference might justify spraying once every 5 days. Underestimation can make one decide, not conservatively but wrongly.

23.25. The Field Side of the Story

At the beginning of this chapter we discussed the problem of an experiment in an area where practically all farmers spray their crops effectively. The presence of unsprayed check plots in the experiment would cause the effectiveness of fungicides to be underestimated, because sprayed plots would be greatly contaminated with spores from unsprayed plots, whereas well sprayed fields would be little contaminated by spores from other well sprayed fields.

But what would happen in an epidemic of a foliage disease if only some farmers sprayed their fields and others left theirs unsprayed? Unless the fields were few and far between, they would interfere with one another, just as plots interfere with one another. There would still be a representational error in an experiment that used sprayed and unsprayed plots mixed up without a precautionary layout; the error would persist so long as sprayed plots received spores from unsprayed plots with more disease, irrespective of whether there was a fallout of spores from fields in the neighborhood or not. But the error would be smaller.

In this chapter, representational error rather than interplot effect is discussed, because to consider only plots is to consider only part of the story. One must know what happens on farms as well. Although an interplot effect is the source of a representational error, it is only one of the factors that determine the error's size. The interfield effect which reflects the current farming pattern is another.

APPENDIX

Logits

Yule (1925) gave a short table of $\log_e [x/(1-x)]$ in his work on the logistic increase of human populations. (See next section.) Berkson (1944), who, among others, applied the logistic function to the problems of bioassay, named $\log_e [x/(1-x)]$ the logit of x. Berkson (1953) published tables of logits and antilogits.

Finney (1952) also used *logit*, but with a somewhat different definition. The same word with two different meanings would be intolerably confusing; and *logit* Berkson has clear priority.

We have avoided *logit* in this book simply because it is unfamiliar, but regard its use in future as inevitable and to be welcomed.

Logistic Increase

Logistic growth proceeds according to the equation

$$\frac{dx}{dt} = bx(a-x)$$

With $a = 1$, to conform with the use of proportions of disease,

$$\frac{dx}{dt} = bx(1-x)$$

This is similar to Eq. (3.2)

$$\frac{dx}{dt} = rx(1-x)$$

The difference is that, by all definitions of logistic increase, b is a constant.

The word logistic was first used 120 years ago by Verhulst (quoted by Yule, 1925) when he inquired how populations increased in France and Belgium. The basic concept of logistic increase is that there is an upper limit to the population a country can sustain. The absolute rate of increase of population at any date is determined both by the population at that date and by "the still unutilized reserves of population-support."

Pearl and Reed (1920) fitted a logistic curve to the increase of population of the United States after 1790. Yule (1925) fitted logistic curves to the increase in England and Wales, the United States, and France.

These workers set out to estimate the constants in the logistic equation best fitted to the observed increase of population. We are content to let r vary. Equation (3.2) defines r at any instant, without describing a method of increase and without implying that r stays constant.

The difference is more easily seen from the fact that *logistic* and *autocatalytic* are synonyms. They are so defined by Kendall and Buckland (1957) in their "Dictionary of Statistical Terms". In an autocatalytic chemical reaction the velocity constant, the equivalent of r here, is, as the name states, a constant in constant conditions of temperature, etc. In an infection process in constant conditions of temperature, humidity, resistance of the host, aggressiveness of the pathogen, etc., r must vary. It must increase as an epidemic progresses (see Section 6.8), because in an infection process there is a latent period which an autocatalytic chemical process does not have.

Infection processes are not logistic and should not be so described. It is admittedly an experimental finding that r often stays nearly constant during much of an epidemic; but it can only blur essential differences if, because of this finding, we try to relate an epidemic with an autocatalytic chemical reaction or other logistic process.

Transformations

In various chapters the proportion x of disease has been transformed to $\log[x/(1-x)]$ or $\log[1/(1-x)]$ according to the circumstances. The following is a summary.

Use $\log[x/(1-x)]$ when the disease is of the "compound interest" type. The pathogen must have multiplied, in the sense that it must have moved from lesion to lesion (or from plant to plant if the disease is systemic). Plot $\log[x/(1-x)]$ against t or $\log x_0$. That is, one uses $\log[x/(1-x)]$ when one wishes to relate compound interest disease to time or to the logarithm of the amount of initial inoculum (or to any factor directly and simply related to the amount).

Use $\log[1/(1-x)]$ when the disease is of the "simple interest" type, or when the pathogen has passed through only one generation during the season. Plot $\log[1/(1-x)]$ against t or x_0. (Note that one plots $\log[1/(1-x)]$ against x_0, not $\log x_0$.)

In this summary replace x_0 by Q where this is appropriate (see Section 13.9).

These rules are probably most useful when $x < 0.5$. When x is greater, $1-x$ becomes increasingly inadequate as a correction factor, and one's knowledge of epidemics becomes increasingly empirical (for reasons discussed in Chapter 6).

TABLE 1

$\log_{10}[x/(1-x)]$

Table 1 gives $\log_{10}[x/(1-x)]$ with three digits after the decimal point. This is accurate enough for dealing with most records of plant disease. A bar is used, as with ordinary logarithms: thus $\bar{2}.490$ instead of -1.510.

For values of x less than 0.001 one may use $\log x$ for $\log[x/(1-x)]$, with enough accuracy for most purposes.

x	.000	.001	.002	.003	.004	.005	.006	.007	.008	.009
.00		$\bar{3}.000$	$\bar{3}.302$	$\bar{3}.478$	$\bar{3}.604$	$\bar{3}.701$	$\bar{3}.781$	$\bar{3}.848$	$\bar{3}.907$	$\bar{3}.958$
.01	$\bar{2}.004$	$\bar{2}.046$	$\bar{2}.084$	$\bar{2}.120$	$\bar{2}.152$	$\bar{2}.183$	$\bar{2}.211$	$\bar{2}.238$	$\bar{2}.263$	$\bar{2}.287$
.02	$\bar{2}.310$	$\bar{2}.331$	$\bar{2}.352$	$\bar{2}.372$	$\bar{2}.391$	$\bar{2}.409$	$\bar{2}.426$	$\bar{2}.443$	$\bar{2}.460$	$\bar{2}.475$
.03	$\bar{2}.490$	$\bar{2}.505$	$\bar{2}.519$	$\bar{2}.533$	$\bar{2}.546$	$\bar{2}.559$	$\bar{2}.572$	$\bar{2}.585$	$\bar{2}.597$	$\bar{2}.608$
.04	$\bar{2}.620$	$\bar{2}.631$	$\bar{2}.642$	$\bar{2}.653$	$\bar{2}.663$	$\bar{2}.673$	$\bar{2}.683$	$\bar{2}.693$	$\bar{2}.703$	$\bar{2}.712$
.05	$\bar{2}.721$	$\bar{2}.730$	$\bar{2}.739$	$\bar{2}.748$	$\bar{2}.756$	$\bar{2}.765$	$\bar{2}.773$	$\bar{2}.781$	$\bar{2}.789$	$\bar{2}.797$
.06	$\bar{2}.805$	$\bar{2}.813$	$\bar{2}.820$	$\bar{2}.828$	$\bar{2}.835$	$\bar{2}.842$	$\bar{2}.849$	$\bar{2}.856$	$\bar{2}.863$	$\bar{2}.870$
.07	$\bar{2}.877$	$\bar{2}.883$	$\bar{2}.890$	$\bar{2}.896$	$\bar{2}.903$	$\bar{2}.909$	$\bar{2}.915$	$\bar{2}.921$	$\bar{2}.927$	$\bar{2}.933$
.08	$\bar{2}.939$	$\bar{2}.945$	$\bar{2}.951$	$\bar{2}.957$	$\bar{2}.962$	$\bar{2}.968$	$\bar{2}.974$	$\bar{2}.979$	$\bar{2}.984$	$\bar{2}.990$
.09	$\bar{2}.995$	$\bar{1}.000$	$\bar{1}.006$	$\bar{1}.011$	$\bar{1}.016$	$\bar{1}.021$	$\bar{1}.026$	$\bar{1}.031$	$\bar{1}.036$	$\bar{1}.041$
.10	$\bar{1}.046$	$\bar{1}.051$	$\bar{1}.055$	$\bar{1}.060$	$\bar{1}.065$	$\bar{1}.069$	$\bar{1}.074$	$\bar{1}.078$	$\bar{1}.083$	$\bar{1}.088$
.11	$\bar{1}.092$	$\bar{1}.096$	$\bar{1}.101$	$\bar{1}.105$	$\bar{1}.109$	$\bar{1}.114$	$\bar{1}.118$	$\bar{1}.122$	$\bar{1}.126$	$\bar{1}.131$
.12	$\bar{1}.135$	$\bar{1}.139$	$\bar{1}.143$	$\bar{1}.147$	$\bar{1}.151$	$\bar{1}.155$	$\bar{1}.159$	$\bar{1}.163$	$\bar{1}.167$	$\bar{1}.171$
.13	$\bar{1}.174$	$\bar{1}.178$	$\bar{1}.182$	$\bar{1}.186$	$\bar{1}.190$	$\bar{1}.193$	$\bar{1}.197$	$\bar{1}.201$	$\bar{1}.204$	$\bar{1}.208$
.14	$\bar{1}.212$	$\bar{1}.215$	$\bar{1}.219$	$\bar{1}.222$	$\bar{1}.226$	$\bar{1}.229$	$\bar{1}.233$	$\bar{1}.236$	$\bar{1}.240$	$\bar{1}.243$
.15	$\bar{1}.247$	$\bar{1}.250$	$\bar{1}.253$	$\bar{1}.257$	$\bar{1}.260$	$\bar{1}.263$	$\bar{1}.267$	$\bar{1}.270$	$\bar{1}.273$	$\bar{1}.277$
.16	$\bar{1}.280$	$\bar{1}.283$	$\bar{1}.286$	$\bar{1}.289$	$\bar{1}.293$	$\bar{1}.296$	$\bar{1}.299$	$\bar{1}.302$	$\bar{1}.305$	$\bar{1}.308$
.17	$\bar{1}.311$	$\bar{1}.314$	$\bar{1}.317$	$\bar{1}.321$	$\bar{1}.324$	$\bar{1}.327$	$\bar{1}.330$	$\bar{1}.333$	$\bar{1}.335$	$\bar{1}.338$
.18	$\bar{1}.341$	$\bar{1}.344$	$\bar{1}.347$	$\bar{1}.350$	$\bar{1}.353$	$\bar{1}.356$	$\bar{1}.359$	$\bar{1}.362$	$\bar{1}.365$	$\bar{1}.367$
.19	$\bar{1}.370$	$\bar{1}.373$	$\bar{1}.376$	$\bar{1}.379$	$\bar{1}.381$	$\bar{1}.384$	$\bar{1}.387$	$\bar{1}.390$	$\bar{1}.392$	$\bar{1}.395$
.20	$\bar{1}.398$	$\bar{1}.401$	$\bar{1}.403$	$\bar{1}.406$	$\bar{1}.409$	$\bar{1}.411$	$\bar{1}.414$	$\bar{1}.417$	$\bar{1}.419$	$\bar{1}.422$
.21	$\bar{1}.425$	$\bar{1}.427$	$\bar{1}.430$	$\bar{1}.432$	$\bar{1}.435$	$\bar{1}.438$	$\bar{1}.440$	$\bar{1}.443$	$\bar{1}.445$	$\bar{1}.448$
.22	$\bar{1}.450$	$\bar{1}.453$	$\bar{1}.455$	$\bar{1}.458$	$\bar{1}.460$	$\bar{1}.463$	$\bar{1}.465$	$\bar{1}.468$	$\bar{1}.470$	$\bar{1}.473$
.23	$\bar{1}.475$	$\bar{1}.478$	$\bar{1}.480$	$\bar{1}.483$	$\bar{1}.485$	$\bar{1}.487$	$\bar{1}.490$	$\bar{1}.492$	$\bar{1}.495$	$\bar{1}.497$
.24	$\bar{1}.499$	$\bar{1}.502$	$\bar{1}.504$	$\bar{1}.506$	$\bar{1}.509$	$\bar{1}.511$	$\bar{1}.514$	$\bar{1}.516$	$\bar{1}.518$	$\bar{1}.521$
.25	$\bar{1}.523$	$\bar{1}.525$	$\bar{1}.527$	$\bar{1}.530$	$\bar{1}.532$	$\bar{1}.534$	$\bar{1}.537$	$\bar{1}.539$	$\bar{1}.541$	$\bar{1}.543$
.26	$\bar{1}.546$	$\bar{1}.548$	$\bar{1}.550$	$\bar{1}.552$	$\bar{1}.555$	$\bar{1}.557$	$\bar{1}.559$	$\bar{1}.561$	$\bar{1}.564$	$\bar{1}.566$
.27	$\bar{1}.568$	$\bar{1}.570$	$\bar{1}.572$	$\bar{1}.575$	$\bar{1}.577$	$\bar{1}.579$	$\bar{1}.581$	$\bar{1}.583$	$\bar{1}.586$	$\bar{1}.588$
.28	$\bar{1}.590$	$\bar{1}.592$	$\bar{1}.594$	$\bar{1}.596$	$\bar{1}.598$	$\bar{1}.601$	$\bar{1}.603$	$\bar{1}.605$	$\bar{1}.607$	$\bar{1}.609$
.29	$\bar{1}.611$	$\bar{1}.613$	$\bar{1}.615$	$\bar{1}.617$	$\bar{1}.620$	$\bar{1}.622$	$\bar{1}.624$	$\bar{1}.626$	$\bar{1}.628$	$\bar{1}.630$

TABLE 1 *(cont.)*

x	.000	.001	.002	.003	.004	.005	.006	.007	.008	.009
.30	$\bar{1}.632$	$\bar{1}.634$	$\bar{1}.636$	$\bar{1}.638$	$\bar{1}.640$	$\bar{1}.642$	$\bar{1}.644$	$\bar{1}.646$	$\bar{1}.648$	$\bar{1}.650$
.31	$\bar{1}.652$	$\bar{1}.655$	$\bar{1}.657$	$\bar{1}.659$	$\bar{1}.661$	$\bar{1}.663$	$\bar{1}.665$	$\bar{1}.667$	$\bar{1}.669$	$\bar{1}.671$
.32	$\bar{1}.673$	$\bar{1}.675$	$\bar{1}.677$	$\bar{1}.679$	$\bar{1}.681$	$\bar{1}.683$	$\bar{1}.685$	$\bar{1}.686$	$\bar{1}.688$	$\bar{1}.690$
.33	$\bar{1}.692$	$\bar{1}.694$	$\bar{1}.696$	$\bar{1}.698$	$\bar{1}.700$	$\bar{1}.702$	$\bar{1}.704$	$\bar{1}.706$	$\bar{1}.708$	$\bar{1}.710$
.34	$\bar{1}.712$	$\bar{1}.714$	$\bar{1}.716$	$\bar{1}.718$	$\bar{1}.720$	$\bar{1}.722$	$\bar{1}.723$	$\bar{1}.725$	$\bar{1}.727$	$\bar{1}.729$
.35	$\bar{1}.731$	$\bar{1}.733$	$\bar{1}.735$	$\bar{1}.737$	$\bar{1}.739$	$\bar{1}.741$	$\bar{1}.743$	$\bar{1}.744$	$\bar{1}.746$	$\bar{1}.748$
.36	$\bar{1}.750$	$\bar{1}.752$	$\bar{1}.754$	$\bar{1}.756$	$\bar{1}.758$	$\bar{1}.760$	$\bar{1}.761$	$\bar{1}.763$	$\bar{1}.765$	$\bar{1}.767$
.37	$\bar{1}.769$	$\bar{1}.771$	$\bar{1}.773$	$\bar{1}.774$	$\bar{1}.776$	$\bar{1}.778$	$\bar{1}.780$	$\bar{1}.782$	$\bar{1}.784$	$\bar{1}.785$
.38	$\bar{1}.787$	$\bar{1}.789$	$\bar{1}.791$	$\bar{1}.793$	$\bar{1}.795$	$\bar{1}.797$	$\bar{1}.798$	$\bar{1}.800$	$\bar{1}.802$	$\bar{1}.804$
.39	$\bar{1}.806$	$\bar{1}.808$	$\bar{1}.809$	$\bar{1}.811$	$\bar{1}.813$	$\bar{1}.815$	$\bar{1}.817$	$\bar{1}.818$	$\bar{1}.820$	$\bar{1}.822$
.40	$\bar{1}.824$	$\bar{1}.826$	$\bar{1}.828$	$\bar{1}.829$	$\bar{1}.831$	$\bar{1}.833$	$\bar{1}.835$	$\bar{1}.837$	$\bar{1}.838$	$\bar{1}.840$
.41	$\bar{1}.842$	$\bar{1}.844$	$\bar{1}.846$	$\bar{1}.847$	$\bar{1}.849$	$\bar{1}.851$	$\bar{1}.853$	$\bar{1}.854$	$\bar{1}.856$	$\bar{1}.858$
.42	$\bar{1}.860$	$\bar{1}.862$	$\bar{1}.863$	$\bar{1}.865$	$\bar{1}.867$	$\bar{1}.869$	$\bar{1}.870$	$\bar{1}.872$	$\bar{1}.874$	$\bar{1}.876$
.43	$\bar{1}.878$	$\bar{1}.879$	$\bar{1}.881$	$\bar{1}.883$	$\bar{1}.885$	$\bar{1}.886$	$\bar{1}.888$	$\bar{1}.890$	$\bar{1}.892$	$\bar{1}.894$
.44	$\bar{1}.895$	$\bar{1}.897$	$\bar{1}.899$	$\bar{1}.901$	$\bar{1}.902$	$\bar{1}.904$	$\bar{1}.906$	$\bar{1}.908$	$\bar{1}.909$	$\bar{1}.911$
.45	$\bar{1}.913$	$\bar{1}.915$	$\bar{1}.916$	$\bar{1}.918$	$\bar{1}.920$	$\bar{1}.922$	$\bar{1}.923$	$\bar{1}.925$	$\bar{1}.927$	$\bar{1}.929$
.46	$\bar{1}.930$	$\bar{1}.932$	$\bar{1}.934$	$\bar{1}.936$	$\bar{1}.937$	$\bar{1}.939$	$\bar{1}.941$	$\bar{1}.943$	$\bar{1}.944$	$\bar{1}.946$
.47	$\bar{1}.948$	$\bar{1}.950$	$\bar{1}.951$	$\bar{1}.953$	$\bar{1}.955$	$\bar{1}.957$	$\bar{1}.958$	$\bar{1}.960$	$\bar{1}.962$	$\bar{1}.963$
.48	$\bar{1}.965$	$\bar{1}.967$	$\bar{1}.969$	$\bar{1}.970$	$\bar{1}.972$	$\bar{1}.974$	$\bar{1}.976$	$\bar{1}.977$	$\bar{1}.979$	$\bar{1}.981$
.49	$\bar{1}.983$	$\bar{1}.984$	$\bar{1}.986$	$\bar{1}.988$	$\bar{1}.990$	$\bar{1}.991$	$\bar{1}.993$	$\bar{1}.995$	$\bar{1}.997$	$\bar{1}.998$
.50	0.000	0.002	0.003	0.005	0.007	0.009	0.010	0.012	0.014	0.016
.51	0.017	0.019	0.021	0.023	0.024	0.026	0.028	0.030	0.031	0.033
.52	0.035	0.037	0.038	0.040	0.042	0.043	0.045	0.047	0.049	0.050
.53	0.052	0.054	0.056	0.057	0.059	0.061	0.063	0.064	0.066	0.068
.54	0.070	0.071	0.073	0.075	0.077	0.078	0.080	0.082	0.084	0.085
.55	0.087	0.089	0.091	0.092	0.094	0.096	0.098	0.099	0.101	0.103
.56	0.105	0.106	0.108	0.110	0.112	0.114	0.115	0.117	0.119	0.121
.57	0.122	0.124	0.126	0.128	0.130	0.131	0.133	0.135	0.137	0.138
.58	0.140	0.142	0.144	0.146	0.147	0.149	0.151	0.153	0.154	0.156
.59	0.158	0.160	0.162	0.163	0.165	0.167	0.169	0.171	0.172	0.174
.60	0.176	0.178	0.180	0.182	0.183	0.185	0.187	0.189	0.191	0.192
.61	0.194	0.196	0.198	0.200	0.202	0.203	0.205	0.207	0.209	0.211
.62	0.213	0.215	0.216	0.218	0.220	0.222	0.224	0.226	0.227	0.229
.63	0.231	0.233	0.235	0.237	0.239	0.240	0.242	0.244	0.246	0.248
.64	0.250	0.252	0.254	0.256	0.257	0.259	0.261	0.263	0.265	0.267

TABLE 1 *(cont.)*

x	.000	.001	.002	.003	.004	.005	.006	.007	.008	.009
.65	0.269	0.271	0.273	0.275	0.277	0.278	0.280	0.282	0.284	0.286
.66	0.288	0.290	0.292	0.294	0.296	0.298	0.300	0.302	0.304	0.306
.67	0.308	0.310	0.312	0.314	0.315	0.317	0.319	0.321	0.323	0.325
.68	0.327	0.329	0.331	0.333	0.335	0.337	0.339	0.341	0.343	0.345
.69	0.348	0.350	0.352	0.354	0.356	0.358	0.360	0.362	0.364	0.366
.70	0.368	0.370	0.372	0.374	0.376	0.378	0.380	0.383	0.385	0.387
.71	0.389	0.391	0.393	0.395	0.397	0.399	0.402	0.404	0.406	0.408
.72	0.410	0.412	0.414	0.417	0.419	0.421	0.423	0.425	0.428	0.430
.73	0.432	0.434	0.436	0.439	0.441	0.443	0.445	0.448	0.450	0.452
.74	0.454	0.457	0.459	0.461	0.463	0.466	0.468	0.470	0.473	0.475
.75	0.477	0.479	0.481	0.484	0.486	0.489	0.491	0.494	0.496	0.498
.76	0.501	0.503	0.505	0.508	0.510	0.513	0.515	0.517	0.520	0.522
.77	0.525	0.527	0.530	0.532	0.535	0.537	0.540	0.542	0.545	0.547
.78	0.550	0.552	0.555	0.557	0.560	0.562	0.565	0.567	0.570	0.573
.79	0.575	0.578	0.581	0.583	0.586	0.589	0.591	0.594	0.597	0.599
.80	0.602	0.605	0.608	0.610	0.613	0.616	0.619	0.621	0.624	0.627
.81	0.630	0.633	0.635	0.638	0.641	0.644	0.647	0.650	0.653	0.656
.82	0.659	0.662	0.665	0.667	0.670	0.673	0.676	0.679	0.683	0.686
.83	0.689	0.692	0.695	0.698	0.701	0.704	0.707	0.711	0.714	0.717
.84	0.720	0.723	0.727	0.730	0.733	0.737	0.740	0.743	0.747	0.750
.85	0.753	0.757	0.760	0.764	0.767	0.771	0.774	0.778	0.781	0.785
.86	0.788	0.792	0.796	0.799	0.803	0.807	0.810	0.814	0.818	0.822
.87	0.826	0.829	0.833	0.837	0.841	0.845	0.849	0.853	0.857	0.861
.88	0.865	0.869	0.874	0.878	0.882	0.886	0.891	0.895	0.899	0.904
.89	0.908	0.912	0.917	0.922	0.926	0.931	0.935	0.940	0.945	0.949
.90	0.954	0.959	0.964	0.969	0.974	0.979	0.984	0.989	0.994	1.000
.91	1.005	1.010	1.016	1.021	1.026	1.032	1.038	1.043	1.049	1.055
.92	1.061	1.067	1.073	1.079	1.085	1.091	1.097	1.104	1.110	1.117
.93	1.123	1.130	1.137	1.144	1.151	1.158	1.165	1.172	1.180	1.187
.94	1.195	1.203	1.211	1.219	1.227	1.235	1.244	1.252	1.261	1.270
.95	1.279	1.288	1.297	1.307	1.317	1.327	1.337	1.347	1.358	1.369
.96	1.380	1.392	1.403	1.415	1.428	1.441	1.454	1.467	1.481	1.495
.97	1.510	1.525	1.541	1.557	1.574	1.591	1.609	1.628	1.648	1.669
.98	1.690	1.713	1.737	1.762	1.789	1.817	1.848	1.880	1.916	1.954
.99	1.996	2.042	2.093	2.152	2.219	2.299	2.396	2.522	2.698	3.000

TABLE 2

Log$_e$ $[x/(1-x)]$ or Logits[a]

(For x less than .50 on left, logit is negative. For x greater than .50 on right, logit is positive)

x	Thousandths, for x in left column											
	0	1	2	3	4	5	6	7	8	9		
.00	—	6.90675	6.21261	5.80614	5.51745	5.29330	5.10998	4.95482	4.82028	4.70149	4.59512	.99
.01	4.59512	4.49880	4.41078	4.32972	4.25460	4.18459	4.11904	4.05740	3.99922	3.94413	3.89182	.98
.02	3.89182	3.84201	3.79447	3.74899	3.70541	3.66356	3.62331	3.58455	3.54715	3.51103	3.47610	.97
.03	3.47610	3.44228	3.40950	3.37769	3.34680	3.31678	3.28757	3.25914	3.23143	3.20441	3.17805	.96
.04	3.17805	3.15232	3.12718	3.10260	3.07857	3.05505	3.03202	3.00947	2.98736	2.96569	2.94444	.95
.05	2.94444	2.92358	2.90311	2.88301	2.86326	2.84385	2.82477	2.80601	2.78756	2.76941	2.75154	.94
.06	2.75154	2.73394	2.71662	2.69955	2.68273	2.66616	2.64982	2.63371	2.61783	2.60215	2.58669	.93
.07	2.58669	2.57143	2.55637	2.54149	2.52681	2.51231	2.49798	2.48382	2.46984	2.45601	2.44235	.92
.08	2.44235	2.42884	2.41548	2.40227	2.38920	2.37627	2.36348	2.35083	2.33830	2.32591	2.31363	.91
.09	2.31363	2.30149	2.28946	2.27754	2.26574	2.25406	2.24248	2.23101	2.21965	2.20839	2.19722	.90
.10	2.19722	2.18616	2.17520	2.16433	2.15355	2.14286	2.13227	2.12176	2.11133	2.10100	2.09074	.89
.11	2.09074	2.08057	2.07047	2.06046	2.05052	2.04066	2.03087	2.02115	2.01151	2.00193	1.99243	.88
.12	1.99243	1.98299	1.97363	1.96432	1.95508	1.94591	1.93680	1.92775	1.91876	1.90983	1.90096	.87
.13	1.90096	1.89215	1.88339	1.87469	1.86605	1.85745	1.84892	1.84043	1.83200	1.82362	1.81529	.86
.14	1.81529	1.80701	1.79878	1.79059	1.78246	1.77437	1.76632	1.75833	1.75037	1.74247	1.73460	.85
.15	1.73460	1.72678	1.71900	1.71126	1.70357	1.69591	1.68830	1.68072	1.67318	1.66569	1.65823	.84
.16	1.65823	1.65081	1.64342	1.63607	1.62876	1.62149	1.61425	1.60704	1.59987	1.59273	1.58563	.83
.17	1.58563	1.57856	1.57152	1.56451	1.55754	1.55060	1.54369	1.53681	1.52996	1.52314	1.51635	.82
.18	1.51635	1.50959	1.50286	1.49615	1.48948	1.48283	1.47621	1.46962	1.46306	1.45652	1.45001	.81
.19	1.45001	1.44353	1.43707	1.43063	1.42423	1.41784	1.41148	1.40515	1.39884	1.39256	1.38629	.80

TABLE 2 *(cont.)*

x	Thousandths, for x in left column											
	0	1	2	3	4	5	6	7	8	9		
.20	1.38629	1.38006	1.37384	1.36765	1.36148	1.35533	1.34921	1.34310	1.33702	1.33096	1.32493	.79
.21	1.32493	1.31891	1.31291	1.30694	1.30098	1.29505	1.28913	1.28324	1.27736	1.27150	1.26567	.78
.22	1.26567	1.25985	1.25405	1.24827	1.24251	1.23676	1.23104	1.22533	1.21964	1.21397	1.20831	.77
.23	1.20831	1.20267	1.19705	1.19145	1.18586	1.18029	1.17474	1.16920	1.16368	1.15817	1.15268	.76
.24	1.15268	1.14720	1.14175	1.13630	1.13087	1.12546	1.12006	1.11468	1.10931	1.10395	1.09861	.75
.25	1.09861	1.09329	1.08797	1.08268	1.07739	1.07212	1.06686	1.06162	1.05639	1.05117	1.04597	.74
.26	1.04597	1.04078	1.03560	1.03043	1.02528	1.02014	1.01501	1.00990	1.00479	0.99970	0.99462	.73
.27	0.99462	0.98955	0.98450	0.97945	0.97442	0.96940	0.96439	0.95939	0.95440	0.94943	0.94446	.72
.28	0.94446	0.93951	0.93456	0.92963	0.92471	0.91979	0.91489	0.91000	0.90512	0.90025	0.89538	.71
.29	0.89538	0.89053	0.88569	0.88086	0.87604	0.87122	0.86642	0.86162	0.85684	0.85206	0.84730	.70
.30	0.84730	0.84254	0.83779	0.83305	0.82832	0.82360	0.81889	0.81418	0.80949	0.80480	0.80012	.69
.31	0.80012	0.79545	0.79079	0.78613	0.78148	0.77685	0.77222	0.76759	0.76298	0.75837	0.75377	.68
.32	0.75377	0.74918	0.74460	0.74002	0.73545	0.73089	0.72633	0.72179	0.71724	0.71271	0.70819	.67
.33	0.70819	0.70367	0.69915	0.69465	0.69015	0.68566	0.68117	0.67669	0.67222	0.66775	0.66329	.66
.34	0.66329	0.65884	0.65439	0.64995	0.64552	0.64109	0.63667	0.63225	0.62784	0.62344	0.61904	.65
.35	0.61904	0.61465	0.61026	0.60588	0.60150	0.59713	0.59277	0.58841	0.58406	0.57971	0.57536	.64
.36	0.57536	0.57103	0.56669	0.56237	0.55804	0.55373	0.54942	0.54511	0.54081	0.53651	0.53222	.63
.37	0.53222	0.52793	0.52365	0.51937	0.51509	0.51083	0.50656	0.50230	0.49805	0.49379	0.48955	.62
.38	0.48955	0.48531	0.48107	0.47683	0.47260	0.46838	0.46416	0.45994	0.45573	0.45152	0.44731	.61
.39	0.44731	0.44311	0.43891	0.43472	0.43053	0.42634	0.42216	0.41798	0.41381	0.40963	0.40547	.60

TABLE 2 (cont.)

x	Thousandths, for x in left column											
	0	1	2	3	4	5	6	7	8	9		
.40	0.40547	0.40130	0.39714	0.39298	0.38883	0.38467	0.38053	0.37638	0.37224	0.36810	0.36397	.59
.41	0.36397	0.35983	0.35570	0.35158	0.34745	0.34333	0.33922	0.33510	0.33099	0.32688	0.32277	.58
.42	0.32277	0.31867	0.31457	0.31047	0.30637	0.30228	0.29819	0.29410	0.29002	0.28593	0.28185	.57
.43	0.28185	0.27777	0.27370	0.26962	0.26555	0.26148	0.25741	0.25335	0.24928	0.24522	0.24116	.56
.44	0.24116	0.23710	0.23305	0.22900	0.22494	0.22089	0.21685	0.21280	0.20875	0.20471	0.20067	.55
.45	0.20067	0.19663	0.19259	0.18856	0.18452	0.18049	0.17646	0.17243	0.16840	0.16437	0.16034	.54
.46	0.16034	0.15632	0.15229	0.14827	0.14425	0.14023	0.13621	0.13219	0.12818	0.12416	0.12014	.53
.47	0.12014	0.11613	0.11212	0.10811	0.10409	0.10008	0.09607	0.09206	0.08806	0.08405	0.08004	.52
.48	0.08004	0.07604	0.07203	0.06803	0.06402	0.06002	0.05601	0.05201	0.04801	0.04401	0.04001	.51
.49	0.04001	0.03600	0.03200	0.02800	0.02400	0.02000	0.01600	0.01200	0.00800	0.00400	0.00000	.50
		9	8	7	6	5	4	3	2	1	0	x
		Thousandths, for x in right column										

[a] Berkson's (1953) table of logits is reproduced by kind permission of Dr. Joseph Berkson, of the Division of Biometry and Medical Statistics of the Mayo Clinic, Rochester, Minnesota and the Editor of the *Journal of the American Statistical Association.*

TABLE 3

$\log_e[1/(1-x)]$

The table begins with $x = 0.01$. For lower values of x use $\log_e[1/(1-x)] = x$.

x	0.000	0.001	0.002	0.003	0.004	0.005	0.006	0.007	0.008	0.009
.01	0.010	0.011	0.012	0.013	0.014	0.015	0.016	0.017	0.018	0.019
.02	0.020	0.021	0.022	0.023	0.024	0.025	0.026	0.027	0.028	0.029
.03	0.030	0.031	0.033	0.034	0.035	0.036	0.037	0.038	0.039	0.040
.04	0.041	0.042	0.043	0.044	0.045	0.046	0.047	0.048	0.049	0.050
.05	0.051	0.052	0.053	0.054	0.056	0.057	0.058	0.059	0.060	0.061
.06	0.062	0.063	0.064	0.065	0.066	0.067	0.068	0.069	0.070	0.071
.07	0.073	0.074	0.075	0.076	0.077	0.078	0.079	0.080	0.081	0.082
.08	0.083	0.084	0.086	0.087	0.088	0.089	0.090	0.091	0.092	0.093
.09	0.094	0.095	0.097	0.098	0.099	0.100	0.101	0.102	0.103	0.104
.10	0.105	0.106	0.108	0.109	0.110	0.111	0.112	0.113	0.114	0.115
.11	0.117	0.118	0.119	0.120	0.121	0.122	0.123	0.124	0.126	0.127
.12	0.128	0.129	0.130	0.131	0.132	0.134	0.135	0.136	0.137	0.138
.13	0.139	0.140	0.142	0.143	0.144	0.145	0.146	0.147	0.148	0.149
.14	0.151	0.152	0.153	0.154	0.155	0.157	0.158	0.159	0.160	0.161
.15	0.163	0.164	0.165	0.166	0.167	0.168	0.170	0.171	0.172	0.173
.16	0.174	0.176	0.177	0.178	0.179	0.180	0.182	0.183	0.184	0.185
.17	0.186	0.188	0.189	0.190	0.191	0.192	0.194	0.195	0.196	0.197
.18	0.198	0.200	0.201	0.202	0.203	0.205	0.206	0.207	0.208	0.209
.19	0.211	0.212	0.213	0.214	0.216	0.217	0.218	0.219	0.221	0.222
.20	0.223	0.224	0.226	0.227	0.228	0.229	0.231	0.232	0.233	0.234
.21	0.236	0.237	0.238	0.240	0.241	0.242	0.243	0.245	0.246	0.247
.22	0.248	0.250	0.251	0.252	0.254	0.255	0.256	0.257	0.259	0.260
.23	0.261	0.263	0.264	0.265	0.267	0.268	0.269	0.270	0.272	0.273
.24	0.274	0.276	0.277	0.278	0.280	0.281	0.282	0.284	0.285	0.286
.25	0.288	0.289	0.290	0.292	0.293	0.294	0.296	0.297	0.298	0.300
.26	0.301	0.302	0.304	0.305	0.306	0.308	0.309	0.311	0.312	0.313
.27	0.315	0.316	0.317	0.319	0.320	0.322	0.323	0.324	0.326	0.327
.28	0.329	0.330	0.331	0.333	0.334	0.336	0.337	0.338	0.340	0.341
.29	0.342	0.344	0.345	0.347	0.348	0.350	0.351	0.352	0.354	0.355
.30	0.357	0.358	0.359	0.361	0.362	0.364	0.365	0.367	0.368	0.370

TABLE 3 ***(cont.)***

x	0.000	0.001	0.002	0.003	0.004	0.005	0.006	0.007	0.008	0.009
.31	0.371	0.372	0.374	0.375	0.377	0.378	0.380	0.381	0.383	0.384
.32	0.386	0.387	0.389	0.390	0.392	0.393	0.394	0.396	0.398	0.399
.33	0.400	0.402	0.403	0.405	0.406	0.408	0.409	0.411	0.413	0.414
.34	0.416	0.417	0.419	0.420	0.422	0.423	0.425	0.426	0.428	0.429
.35	0.431	0.432	0.434	0.435	0.437	0.439	0.440	0.442	0.443	0.445
.36	0.446	0.448	0.449	0.451	0.453	0.454	0.456	0.457	0.459	0.460
.37	0.462	0.464	0.465	0.467	0.468	0.470	0.472	0.473	0.475	0.476
.38	0.478	0.480	0.481	0.483	0.485	0.486	0.488	0.489	0.491	0.493
.39	0.494	0.496	0.498	0.499	0.501	0.503	0.504	0.506	0.508	0.509
.40	0.511	0.513	0.514	0.516	0.518	0.519	0.521	0.523	0.524	0.526
.41	0.528	0.529	0.531	0.533	0.534	0.536	0.538	0.540	0.541	0.543
.42	0.545	0.546	0.548	0.550	0.552	0.553	0.555	0.557	0.559	0.560
.43	0.562	0.564	0.566	0.567	0.569	0.571	0.573	0.574	0.576	0.578
.44	0.580	0.582	0.583	0.585	0.587	0.589	0.591	0.592	0.594	0.596
.45	0.598	0.600	0.601	0.603	0.605	0.607	0.609	0.611	0.612	0.614
.46	0.616	0.618	0.620	0.622	0.624	0.626	0.627	0.629	0.631	0.633
.47	0.635	0.637	0.639	0.641	0.642	0.644	0.646	0.648	0.650	0.652
.48	0.654	0.656	0.658	0.660	0.662	0.664	0.666	0.667	0.669	0.671
.49	0.673	0.675	0.677	0.679	0.681	0.683	0.685	0.687	0.689	0.691
.50	0.693	0.695	0.697	0.699	0.701	0.703	0.705	0.707	0.709	0.711
.51	0.713	0.715	0.717	0.719	0.721	0.724	0.726	0.728	0.730	0.732
.52	0.734	0.736	0.738	0.740	0.742	0.744	0.747	0.749	0.751	0.753
.53	0.755	0.757	0.759	0.761	0.764	0.766	0.768	0.770	0.772	0.774
.54	0.777	0.779	0.781	0.783	0.785	0.787	0.790	0.792	0.794	0.796
.55	0.799	0.801	0.803	0.805	0.807	0.810	0.812	0.814	0.816	0.819
.56	0.821	0.823	0.826	0.828	0.830	0.832	0.835	0.837	0.839	0.842
.57	0.844	0.846	0.849	0.851	0.853	0.856	0.858	0.860	0.863	0.865
.58	0.868	0.870	0.872	0.875	0.877	0.879	0.882	0.884	0.887	0.889
.59	0.892	0.894	0.896	0.899	0.901	0.904	0.906	0.909	0.911	0.914
.60	0.916	0.918	0.921	0.924	0.926	0.929	0.931	0.934	0.937	0.939
.61	0.942	0.944	0.947	0.949	0.952	0.955	0.957	0.960	0.962	0.965
.62	0.968	0.970	0.973	0.976	0.978	0.981	0.983	0.986	0.989	0.992
.63	0.994	0.997	1.000	1.002	1.005	1.008	1.010	1.013	1.016	1.019
.64	1.022	1.024	1.027	1.030	1.033	1.036	1.038	1.041	1.044	1.047
.65	1.050	1.053	1.056	1.058	1.061	1.064	1.067	1.070	1.073	1.076

TABLE 3 *(cont.)*

x	0.000	0.001	0.002	0.003	0.004	0.005	0.006	0.007	0.008	0.009
.66	1.079	1.082	1.085	1.088	1.091	1.094	1.097	1.100	1.103	1.106
.67	1.109	1.112	1.115	1.118	1.121	1.124	1.127	1.130	1.133	1.136
.68	1.139	1.142	1.146	1.149	1.152	1.155	1.158	1.162	1.165	1.168
.69	1.171	1.174	1.178	1.181	1.184	1.187	1.191	1.194	1.197	1.201
.70	1.204	1.207	1.211	1.214	1.217	1.221	1.224	1.228	1.231	1.234
.71	1.238	1.241	1.245	1.248	1.252	1.255	1.259	1.262	1.266	1.269
.72	1.273	1.277	1.280	1.284	1.287	1.291	1.295	1.298	1.302	1.306
.73	1.309	1.313	1.317	1.321	1.324	1.328	1.332	1.336	1.339	1.343
.74	1.347	1.351	1.355	1.359	1.363	1.367	1.370	1.374	1.378	1.382
.75	1.386	1.390	1.394	1.398	1.402	1.406	1.411	1.415	1.419	1.423
.76	1.427	1.431	1.435	1.440	1.444	1.448	1.452	1.457	1.461	1.465
.77	1.470	1.474	1.478	1.483	1.487	1.492	1.496	1.501	1.505	1.510
.78	1.514	1.519	1.523	1.528	1.532	1.537	1.542	1.546	1.551	1.556
.79	1.561	1.565	1.570	1.575	1.580	1.585	1.590	1.595	1.599	1.604
.80	1.609	1.614	1.619	1.625	1.630	1.635	1.640	1.645	1.650	1.655
.81	1.661	1.666	1.671	1.677	1.682	1.687	1.693	1.698	1.704	1.709
.82	1.715	1.720	1.726	1.732	1.737	1.743	1.749	1.754	1.760	1.766
.83	1.772	1.778	1.784	1.790	1.796	1.802	1.808	1.814	1.820	1.826
.84	1.833	1.839	1.845	1.852	1.858	1.864	1.871	1.877	1.884	1.890
.85	1.897	1.904	1.911	1.917	1.924	1.931	1.938	1.945	1.952	1.959
.86	1.966	1.973	1.981	1.988	1.995	2.002	2.010	2.017	2.025	2.033
.87	2.040	2.048	2.056	2.064	2.072	2.079	2.088	2.096	2.104	2.112
.88	2.120	2.129	2.137	2.146	2.154	2.163	2.172	2.180	2.189	2.198
.89	2.207	1.216	2.226	2.235	2.244	2.254	2.264	2.273	2.283	2.293
.90	2.303	2.313	2.323	2.333	2.343	2.354	2.364	2.375	2.386	2.397
.91	2.408	2.419	2.430	2.442	2.453	2.465	2.477	2.489	2.501	2.513
.92	2.526	2.538	2.551	2.564	2.577	2.590	2.604	2.617	2.631	2.645
.93	2.659	2.674	2.688	2.703	2.718	2.733	2.749	2.765	2.781	2.797
.94	2.813	2.830	2.847	2.865	2.882	2.900	2.919	2.937	2.957	2.976
.95	2.996	3.016	3.037	3.058	3.079	3.101	3.124	3.147	3.170	3.194
.96	3.219	3.244	3.270	3.297	3.324	3.352	3.381	3.411	3.442	3.474
.97	3.507	3.541	3.576	3.612	3.650	3.689	3.730	3.772	3.817	3.863
.98	3.912	3.963	4.017	4.075	4.135	4.200	4.269	4.343	4.423	4.510
.99	4.605	4.710	4.828	4.962	5.116	5.298	5.521	5.809	6.215	6.908

TABLE 4

e^a

The table is useful for evaluating e^{rt}, e^{pr}, and $e^{(i+p)rl}$. Solving Eq. (8.5) demands a wide range of values of a.

For values not in the table use $\log_{10} e^a = 0.4343a$.

a	.00	.01	.02	.03	.04	.05	.06	.07	.08	.09
0.0	1.000	1.010	1.020	1.030	1.041	1.051	1.062	1.073	1.083	1.094
0.1	1.105	1.116	1.127	1.139	1.150	1.162	1.174	1.185	1.197	1.209
0.2	1.221	1.234	1.246	1.259	1.271	1.284	1.297	1.310	1.323	1.336
0.3	1.350	1.363	1.377	1.391	1.405	1.419	1.433	1.448	1.462	1.477
0.4	1.492	1.507	1.522	1.537	1.553	1.568	1.584	1.600	1.616	1.632
0.5	1.647	1.665	1.682	1.699	1.716	1.733	1.751	1.768	1.786	1.804
0.6	1.822	1.840	1.859	1.878	1.896	1.916	1.935	1.954	1.974	1.994
0.7	2.014	2.034	2.054	2.075	2.096	2.117	2.138	2.160	2.181	2.203
0.8	2.226	2.248	2.270	2.293	2.316	2.340	2.363	2.387	2.411	2.435
0.9	2.460	2.484	2.509	2.535	2.560	2.586	2.612	2.638	2.664	2.691
1.0	2.718	2.746	2.773	2.801	2.829	2.858	2.886	2.915	2.945	2.974
1.1	3.004	3.034	3.065	3.096	3.127	3.158	3.190	3.222	3.254	3.287
1.2	3.320	3.353	3.387	3.421	3.456	3.490	3.525	3.561	3.597	3.633
1.3	3.670	3.706	3.743	3.781	3.819	3.857	3.896	3.935	3.975	4.015
1.4	4.055	4.096	4.137	4.179	4.221	4.263	4.306	4.349	4.393	4.437
1.5	4.482	4.527	4.572	4.618	4.665	4.711	4.759	4.807	4.855	4.904
1.6	4.953	5.003	5.053	5.104	5.155	5.207	5.259	5.312	5.366	5.419
1.7	5.474	5.529	5.585	5.641	5.697	5.755	5.812	5.871	5.930	5.989
1.8	6.050	6.110	6.172	6.234	6.297	6.360	6.424	6.488	6.553	6.619
1.9	6.686	6.753	6.821	6.890	6.959	7.029	7.099	7.171	7.243	7.316
2.0	7.389	7.463	7.538	7.614	7.691	7.768	7.846	7.925	8.004	8.085
2.1	8.166	8.248	8.331	8.415	8.499	8.585	8.671	8.758	8.846	8.935
2.2	9.025	9.116	9.207	9.300	9.393	9.488	9.583	9.679	9.777	9.875
2.3	9.974	10.07	10.18	10.28	10.38	10.49	10.59	10.70	10.80	10.91
2.4	11.02	11.13	11.25	11.36	11.47	11.59	11.70	11.82	11.94	12.06
2.5	12.18	12.30	12.43	12.55	12.68	12.81	12.94	13.07	13.20	13.33
2.6	13.46	13.60	13.74	13.87	14.01	14.15	14.30	14.44	14.59	14.73
2.7	14.88	15.03	15.18	15.33	15.49	15.64	15.80	15.96	16.12	16.28
2.8	16.44	16.61	16.78	16.95	17.12	17.29	17.46	17.64	17.81	17.99
2.9	18.17	18.36	18.54	18.73	18.92	19.11	19.30	19.49	19.69	19.89

TABLE 4 *(cont.)*

a	.00	.01	.02	.03	.04	.05	.06	.07	.08	.09
3.0	20.09	20.29	20.49	20.70	20.91	21.12	21.33	21.54	21.76	21.98
3.1	22.20	22.42	22.65	22.87	23.10	23.34	23.57	23.81	24.05	24.29
3.2	24.53	24.78	25.03	25.28	25.53	25.79	26.05	26.31	26.58	26.84
3.3	27.11	27.39	27.66	27.94	28.22	28.50	28.79	29.08	29.37	29.67
3.4	29.96	30.27	30.57	30.88	31.19	31.50	31.82	32.14	32.46	32.79
3.5	33.12	33.45	33.78	34.12	34.47	34.81	35.16	35.52	35.87	36.23
3.6	36.60	36.97	37.34	37.71	38.09	38.47	38.86	39.25	39.65	40.04
3.7	40.45	40.85	41.26	41.68	42.10	42.52	42.95	43.38	43.82	44.26
3.8	44.70	45.15	45.60	46.06	46.53	46.99	47.47	47.94	48.42	48.91
3.9	49.40	49.90	50.40	50.91	51.42	51.94	52.46	52.98	53.52	54.05
4.0	54.60	55.15	55.70	56.26	56.83	57.40	57.97	58.56	59.15	59.74
4.1	60.34	60.95	61.56	62.18	62.80	63.43	64.07	64.72	65.37	66.02
4.2	66.69	67.36	68.03	68.72	69.41	70.11	70.81	71.52	72.24	72.97
4.3	73.70	74.44	75.19	75.94	76.71	77.48	78.26	79.04	79.84	80.64
4.4	81.45	82.27	83.10	83.93	84.77	85.63	86.49	87.36	88.23	89.12
4.5	90.02	90.92	91.84	92.76	93.69	94.63	95.58	96.54	97.51	98.49
4.6	99.48	100.5	101.5	102.5	103.5	104.6	105.6	106.7	107.8	108.9
4.7	109.9	111.1	112.2	113.3	114.4	115.6	116.7	117.9	119.1	120.3
4.8	121.5	122.7	124.0	125.2	126.5	127.4	129.0	130.3	131.6	133.0
4.9	134.3	135.6	137.0	138.4	139.8	141.2	142.6	144.0	145.5	146.9
5.0	148.4	149.9	151.4	152.9	154.5	156.0	157.6	159.2	160.8	162.4
5.1	164.0	165.7	167.3	169.0	170.7	172.4	174.2	175.9	177.7	179.5
5.2	181.3	183.1	184.9	186.8	188.7	190.6	192.5	194.4	196.4	198.3
5.3	200.3	202.4	204.4	206.4	208.5	210.6	212.7	214.9	217.0	219.2
5.4	221.4	223.6	225.9	228.1	230.4	232.8	235.1	237.5	239.8	242.3
5.5	244.7	247.2	249.6	252.1	254.7	257.2	259.8	262.4	265.1	267.7
5.6	270.4	273.1	275.9	278.7	281.5	284.3	287.1	290.0	292.9	295.9
5.7	298.9	301.9	304.9	308.0	311.1	314.2	317.3	320.5	323.8	327.0
5.8	330.3	333.6	337.0	340.4	343.8	347.2	350.7	354.2	357.8	361.4
5.9	365.0	368.7	372.4	376.2	379.9	383.7	387.6	391.5	395.4	399.4
6.0	403.4	407.5	411.6	415.7	419.9	424.1	428.4	432.7	437.1	441.4
6.1	445.9	450.3	454.9	459.4	464.1	468.7	473.4	478.2	483.0	487.8
6.2	492.7	497.7	502.7	507.8	512.9	518.0	523.2	528.4	533.8	539.2
6.3	544.6	550.0	555.6	561.2	566.8	572.5	578.2	584.1	589.9	595.9
6.4	601.8	607.9	614.0	620.2	626.4	632.7	639.1	645.5	652.0	658.5

TABLE 4 *(cont.)*

a	.00	.01	.02	.03	.04	.05	.06	.07	.08	.09
6.5	665.1	671.8	678.6	685.4	692.3	699.2	706.3	713.4	720.5	727.8
6.6	735.1	742.5	749.9	757.5	765.1	772.8	780.6	788.4	796.3	804.3
6.7	812.4	820.6	828.8	837.1	845.6	854.1	862.6	871.3	880.1	888.9
6.8	897.8	906.9	916.0	925.2	934.5	943.9	953.4	962.9	972.6	982.4
6.9	992.3	1002	1012	1022	1033	1043	1054	1064	1075	1086
7.0	1097	1108	1119	1130	1141	1153	1164	1176	1188	1200
7.1	1212	1224	1236	1249	1261	1274	1287	1300	1313	1326
7.2	1339	1353	1366	1384	1394	1408	1422	1437	1451	1466
7.3	1480	1495	1510	1525	1541	1556	1572	1588	1604	1620
7.4	1636	1652	1669	1686	1703	1720	1737	1755	1772	1790
7.5	1808	1826	1845	1863	1882	1901	1920	1939	1959	1978
7.6	1998	2018	2039	2059	2080	2101	2122	2143	2165	2186
7.7	2208	2231	2253	2276	2298	2322	2345	2368	2392	2416
7.8	2441	2465	2490	2515	2540	2566	2592	2618	2644	2670
7.9	2697	2724	2752	2779	2807	2836	2864	2893	2922	2951
8.0	2981	3011	3041	3072	3103	3134	3165	3197	3229	3262
8.1	3294	3328	3361	3395	3429	3463	3498	3533	3569	3605
8.2	3641	3678	3715	3752	3790	3828	3866	3905	3944	3984
8.3	4024	4064	4105	4146	4188	4230	4273	4316	4359	4403
8.4	4447	4492	4537	4583	4629	4675	4722	4770	4817	4866
8.5	4915	4964	5014	5064	5115	5167	5219	5271	5324	5378
8.6	5432	5486	5541	5597	5653	5710	5767	5826	5884	5943
8.7	6003	6063	6124	6186	6248	6311	6374	6438	6503	6568
8.8	6634	6701	6768	6836	6905	6974	7044	7115	7187	7259
8.9	7332	7406	7480	7555	7631	7708	7785	7864	7943	8022

TABLE 5

ae^a

With a standing for pr_l, this table greatly simplifies solutions of Eq. (5.9), when one is given p and R.

For reasons given in Section 5.9, it has been thought unnecessary to take a higher than 5.99.

a	.00	.01	.02	.03	.04	.05	.06	.07	.08	.09
0.0	—	0.0101	0.0204	0.0309	0.0416	0.0526	0.0637	0.0751	0.0867	0.0985
0.1	0.1105	0.1228	0.1353	0.1480	0.1610	0.1743	0.1878	0.2015	0.2155	0.2297
0.2	0.2443	0.2591	0.2741	0.2895	0.3051	0.3210	0.3372	0.3537	0.3705	0.3876
0.3	0.4050	0.4227	0.4407	0.4590	0.4777	0.4967	0.5160	0.5356	0.5557	0.5760
0.4	0.5967	0.6178	0.6392	0.6610	0.6832	0.7057	0.7287	0.7520	0.7757	0.7998
0.5	0.8244	0.8493	0.8746	0.9004	0.9266	0.9533	0.9804	1.008	1.036	1.064
0.6	1.093	1.123	1.153	1.183	1.214	1.245	1.277	1.309	1.342	1.376
0.7	1.410	1.444	1.479	1.515	1.551	1.588	1.625	1.633	1.702	1.741
0.8	1.780	1.821	1.862	1.903	1.946	1.989	2.032	2.077	2.122	2.167
0.9	2.214	2.261	2.309	2.357	2.406	2.456	2.507	2.559	2.611	2.664
1.0	2.718	2.773	2.829	2.885	2.942	3.000	3.060	3.119	3.180	3.242
1.1	3.305	3.368	3.433	3.498	3.565	3.632	3.700	3.770	3.840	3.912
1.2	3.985	4.058	4.132	4.208	4.285	4.363	4.442	4.522	4.604	4.686
1.3	4.771	4.855	4.941	5.029	5.117	5.207	5.299	5.391	5.485	5.581
1.4	5.677	5.775	5.875	5.976	6.078	6.182	6.287	6.393	6.501	6.611
1.5	6.723	6.835	6.950	7.066	7.183	7.303	7.424	7.546	7.671	7.797
1.6	7.925	8.055	8.186	8.319	8.455	8.592	8.730	8.871	9.014	9.159
1.7	9.306	9.455	9.605	9.758	9.913	10.07	10.23	10.39	10.56	10.72
1.8	10.89	11.06	11.23	11.41	11.59	11.77	11.95	12.13	12.32	12.51
1.9	12.70	12.90	13.10	13.30	13.50	13.71	13.91	14.13	14.34	14.56
2.0	14.78	15.00	15.23	15.46	15.69	15.92	16.16	16.40	16.65	16.90
2.1	17.15	17.40	17.66	17.92	18.19	18.46	18.73	19.01	19.28	19.57
2.2	19.86	20.15	20.44	20.74	21.04	21.35	21.66	21.97	22.29	22.61
2.3	22.94	23.27	23.61	23.95	24.29	24.64	24.99	25.35	25.72	26.08
2.4	26.46	26.83	27.22	27.60	27.99	28.39	28.79	29.20	29.61	30.03
2.5	30.46	30.89	31.32	31.76	32.21	32.66	33.12	33.58	34.05	34.52
2.6	35.01	35.49	35.99	36.49	36.99	37.51	38.03	38.55	39.23	39.63
2.7	40.18	40.73	41.29	41.86	42.43	43.02	43.61	44.21	44.81	45.42
2.8	46.05	46.67	47.31	47.95	48.61	49.27	49.94	50.62	51.30	52.00
2.9	52.70	53.42	54.14	54.87	55.61	56.36	57.12	57.89	58.67	59.46

TABLE 5 *(cont.)*

a	.00	.01	.02	.03	.04	.05	.06	.07	.08	.09
3.0	60.26	61.06	61.88	62.71	63.55	64.40	65.26	66.13	67.01	67.91
3.1	68.81	69.73	70.66	71.60	72.55	73.51	74.48	75.47	76.47	77.48
3.2	78.51	79.54	80.59	81.65	82.73	83.82	84.92	86.04	87.17	88.31
3.3	89.47	90.64	91.83	93.03	94.25	95.49	96.73	98.00	99.27	100.6
3.4	101.9	103.2	104.5	105.9	107.3	108.7	110.1	111.5	113.0	114.4
3.5	115.9	117.4	118.9	120.5	122.0	123.6	125.2	126.8	128.4	130.1
3.6	131.8	133.4	135.2	136.9	138.7	140.4	142.2	144.1	145.9	147.8
3.7	149.7	151.6	153.5	155.5	157.4	159.5	161.5	163.5	165.6	167.7
3.8	169.9	172.0	174.2	176.4	178.7	180.9	183.2	185.5	187.9	190.3
3.9	192.7	195.1	197.6	200.1	202.6	205.1	207.7	210.4	213.0	215.7
4.0	218.4	221.1	223.9	226.7	229.6	232.5	235.4	238.3	241.3	244.3
4.1	247.4	250.5	253.6	256.8	260.0	263.3	266.5	269.9	273.2	276.6
4.2	280.1	283.6	287.1	290.6	294.3	297.9	301.7	305.4	309.2	313.0
4.3	316.9	320.8	324.8	328.8	332.9	337.0	341.2	345.4	349.7	354.0
4.4	358.4	362.8	367.3	371.8	376.4	381.0	385.7	390.5	395.3	400.1
4.5	405.1	410.0	415.1	420.2	425.4	430.6	435.8	441.2	446.6	452.1
4.6	457.6	463.2	468.9	474.6	480.4	486.3	492.3	498.3	504.4	510.5
4.7	516.8	523.0	529.4	535.9	542.4	549.0	555.7	562.5	569.3	576.2
4.8	583.3	590.3	597.5	604.8	612.1	619.5	627.0	634.7	642.4	650.1
4.9	658.0	666.0	674.0	682.2	690.5	698.8	707.2	715.8	724.4	733.2
5.0	742.1	751.0	760.1	769.2	778.5	787.9	797.4	807.0	816.7	826.6
5.1	836.5	846.6	856.7	867.1	877.5	888.0	898.7	909.5	920.4	931.4
5.2	942.6	953.4	965.3	976.9	988.6	1000	1012	1025	1037	1049
5.3	1062	1074	1087	1100	1113	1127	1140	1154	1168	1181
5.4	1196	1210	1224	1239	1254	1269	1284	1299	1314	1330
5.5	1346	1362	1378	1394	1411	1428	1445	1462	1479	1497
5.6	1514	1532	1550	1569	1587	1606	1625	1644	1664	1684
5.7	1704	1724	1744	1765	1785	1807	1828	1850	1871	1893
5.8	1916	1938	1961	1984	2008	2031	2055	2079	2104	2129
5.9	2154	2179	2205	2231	2257	2283	2310	2337	2365	2392

References

Akeley, R. V., Stevenson, F. J., and Cunningham, C. E. 1955. Potato variety yields, total solids, and cooking quality as affected by the date of vine killing. *Am. Potato J.* **32,** 303–313.

Albert, J. J., and Lewis, F. H. 1962. Effect of repeated applications of dodine and of captan on apple scab foliage lesions. *Plant Disease Reptr.* **46,** 163–167.

Alexander, L. J., and Tucker, C. M. 1945. Physiologic specialization in the tomato wilt fungus *Fusarium oxysporum* f. *lycopersici*. *J. Agr. Research* **70,** 303–313.

Allen, P. J. 1955. The role of self-inhibitor in the germination of rust uredospores. *Phytopathology* **45,** 259–266.

Allen, P. J. 1957. Properties of a volatile fraction from uredospores of *Puccinia graminis* var. *tritici* affecting their germination and development. I. Biological activity. *Plant Physiol.* **32,** 385–389.

Allen, R. F. 1926. A cytological study of *Puccinia triticina* physiologic form II on Little Club wheat. *J. Agr. Research* **33,** 201–222.

Anonymous. 1947. The measurement of potato blight. *Brit. Mycol. Soc. Trans.* **31,** 140–141.

Anonymous. 1953. Some further definitions of terms used in plant pathology. *Brit. Mycol. Soc. Trans.* **36,** 267.

Anonymous. 1954. Verslag van de enquete over het optreden van de aartappelziekte, *Phytophthora infestans* (Mont.) de Bary in 1953. *Jaarboek Plantenziektenkundige Dienst Wageningen* **1953,** 34–53.

Anonymous. 1958. West African Maize Research Institute. Review of research for the period January 1955–December 1957. Published by the Federal Department of Agricultural Research Ibadan, Nigeria. 50 pp.

Asai, G. N. 1960. Intra- and interregional movement of uredospores of black stem rust in the Upper Mississippi River Valley. *Phytopathology* **50,** 535–541.

Bailey, N. T. J. 1957. "The Mathematical Theory of Epidemics", 194 pp. Charles Griffin, London.

Bald, J. G. 1937. Investigations on "spotted wilt" of tomatoes. III. Infection in field plots. *Bull. Council Sci. Ind. Research Australia* **106,** 32 pp.

Beaumont, A. 1954. Tomato leaf mould: Spraying trials in Lancashire and Yorkshire, 1949–52. *Plant Pathol.* **3,** 21–25.

Bell, F. H. 1951. Distribution of hyphae of several plant-pathogenic fungi in leaf tissue (Abstr.). *Phytopathology* **41,** 3.

Bennett, C. W., and Costa, A. S. 1949. Tristeza disease of citrus. *J. Agr. Research* **78,** 207–237.

Benstead, R. J. 1951. Cocoa re-establishment. *Proc. Cocoa Conf. London* pp. 111–115.

Benstead, R. J. 1953. Swollen shoot disease. *Proc. West African Intern. Cacao Conf. Tafo* pp. 25–28.

Berkson, J. 1944. Application of the logistic function to bio-assay. *J. Am. Statist. Assoc.* **39,** 357–365.

Berkson, J. 1953. A statistically precise and relatively simple method of estimating the bio-assay with quantal response, based on the logistic function. *J. Am. Statist. Assoc.* **48,** 565–599.

Biffen, R. H. 1905. Mendel's laws of inheritance and wheat breeding. *J. Agr. Sci.* **1,** 4–48.

Björling, K., and Sellgren, K. A. 1957. Protection and its connection with redistribution of different droplet sizes in sprays against *Phytophthora infestans*. *Kgl. Lantbruks Högskol. Ann.* **23**, 291–308.

Black, W. 1957. Incidence of physiological races of *Phytophthora infestans* in various countries. *Scot. Soc. Research Plant-Breeding Ann. Rept.* **1957**, 43–49.

Black, W. 1960. Races of *Phytophthora infestans* and resistance problems in potatoes. *Scot. Plant Breeding Sta. Ann. Rept.* **1960**, 29–38.

Black, W., Mastenbroek, C., Mills, W. R., and Petersen, L. C. 1953. A proposal for an international nomenclature of races of *Phytophthora infestans* and of genes controlling immunity in *Solanum demissum* derivatives. *Euphytica* **2**, 173–178.

Bliss, C. I. 1934. The method of probits. *Science* **79**, 409–410.

Blodgett, F. M. 1941. A method for the determination of losses due to diseased or missing plants. *Am. Potato J.* **18**, 132–135.

Bohn, G. W., and Tucker, C. M. 1939. Immunity to Fusarium wilt in the tomato. *Science* **89**, 603–604.

Bohn, G. W., and Tucker, C. M. 1940. Studies on Fusarium wilt of the tomato: I. Immunity in *Lycopersicon pimpinellifolium* Mill. and its inheritance in hybrids. *Missouri Univ. Agr. Expt. Sta. Research Bull.* **311**, 82 pp.

Bond, T. E. T. 1938. Infection experiments with *Cladosporium fulvum* Cooke and related species. *Ann. Appl. Biol.* **25**, 277–307.

Bonde, R., and Schultz, E. S. 1943. Potato refuse piles as a factor in the dissemination of late blight. *Maine Agr. Expt. Sta. Bull.* **416**, 230–246.

Bonde, R., and Schultz, E. S. 1944. Potato refuse piles and late-blight epidemics. *Maine Agr. Expt. Sta. Bull.* **426**, 233–234.

Bonde, R., Hyre, R. A., and Johnson, B. 1957. Forecasting late blight in Aroostook County, Maine, in 1957. *Plant Disease Reptr.* **41**, 936–938.

Borlaug, N. E. 1959. The use of multilineal or composite varieties to control airborne epidemic diseases of self-pollinated crop plants. *Proc. 1st Intern. Wheat Genetics Symposium 1958*, pp. 12–27.

Boyce, J. S. 1957. Oak wilt spread and damage in the Southern Appalachians. *J. Forestry* **55**, 499–505.

Brandes, G. A., Bonde, R., Cetas, R. C., Samson, R. W., and Rich, A. E. 1959. A review of uniform dosage and timing experiments comparing maneb and zineb fungicides on potatoes in Maine, New York, Indiana, and New Hampshire 1958. *Plant Disease Reptr.* **43**, 201–212.

Broadbent, L. 1957. "Investigations of Virus Diseases of Brassica Crops", 94 pp. Cambridge Univ. Press, London and New York.

Broadbent, L., Heathcote, G. D., and Burt, P. E. 1960. Field trials on the retention of potato stocks in England. *Europ. Potato J.* **3**, 251–262.

Bromfield, K. R., Underwood, J. F., Peet, C. E., Grissinger, E. H., and Kingsolver, C. H. 1959. Epidemiology of stem rust of wheat: IV. The use of rods as spore collecting devices in a study of the dissemination of stem rust of wheat uredospores. *Plant Disease Reptr.* **43**, 1160–1168.

Cammack, R. H. 1958. Factors affecting infection gradients from a point source of *Puccinia polysora* in a plot of *Zea mays*. *Ann. Appl. Biol.* **46**, 186–197.

Cammack, R. H. 1961. *Puccinia polysora*: a review of some factors affecting the epiphytotic in West Africa. *Rept. 6th Commonwealth Mycol. Conf. 1960*, pp. 134–138.

Carleton, M. A. 1899. Cereal rusts of the United States: a physiological investigation. *U.S. Dept. Agr. Div. Veg. Physiol. and Pathol. Bull.* **16**, 73 pp.

Chester, K. S. 1946. "The Cereal Rusts", 269 pp. Chronica Botanica, Waltham, Massachusetts.

Chester, K. S. 1950. Plant disease losses: their appraisal and interpretation. *Plant Disease Reptr., Suppl.* **193**, 190–362.

Christ, R. A. 1957. Control plots in experiments with fungicides. *Commonwealth Phytopathol. News* **3**, 54 and 62.

Cockerham, G. 1958. Observations on the spread of virus X. *Proc. 3rd Conf. Potato Virus Diseases, Lisse-Wageningen 1957*, pp.144–148.

Colhoun, J. 1961. Spore load, light intensity and plant nutrition as factors influencing the incidence of club root of brassicae. *Brit. Mycol. Soc. Trans.* **44**, 593–600.

Colwell, R. N. 1956. Determining the prevalence of certain cereal crop diseases by means of aerial photography. *Hilgardia* **26**, 223–286.

Cornwell, P. B. 1958. Movements of the vectors of virus diseases of cacao in Ghana. I. Canopy movement in and between trees. *Bull. Entomol. Research* **49**, 613–630.

Cox, A. E., and Large, E. C. 1960. Potato blight epidemics throughout the world. *U.S. Dept. Agr., Agr. Handbook* **174**, 230 pp.

Cox, D. R. 1958. "Planning of Experiments", 308 pp. Wiley, New York.

Craigie, J. H. 1944. Increase in production and value of the wheat crops in Manitoba and Eastern Saskatchewan as a result of the introduction of rust resistant wheat varieties. *Sci. Agr.* **25**, 51–64.

Craigie, J. H. 1945. Epidemiology of stem rust in Western Canada. *Sci. Agr.* **25**, 285–401.

Crozier, W. 1934. Studies in the biology of *Phytophthora infestans* (Mont.) de Bary. *Cornell Univ. Agr. Expt. Sta. Mem.* **155**, 40 pp.

Cummins, G. B. 1941. Identity and distribution of the three rusts of corn. *Phytopathology* **31**, 856–857.

Dale, W. T. 1953. Further notes on the spread of virus in a field of clonal cacao in Trinidad. *Cocoa Research* **1945–51**, *Imp. Coll. Trop. Agr.* 130. (Quoted by Thresh, 1958b.)

Davidson, W. D. 1926. Production of healthy stocks of seed potatoes. *J. Dept. Lands Agr. Ireland* **25**, 281–283.

Davidson, W. D. 1928. The rejuvenation of the Champion potato. *Econ. Proc. Roy. Dublin Soc.* **2**, 319–330.

de Lint, M. M., and Meijers, C. P. 1956. Resultaten van de enquete over het optreden van de aartappelziekte. *Jaarboek 1955 Plantenziektenkundige Dienst, Wageningen* pp. 116–133.

de Lint, M. M., and Meijers, C. P. 1958. Verslag van de Phytophthora-bestrydingsproeven 1957. *Proefsta. akker-en weidebouw, Wageningen*, 81 pp.

de Lint, M. M., and Meijers, C. P. 1959. Verslagen van series 609 en 610 van de Phytophthora-bestrydingsproeven 1958. *Proefsta. akker-en weidebouw, Wageningen*, 42 pp.

Durrell, L. W., and Parker, J. H. 1920. Comparative resistance of varieties of oats to crown and stem rusts. *Iowa Agr. Expt. Sta. Research Bull.* **62**, 60 pp.

Eide, C. J., Bonde, R., Gallegly, M. E., Graham, K. M., Mills, W. R., Niederhauser, J., and Wallin, J. R. 1959. Report of the late blight investigations committee. *Am. Potato J.* **36**, 421–423.

Finney, D. J. 1952. "Probit Analysis", 318 pp. Cambridge Univ. Press, London and New York.

Flor, H. H. 1953. Epidemiology of flax rust in the North Central States. *Phytopatholgy* **43**, 624–628.

Flor, H. H. 1959. Genetic controls and host parasite interactions in rust diseases. *In* "Plant Pathology, Problems and Progress 1908–1958" (Holton, C. S., Fischer, G. W., Fulton, R. W., Hart, H., and McCallan, S. E. A., eds.), pp. 137–144. Univ. of Wisconsin Press, Madison, Wisconsin.

Flor, H. H., Gaines, E. F., and Smith, W. K. 1932. The effect of bunt on yield of wheat. *J. Amer. Soc. Agron.* **24**, 778–784.

Forsyth, F. R., and Peturson, B. 1960. Control of leaf and stem rust of wheat by zineb and inorganic nickel salts. *Plant Disease Reptr.* **44**, 208–211.

Gallegly, M. E., and Galindo, J. 1958. Mating types and oospores of *Phytophthora infestans* in nature in Mexico. *Phytopathology* **48,** 274–277.

Gassner, G., and Straib, W. 1936. Unterzuchungen zur Bestimmung der Ernteverluste des Weizens durch Gelb- und Schwartzrostbefall. *Phytopath. Z.* **9,** 479–505.

Gerdemann, J. W., and Finley, A. M. 1951. The pathogenicity of races 1 and 2 of *Fusarium oxysporum* f. *Lycopersici*. *Phytopathology* **41,** 238–244.

Goulden, C. H., and Greaney, F. J. 1930. The relation between stem rust infection and the yield of wheat. *Sci. Agr.* **10,** 405–410.

Graham, K. M., Neiderhauser, J. S., and Romero, S. 1959. Observations on races of *Phytophthora infestans* in Mexico during 1956–1957. *Am. Potato J.* **36,** 196–203.

Graham, K. M., Dionne, L. A., and Hodgson, W. A. 1961. Mutability of *Phytophthora infestans* on blight resistant selections of potato and tomatoes. *Phytopathology* **51,** 264–265.

Grainger, J. 1956. Host nutrition and attack by fungal parasites. *Phytopathology* **46,** 445–456.

Grainger, J. 1962. The host plant as a habitat for fungal and bacterial parasites. *Phytopathology* **62,** 140–150.

Greaney, F. J., Woodward, J. C., and Whiteside, A. G. O. 1941. The effect of stem rust on the yield, quality, chemical composition, and milling and baking qualities of Marquis wheat. *Sci. Agr.* **22,** 40–60.

Gregory, P. H. 1945. The dispersal of air-borne spores. *Brit. Mycol. Soc. Trans.* **28,** 26–72.

Gregory, P. H. 1948. The multiple-infection transformation. *Ann. Appl. Biol.* **35,** 412–417.

Gregory, P. H. 1961. "The Microbiology of the Atmosphere", 251 pp. Leonard Hill, London.

Harrison, A. L. 1947. The relation of weather to epiphytotics of late blight on tomatoes. *Phytopathology* **37,** 533–538.

Hart, H. 1931. Morphologic and physiologic studies on stem-rust resistance in cereals. *U.S. Dept. Agr. Tech. Bull.* **266,** 76 pp.

Haymaker, H. H. 1928. Pathogenicity of two strains of the tomato-wilt fungus, *Fusarium lycopersici* Sacc. *J. Agr. Research* **36,** 675–695.

Heald, F. D. 1921. The relation of spore load to the per cent of stinking smut appearing in the crop. *Phytopathology* **11,** 269–278.

Heald, F. D., and Gaines, E. F. 1930. The control of bunt or stinking smut of wheat. *Wash. State Coll. Agr. Expt. Sta. Bull.* **241,** 30 pp.

Heathcote, G. D., and Broadbent, L. 1961. Local spread of potato leaf roll and Y viruses. *Europ. Potato J.* **4,** 138–143.

Heuberger, J. W., and Jones, R. K. 1962. Apple scab: II. Effect of serial applications of fungicides on leaf lesions on previously unsprayed trees. *Plant Disease Reptr.* **46,** 159–162.

Hewitt, W. B., Frazier, N. W., Freitag, J. H., and Winkler, A. J. 1949. Pierce's disease investigations. *Hilgardia* **19,** 207–264.

Hirst, J. M. 1953. Changes in atmospheric spore content: diurnal periodicity and the effect of weather. *Brit. Mycol. Soc. Trans.* **36,** 375–392.

Hirst, J. M. 1955. The early history of a potato blight epidemic. *Plant Pathol.* **4,** 44–50.

Hirst, J. M. 1958. New methods for studying plant epidemics. *Outlook on Agr.* **2,** 16–26.

Hirst, J. M. 1959. Spore liberation and dispersal. *In* "Plant Pathology Problems and Progress 1908–1958" (Holton, C. S., Fischer, G. W., Fulton, R. W., Hart, H., and McCallan, S. E. A., eds.), pp. 529–538. Univ. of Wisconsin Press, Madison, Wisconsin.

Hirst, J. M. 1961. The aerobiology of *Puccinia graminis* uredospores (Abstr.). *Brit. Mycol. Soc. Trans.* **44,** 138.

Hirst, J. M., and Stedman, O. J. 1960a. The epidemiology of *Phytophthora infestans*. I. Climate, ecoclimate and the phenology of disease outbreak. *Ann. Appl. Biol.* **48**, 471–488.

Hirst, J. M., and Stedman, O. J. 1960b. The epidemiology of *Phytophthora infestans*. II. The source of inoculum. *Ann. Appl. Biol.* **48**, 489–517.

Hirst, J. M., and Stedman, O. J. 1962. The epidemiology of *Phytophthora infestans*. III. Spraying trials 1952–1958. *Plant Pathol.* **11**, 7–13.

Hodgson, W. A. 1961. Laboratory testing of the potato for partial resistance to *Phytophthora infestans*. *Am. Potato J.* **38**, 259–264.

Hogen Esch, J. A., and Zingstra, H. 1957. "Geniteurslijst voor aardappelrassen 1957", 147 pp. Commissie ter bevordering van het kweken en het onderzoek van nieuwe aardappelrassen, Wageningen, Netherlands.

Holmes, F. O. 1959. Discussion *in* "Plant Pathology, Problems and Progress 1908–1958" (Holton, C. S., Fischer, G. W., Fulton, R. W., Hart, H., and McCallan, S. E. A., eds.), pp. 521–523. Univ. of Wisconsin Press, Madison, Wisconsin.

Holmes, F. O. 1960. Inheritance on tobacco of an improved resistance to infection by tobacco mosaic virus. *Virology* **12**, 59–67.

Holmes, F. O. 1961. Concomitant inheritance of resistance to several virus diseases in tobacco. *Virology* **13**, 409–413.

Holton, C. S. 1947. Host selectivity as a factor in the establishment of physiological races of *Tilletia caries* and *T. foetida* produced by hybridization. *Phytopathology* **37**, 817–821.

Hooker, A. L., and le Roux, P. M. 1957. Sources of protoplasmic resistance to *Puccinia sorghi* in corn. *Phytopathology* **47**, 187–191.

Hooker, A. L., and Russel, W. A. 1962. Inheritance of resistance to *Puccinia sorghi* in six corn inbred lines. *Phytopathology* **52**, 122–128.

Hooker, W. J. 1956. Foliage fungicides for potatoes in Iowa. *Am. Potato J.* **33**, 47–52.

Horsfall, J. G., and Barratt, R. W. 1945. An improved grading system for measuring plant diseases. *Phytopathology* (Abstr.) **35**, 655.

Hutchins, L. M., Bodine, E. W., and Thornberry, H. H. 1937. Peach mosaic: its identification and control. *U.S. Dept. Agr. Circ.* **427**, 48 pp.

Hyre, R. A. 1950. Spore traps as an aid in forecasting several downy mildew type diseases. *Plant Disease Reptr. Suppl.* **190**, 14–18.

Jensen, N. F. 1952. Intra-varietal diversification in oat breeding. *Agron. J.* **44**, 30–34.

Johnson, T. 1954. Selfing studies with physiologic races of wheat stem rust, *Puccinia graminis tritici*. *Can. J. Botany* **32**, 506–522.

Johnson, T., and Newton, M. 1940. Mendelian inheritance of certain pathogenic characters of *Puccinia graminis tritici*. *Can. J. Research* **C18**, 599–611.

Johnson, T., and Newton, M. 1941. The predominance of race 56 in relation to stem-rust resistance of Ceres wheat. *Sci. Agr.* **22**, 152–156.

Joyce, R. J. V. 1956. Insect mobility and design of field experiments. *Nature,* **177**, 282–283.

Joyce, R. J. V., and Roberts, P. 1959. The determination of the size of plot suitable for cotton spraying experiments in the Sudan Gezira. *Ann. Appl. Biol.* **47**, 287–305.

Kammermann, N. 1950. Undersökningar rörande Potatisbladmöglet, *Phytophthora infestans* (Mont.) de By. I. Metodologisk undersökning angående prövningen av potatisblastens resistens mot bladmöglet. *Medd. Vaxtsdyddsanst.* **57**, 41 pp.

Kendall, M. G., and Buckland, W. R. 1957. "A Dictionary of Statistical Terms", 493 pp. Oliver and Boyd, Edinburgh.

Kendrick, E. L., and Holton, C. S. 1961. Racial population dynamics in *Tilletia caries* and *T. foetida* as influenced by wheat varietal populations in the Pacific Northwest. *Plant Disease Reptr.* **45**, 5–9.

KenKnight, G. 1961. Epidemiology of peach rosette virus in *Prunus augustifolia*. *Plant Disease Reptr.* **45**, 304–305.

Kiely, T. B. 1950. Control and epiphytology of black spot of citrus. *N. S. Wales Dept. Agr. Sci. Bull.* **71**, 88 pp.

Kingsolver, C. H., Schmitt, C. G., Peet, C. E., and Bromfield, K. R. 1959. Epidemiology of stem rust: II (Relation of quantity of inoculum and growth stage of wheat and rye at infection to yield reduction by stem rust). *Plant Disease Reptr.* **43**, 855–862.

Kirby, R. S., and Archer, W. A. 1927. Diseases of cereal and forage crops in the United States in 1926. *Plant Disease Reptr. Suppl.* **53**, 110–208.

Kirkpatrick, H. C., and Blodgett, F. M. 1943. Yield losses caused by leaf roll of potatoes. *Am. Potato J.* **20**, 53–56.

Kirste [initials not stated]. 1958. Ergebnisse von Krautfäule-Spritzversuchen. *Kartoffelbau* **9**, 114–115.

Knutson, K. W., and Eide, C. J. 1961. Parasitic aggressiveness in *Phytophthora infestans*. *Phytopathology* **51**, 286–290.

Kotze, J. M. 1963. Studies on the Black Spot Disease of Citrus Caused by *Guignardia citricarpa* Kiely, with Particular Reference to Its Epiphytology and Control at Letaba. Thesis, University of Pretoria, 147 pp.

Lambert, E. B. 1929. The relation of weather to the development of stem rust in the Mississippi Valley. *Phytopathology* **19**, 1–71.

Lapwood, D. H. 1961a. Potato haulm resistance to *Phytophthora infestans*. I. Field assessment of resistance. *Ann. Appl. Biol.* **49**, 140–151.

Lapwood, D. H. 1961b. Potato haulm resistance to *Phytophthora infestans*. II. Lesion production and sporulation. *Ann. Appl. Biol.* **49**, 316–330.

Lapwood, D. H. 1961c. Laboratory assessment of the susceptibility of potato haulm to blight (*Phytophthora infestans*). *Europ. Potato J.* **4**, 117–128.

Lapwood, D. H. 1962. Haulm and tuber resistance to blight (*Phytophthora infestans*). *Rept. Rothamsted Expt. Sta. for* **1961**, 116–117.

Large, E. C. 1945. Field trials of copper fungicides for the control of potato blight. I. Foliage protection and yield. *Ann. Appl. Biol.* **32**, 319–329.

Large, E. C. 1952. The interpretation of progress curves for potato blight and other plant diseases. *Plant Pathol.* **1**, 109–117.

Large, E. C., and Taylor, G. G. 1953. The distribution of spray deposits in low-volume potato spraying. *Plant Pathol.* **2**, 93–98.

Large, E. C., Beer, W. J., and Patterson, J. B. E. 1946. Field trials for the control of potato blight. II. Spray retention. *Ann. Appl. Biol.* **33**, 54–63.

Leukel, R. W. 1937. Studies on bunt, or stinking smut, of wheat and its control. *U.S. Dept. Agr. Tech. Bull.* **582**, 48 pp.

Levine, M. N., and Stakman, E. C. 1918. A third biologic form of *Puccinia graminis*. *J. Agr. Research* **13**, 651–654.

Loegering, W. Q. 1951. Survival of races of wheat stem rust in mixtures. *Phytopathology* **41**, 56–65.

Loegering, W. Q., McKinney, H. H., Harmon, D. L., and Clark, W. A. 1961. A long term experiment for the preservation of uredospores of *Puccinia graminis tritici* in liquid nitrogen. *Plant Disease Reptr.* **45**, 384–385.

McCallan, S. E. A. 1958. Effectiveness of various formulations of Bordeaux mixture in controlling early and late blight of tomato in greenhouse tests. *Contribs. Boyce Thompson Inst.* **19**, 157–167.

McCallan, S. E. A., and Wellman, R. H. 1943. A greenhouse method of evaluating fungicides by means of tomato foliage diseases. *Contribs. Boyce Thompson Inst.* **13**, 93–141.

Mains, E. B. 1931. Inheritance of resistance to rust, *Puccinia sorghi*, in maize. *J. Agr. Research* **43**, 419–430.

Miller, J. D., and Stewart, D. M. 1961. Reaction of wheat seedlings to new isolates of wheat stem rust. *Plant Disease Reptr.* **45**, 657–658.

Mills, W. R., and Peterson, L. C. 1952. The development of races of *Phytophthora infestans* (Mont.) de Bary on potato hybrids (Abstr.). *Phytopathology* **42**, 26.

Mooi, J. C. 1962. Onderzoek inzake de "veldresistentie" van het loof van aardappelen tegen aantasting door *Phytophthora infestans* (Mont.) de Bary. *Inst. Plantenziektenkundig Onderzoek Jaarverslag* **1961**, 39–40.

Müller, K. O. 1951. Über die Herkunft der W-Rassen, ihre Entwicklungsgeschichte und ihre bisherige Nutzung in der praktischen Kartoffelzüchtung. *Z. Pflanzenzücht* **29**, 366–387.

Müller, K. O. 1953. The nature of resistance of the potato plant to blight—*Phytophthora infestans*. *J. Natl. Inst. Agr. Botany* **6**, 346–360.

Müller, K. O., and Munro, J. 1951. The reaction of virus-infected potato plants to *Phytophthora infestans*. *Ann. Appl. Biol.* **38**, 765–773.

Niederhauser, J. S., Cervantes, J., and Servín, L. 1954. Late blight in Mexico and its implications. *Phytopathology* **44**, 406–408.

Nutman, F. J., and Sheffield, F. M. L. 1949. Studies of the clove tree. I. Sudden death disease and its epidemiology. *Ann. Appl. Biol.* **36**, 419–439.

Nutman, F. J., Sheffield, F. M. L., Swainson, O. S., and Winter, D. W. 1951. The sudden death disease of cloves and its economic and agricultural importance. *Empire J. Exptl. Agr.* **19**, 145–159.

O'Connor, C. 1933. Potato breeding and resistance to blight. *Gardeners' Chronicle* **93**, 104.

Pearl, R., and Reed, L. J. 1920. On the rate of growth of the population of the United States since 1790 and its mathematical representation. *Proc. Natl. Acad. Sci. U.S.* **6**, 275–288.

Percival, J. 1921. "The Wheat Plant", 463 pp. Duckworth, London.

Petersen, L. J. 1959. Relations between inoculum density and infection of wheat by uredospores of *Puccinia graminis* var. *tritici*. *Phytopathology* **49**, 607–614.

Peturson, B., Forsyth, F. R., and Lyon, C. B. 1958. Chemical control of cereal rusts. II. Control of leaf rust of wheat with experimental chemicals under field conditions. *Phytopathology* **48**, 655–657.

Posnette, A. F. 1943. Control measures against swollen shoot virus disease of cacao. *Trop. Agr.* (*Trinidad*) **20**, 116–123.

Pugsley, A. T. 1959. Discussion on a paper by D. R. Knott. *Proc. 1st Intern. Wheat Genetics Symposium, Winnipeg* 1958, p. 38.

Radley, R. W., Taha, M. A., and Bremner, P. M. 1961. Tuber bulking in the potato crop. *Nature* **191**, 782–783.

Roane, C. W., Stakman, E. C., Loegering, W. Q., Stewart, D. M., and Watson, W. M. 1960. Survival of physiologic races of *Puccinia graminis* var. *tritici* on wheat near barberry bushes. *Phytopathology* **50**, 40–44.

Roberts, P. 1960. The analysis of an experiment on interplot effects. *Incorp. Statistician* **10**, 59–65.

Rowell, J. B. 1957. Oil inoculation of wheat with spores of *Puccinia graminis* var. *tritici*. *Phytopathology* **47**, 689–690.

Rowell, J. B. 1959. Problems in the epidemics affecting chemical control of cereal rusts. *Proc. Conf. Chem. Control Cereal Rusts 1959, St. Paul, Minnesota*, pp. 25–27.

Rowell, J. B., and Olien, C. R. 1957. Controlled inoculation of wheat-seedlings with uredospores of *Puccinia graminis* var. *tritici*. *Phytopathology* **47**, 650–655.

Rudorf, W., and Schaper, P. 1951. Grundlagen und Ergebnisse der Züchtung krautfäuleresistenter Kartoffelsorten. *Z. Pflanzenzücht.* **30**, 29–88.

Russell, W. A., and Hooker, A. L. 1959. Inheritance of resistance in corn to rust, *Puccinia sorghi* Schw., and genetic relationships among different sources of resistance. *Agron. J.* **51**, 21–24.

Samborski, D. J., and Peturson, B. 1960. Effect of leaf rust on the yield of resistant wheats. *Can. J. Plant Sci.* **40**, 620–622.

Schaper, P. 1951. Die Bedeutung der Inkubationzeit für die Züchtung krautfäuleresistenter Kartoffelsorten. *Z. Pflanzenzücht.* **30**, 292–299.

Schick, R. 1932. Über das Verhalten von *Solanum demissum*, *Solanum tuberosum* und ihren Bastarden gegenüber verschiedenen Herkünften von *Phytophthora infestans*. *Züchter* **4**, 233–237.

Schick, R., Möller, K. H., Haussdörfer, M., and Schick, E. 1958a. Die Widerstandsfähigkeit von Kartoffelsorten gegenüber der durch *Phytophthora infestans* (Mont.) de Bary hervorgerufenen Krautfäule. *Züchter* **28**, 99–105.

Schick, R., Schick, E., and Haussdörfer, M. 1958b. Ein Beitrag zur physiologischen Spezialisierung von *Phytophthora infestans*. *Phytopathol. Z.* **31**, 225–236.

Schmitt, C. G., Kingsolver, C. H., and Underwood, J. F. 1959. Epidemiology of stem rust of wheat: I. Wheat stem rust development from inoculation foci of different concentration and spatial arrangement. *Plant Disease Reptr.* **43**, 601–606.

Schroeder, W. T. 1959. Early blight of tomato. Results of 1958 fungicide tests, published by the American Phytopathological Society 1959, p. 43.

Schrödter, H. 1960. Dispersal by air and water—the flight and landing. *In* "Plant Pathology" (Horsfall, J. G., and Dimond, A. E., eds.), Vol. 3, pp. 169–227, Academic Press, New York.

Slinkard, A. E., and Elliott, F. C. 1954. The effect of bunt incidence on the yield of wheat in Eastern Washington. *Agron. J.* **46**, 439–441.

Smith, R. S. 1961. Uredospore dissemination and infection (Abstr.). *Brit. Mycol. Soc. Trans.* **44**, 136.

Smoot, J. J., Gough, F. J., Lamey, H. A., Eichenmuller, J. J., and Gallegly, M. E. 1958. Production and germination of oospores of *Phytophthora infestans*. *Phytopathology* **48**, 165–171.

Stakman, E. C., and Christensen, J. J. 1960. The problem of breeding resistant varieties. *In* "Plant Pathology" (Horsfall, J. G., and Dimond, A. E., eds.), Vol. 3, pp. 567–624. Academic Press, New York.

Stakman, E. C., and Fletcher, D. G. 1930. The common barberry and black stem rust. *U.S. Dept. Agr. Farmers' Bull. 1544*, 28 pp.

Stakman, E. C., and Harrar, J. G. 1957. "Principles of Plant Pathology", 581 pp. Ronald Press, New York.

Stakman, E. C., and Lambert, E. B. 1928. The relation of temperature during the growing season in the spring wheat area of the United States to the occurrence of stem-rust epidemics. *Phytopathology* **18**, 369–374.

Stakman, E. C., and Piemeisel, F. J. 1917a. Biologic forms of *Puccinia graminis* on cereals and grasses. *J. Agr. Research* **10**, 429–495.

Stakman, E. C., and Piemeisel, F. J. 1917b. A new strain of *Puccinia graminis*. *Phytopathology* **7**, 73.

Stall, R. E. 1961a. Development of Fusarium wilt on resistant varieties of tomato caused by a strain different from race 1 isolates of *Fusarium oxysporum* f. *lycopersici*. *Plant Disease Reptr.* **45**, 12–15.

Stall, R. E. 1961b. Development in Florida of a different pathogenic race of the Fusarium wilt of tomato organism. *Proc. Florida State Hort. Soc.* **74**, 175–177.

Steven, W. F. 1936. A new disease of cocoa in the Gold Coast. *Gold Coast Farmer* **5**, 122, 144.

Stevens, N. E., and Scott, W. O. 1950. How long will present spring oat varieties last in the central corn belt? *Agron. J.* **42**, 307–309.

Stevenson, F. J., and Jones, H. A. 1953. Some sources of resistance in crop plants. *Yearbook Agr.* (*U.S. Dept. Agr.*) *1953*, 192–216.

Stevenson, F. J., Akeley, R. V., and Webb, R. E. 1955. Reactions of potato varieties to late blight and insect injury as reflected in yields and percentage solids. *Am. Potato J.* **32**, 215–221.

Strickland, A. H. 1951. The entomology of swollen shoot of cacao. II. The bionomics and ecology of the species involved. *Bull. Entomol. Research* **42**, 65.

Thompson, W. R. 1924. La théorie mathématique de l'action des parasites entomophages et le facteur du hasard. *Ann. fac. Sci. Marseille* [2] **2,** 69–89.

Thresh, J. M. 1958a. The control of cacao swollen shoot disease in West Africa. *West African Cocoa Research Inst. Tech. Bull.* **4,** 36 pp.

Thresh, J. M. 1958b. The spread of virus disease in cacao. *West African Cocoa Research Inst. Tech. Bull.* **5,** 36 pp.

Toxopeus, H. J. 1956. Reflections on the origin of new physiologic races of *Phytophthora infestans* and the breeding for resistance in potatoes. *Euphytica* **5,** 221–237.

Troutman, J. L., and Fulton, R. W. 1958. Resistance in tobacco to cucumber mosaic virus. *Virology* **6,** 303–316.

Tuthill, C. S., and Decker, P., 1941. Losses in yield caused by leaf roll in potatoes. *Am. Potato J.* **18,** 136–139.

Ullstrup, A. J. 1961. Corn diseases in the United States and their control. *U.S. Dept. Agr., Agr. Handbook No.* **199,** 29 pp.

Underwood, J. F., Kingsolver, C. H., Peet, C. E., and Bromfield, K. R. 1959. Epidemiology of stem rust of wheat. III. Measurements of increase and spread. *Plant Disease Reptr.* **43,** 1154–1159.

van der Plank, J. E. 1949. Vulnerability and resistance to the harmful plant viruses: a study of why the viruses are where they are. *S. African J. Sci.* **46,** 58–66.

van der Plank, J. E. 1960. Analysis of epidemics. *In* "Plant Pathology" (Horsfall, J. G., and Dimond, A. E., eds.), Vol. 3, pp. 229–289. Academic Press, New York.

van der Plank, J. E. 1961a. Errors due to spore dispersion in field experiments with epidemic disease. *Rept. 6th Commonwealth Mycol. Conf. London* 1960, pp. 79–83.

van der Plank, J. E. 1961b. Dispersal of air-borne fungus spores from plots and fields. *S. African J. Agr. Sci.* **4,** 431–433.

van der Plank, J. E. 1961c. Estimation of the logarithmic infection-rate in plant disease. *S. African J. Agr. Sci.* **4,** 635–637.

van der Zaag, D. E. 1956. Overwintering en epidemiologie van *Phytophthora infestans,* tevens enige nieuwe bestrijdingsmogelijkheden. *Tijdschr. Plantenziekten* **62,** 89–156.

van der Zaag, D. E. 1959. Some observations on breeding for resistance to *Phytophthora infestans. Europ. Potato J.* **2,** 278–286.

Vowinckel, O. 1926. Die Anfälligkeit deutscher Kartoffelsorten gegenüber *Phytophthora infestans* (Mont.) de By, unter besonderer Berücksichtigung der Untersuchungsmethoden. *Arb. biol. Reichsanstalt Land- u. Forstwirtsch. Berlin-Dahlem* **14,** 588–641.

Wager, V. A. 1953. The black spot disease of citrus in South Africa. *S. African Dept. Agr. Sci. Bull.* **303,** 52 pp.

Waggoner, P. E. 1952. Distribution of potato late-blight around inoculum sources. *Phytopathology* **42,** 323–328.

Waldron, L. R. 1935. Stem rust epidemics and wheat breeding. *N. Dakota Agr. Expt. Sta. Cir.* **57,** 12 pp.

Wardlaw, C. W. 1961. "Banana Diseases," 648 pp. Longmans, Green, New York.

Ware, J. O., and Young, V. H. 1934. Control of cotton wilt and "rust". *Arkansas Univ. (Fayetteville) Agr. Expt. Sta. Bull.* **308,** 23 pp.

Waterhouse, W. L., and Watson, I. A. 1941. A note on determinations of physiologic specialisation in flax rust. *J. Proc. Roy. Soc. N. S. Wales* **75,** 115–117.

Watson, I. A., and Singh, D. 1952. The future for rust resistant wheat in Australia. *J. Australian Inst. Agr. Sci.* **18,** 190–197.

Watts Padwick, G. 1956. Losses caused by plant diseases in the colonies. *Commonwealth Mycol. Inst. Phytopathol. Paper* **1,** 60 pp.

Webb, R. E., and Bonde, R. 1956. Physiological races of late blight fungus from potato dump-heap plants in Maine in 1955. *Am. Potato J.* **33**, 53–55.

Wolf, F. A. 1935. "Tobacco Diseases and Decays," 454 pp. Duke Univ. Press, Durham, North Carolina.

Wolfenbarger, D. O. 1946. Dispersion of small organisms, distance dispersion rates of bacteria, spores, seeds, pollen, and insects; incidence rates of diseases and injuries. *Am. Midland Naturalist* **35**, 1–152.

Wolfenbarger, D. O. 1959. Dispersion of small organisms. Incidence of viruses and pollen: dispersion of fungus spores and insects. *Lloydia* **22**, 1–106.

Young, V. H. 1938. Control of cotton wilt and "rust", or potash hunger, by the use of potash-containing fertilizers. *Arkansas Univ.* (*Fayetteville*) *Agr. Expt. Sta. Bull.* **358**, 26 pp.

Yule, G. U. 1925. The growth of population and the factors which control it. *J. Roy. Statist. Soc.* **88**, 1–58.

Zadoks, J. C. 1961. Yellow rust on wheat. Thesis, University of Amsterdam. 188 pp. Also in *Tijdschr. Plantenziekten* **67**, 69–256.

Author Index

Numbers in italics show the page on which the complete reference is listed.

Subject Index